Aymeric Xu
From Culturalist Nationalism to Conservatism

Transformations of Modern China

Editor
Daniel Leese, Eugenia Lean, Alexander C. Cook,
Nikola Spakowski, and Dong Guoqiang,
Jennifer Altehenger, Daniel Leese, Nicola Spakowski,
and Sebastian Veg

Volume 4

Aymeric Xu

From Culturalist Nationalism to Conservatism

Origins and Diversification of Conservative Ideas
in Republican China

DE GRUYTER
OLDENBOURG

ISBN 978-3-11-112220-5
e-ISBN (PDF) 978-3-11-074018-9
e-ISBN (EPUB) 978-3-11-074026-4

Library of Congress Control Number: 2021933950

Bibliographic information published by the Deutsche Nationalbibliothek
The Deutsche Nationalbibliothek lists this publication in the Deutsche Nationalbibliografie;
detailed bibliographic data are available on the Internet at http://dnb.dnb.de.

© 2022 Walter de Gruyter GmbH, Berlin/Boston
This volume is text- and page-identical with the hardback published in 2021.
Cover credit: Photograph of the 14 participants at the opening of the library
of the Society for the Preservation of National learning on October 12th, 1906.
From Guocui xuebao, Vol. 1 (1906), 92.
Printing and binding: CPI books GmbH, Leck

www.degruyter.com

Acknowledgements

This book is the revised and translated version of my PhD dissertation at École des Hautes Études en Sciences Sociales (EHESS), Paris. First and foremost, I thank my supervisor Sebastian Veg for his guidance, encouragement, and open-mindedness that allowed me to elaborate freely on a project about which I am passionate. My deepest gratitude also goes to Nicolas Zufferey, Clarisse Berthezène, Joachim Kurtz, and Yves Chevrier. I was helped enormously by their insightful advice, ideas, critiques, and feedback, which helped me to transform my PhD dissertation into a monograph.

A considerable number of sources that I have used in this book were found at the Institute of Advanced Studies on Asia, University of Tokyo, and the Institute of Modern History, Academia Sinica. I would like to thank Haneda Massahi and Huang Ko-wu who arranged my research stays in Tokyo and Taipei. I thank these two institutes for their generous hospitality. Staff members of the libraries of the University of Tokyo, Academia Sinica, and Tōyō Bunko showed extraordinary professionalism, which greatly facilitated my research there.

My thanks also go to the Centre d'études sur la Chine moderne et contemporaine, Institut d'Asie Orientale and EHESS, who accorded me generous funding to effectuate my research overseas. I am particularly grateful to the Department of History of the EHESS who offered me a doctoral contract. This was instrumental in allowing me to concentrate on my research, since the increased financial burden on students in social sciences is especially problematic and heavy.

A summary of this book has been published in *Journal of Chinese History*, Volume 4, Issue 1, January 2020, pp. 135–159. Copyright: © Cambridge University Press 2019. Portions of Chapter 3 first appeared in *Twentieth-Century China*, Volume 45, Issue 3, October 2020, pages 331–350. Published by Johns Hopkins University Press. Copyright: © 2018 Twentieth Century China Journal, Inc. Reprinted with permission.

Note on Romanization

This book adopts the *pinyin* system of transliteration for Chinese names, places, and terms. However, some long-accepted Wade-Giles versions are retained, most notably Sun Yat-sen, Chiang Kai-shek, and Kuomintang. The Wade-Giles system is also used for transliterating Taiwanese names and places, as well as the names of those who emigrated out of China after 1949, and transliterated their names by this system of romanization. Personal names from Hong Kong are transliterated according to Cantonese. Terms specific to the Shanghainese language are transliterated by using the General Method of Romanization of the Wu language (通用吳語拼音). Occasionally, some personal names are provided in the way they were used by Chinese intellectuals, followed by *pinyin*. Japanese names and terms are transliterated according to the system of Hepburn romanization.

Contents

List of Illustrations —— IX

Introduction —— 1
    The Birth of Modern Conservatism in China —— 5
    "What will you Conserve?" —— 10
    Homogenization of the Cultural Nation and the Political Nation —— 18
    From Culturalist Nationalism to Nationalist Conservatism —— 23
    Chapter Overview —— 28

1 Conservatism: An Idea Rooted in the Present —— 34
    Traditionalism —— 36
    Monarchism and Cultural Radicalism —— 44
    Political and Cultural Radicalism —— 53
    Reactionaryism —— 68
    Conservatism: Preserving the Present —— 72

2 Culturalist Nationalism from the 1890s to the 1900s —— 74
    The Society for Political Education and National Essence in Japan —— 79
    National Essence in the Late Qing Political and Ideological Context —— 84
    Sino-Babylonianism and the Western Origin of the Chinese Nation —— 89
    The First Revolution of Social Liberation —— 92
    Cultural and Ethnical Homogenization of the Chinese Society —— 102
    Origins of Political Reforms —— 110
    Ideas and Counter Ideas —— 117

3 From Culturalist Nationalism to Nationalist Conservatism —— 119
    The Polemic of the Vernacular Language —— 122
    Classical Chinese and Depoliticized Literature —— 125
    The Stigmatization of Conservatism —— 129
    Conservatives' Responses —— 135
    The First World War and the Diversification of Conservative Thought —— 140

4 **Liberal Conservatism and Anti-modern Conservatism (mid-1910s–1930s)** —— 143
   Liberal Conservatism and the Second Revolution of Social Liberation —— 147
   Liberal Conservatism and Politics —— 155
   Antimodern or Antimodern Conservatism? —— 164
   Antimodern Conservative Theorizers —— 167
   Antimodern Conservative Achievers —— 174
   Different Views, Same Dilemma —— 179

5 **Philosophical and Authoritarian Conservatism (1920s–1940s)** —— 180
   Wu Mi and the Critical Review Group —— 182
   The Critical Review Group and the New Culture Movement —— 184
   Babbitt In and Out of China —— 188
   Philosophical Conservatism and Democracy —— 193
   Philosophical Conservatism and Science —— 198
   The Rise of Authoritarian Conservatism —— 201
   Between Order and Liberty: The Case of Zhang Junmai —— 205
   Guofeng banyuekan and Authoritarian Conservatism —— 207
   Authoritarian Conservatism during the Second Sino-Japanese War —— 209
   The Renewal of Confucianism —— 211

Conclusion —— 213
   An Impossible Conservatism? —— 216
   Afterlives of Conservatism of Republican China —— 221
   Continual Influence of Western Conservatism in China —— 225

Annex —— 228

Bibliography —— 230

Glossary —— 250

Index of Persons —— 257

Index of Subjects —— 260

List of Illustrations

Figure 1.1: Devils [i. e. foreigners] worshipping the hog [i. e. jesus]
Figure 1.2: Photograph of the First Assembly of the Southern Society
Figure 1.3: Photograph of the 14 participants at the opening of the library of the Society for the Preservation of National Learnings on October 12, 1906
Figure 2.1: Portrait of Wang Chuanshan
Figure 2.2: Portrait of the Ancestor of China, the Yellow Emperor
Figure 2.3: The Great Nationalist Men in the World, the Yellow Emperor
Figure 4.1: Who is prostitute? Who is student?
Figure 4.2: Liang Shuming and his colleagues in Zouping
Figure 5.1: Portraits of Confucius and Socrates

Introduction

The beginning of the twenty-first century has witnessed the rise of prominent conservative and radical right-wing ideologies across the Western world. Whether one finds this situation pleasant or not, it shows that conservatism has a long-established intellectual tradition and is imposing itself as a vibrant political force in the West. The fate of modern Chinese conservatism, however, is quite different. Conservatism in Republican China (1912–1949) suffers from ill repute; Chinese conservatism, which insisted upon the time-honored cultural and political values of tradition, is too often judged as inconsequential, incapable of producing the results needed to tackle political crises, or even detrimental to the nation's developments.

The negative representation of Chinese conservatives is owed to many of their "progressive" and "radical" contemporaries of dominant discursive power and was readily incorporated into a number of scholarly works. The two dominant paradigms of modern Chinese historiography also have not treated Chinese conservatism favorably. The revolutionary historiography, which prevailed for more than three decades after the Chinese Communist Party (CCP) seized mainland China and expelled the Kuomintang (the Chinese Nationalist Party, KMT) to Taiwan in 1949, based its conceptual framework on historical materialism. Cultural and political revolutions that took place during the first half of the twentieth century link together to form a historical telos that favors the CCP's revolutionary ideology. Under this paradigm, revolution dominated public life in modern China and constitutes the overarching themes of modern Chinese history. Specifically, these themes are the Revolution of 1911 (*Xinhai geming*) that overthrew the imperial system and established a Chinese nation-state; the May Fourth (*Wusi yundong*) and New Culture Movements (*Xinwenhua yundong*) of the mid-1910s to 1920s that replaced "feudal" Chinese traditional culture with Western democratic and scientific enlightenment; and the Communist Revolution from 1949 onwards that "compensated" for the KMT's abortive revolutions to bring about China's long-awaited national revival by the drastic steering of Chinese society, its economy, and politics in a socialist direction.

By contrast, the second dominant historiographical theme takes modernization as the dominant trend that permeated modern China. Since the middle of the nineteenth century, China's socio-political structure and her technological backwardness rendered the country extremely vulnerable to the advances of colonial and imperialist profiteering. Inequitable treaties were imposed on the Qing government, which opened up the path for major Western powers and the Empire of Japan to penetrate China's economy and wrestle for special privileges.

The modernization paradigm emphasizes China's process of reconstruction which began with the Self-Strengthening Movement (*Ziqiang yundong*, ca. 1861–1895), the Hundred Days' Reform (*Bairi weixin*, 1891), and the New Policies (*Xinzheng*, 1901–1911). These were launched by the Manchu dynasty to facilitate the adoption of Western technology, industry, armaments, and institutional models. Underscoring China's path to social and political modernization, this historiographical perspective portrays revolutions as counterproductive and damaging to China's interests.

Despite the contrasting visions of the two historiographies with regard to revolution, both paradigms are wedded to the concept of linear progress, creating a dichotomy between tradition and modernity. Traditional culture and sociopolitical organizations are considered to be a relic of the past that need to be unrooted to make way for modernity. Consequently, conservative political and cultural agendas during the Republican era are often regarded with suspicion. They are viewed as either endorsing Qing loyalists' attempts at imperial restoration, or supporting anti-Westernization during the New Culture Movement, as well as the KMT's authoritarian rule from 1927 to 1949. With its politico-cultural foundations undermined, Chinese conservatism suffered from being in an inherently defensive position when pitted against liberalism and socialism – the other two ideologies that formed the political and intellectual discussion during the Republican era.

However, such a dismissal of conservatism ignores the conservatives' rational convictions and commitment to traditional values and undermines their role in China's quest for survival and reform. Conservatism in this way is understood purely as a reactionary ideology, which resorts to traditional culture in order to obstruct reforms and preserve the political status quo. It is thus indeed difficult to refute the argument that Chinese conservatism was obsolete at a time when major social and political transformations were called for to eliminate China's backwardness in regard to the outside world. Nevertheless, the diversity of Chinese conservative ideological convictions, political views, and cultural ideas opens up a far more textured and nuanced historical landscape. Recent years have witnessed the rise of a historiographical trend that challenges this progressive narrative. Elisabeth Forster's monograph on the New Culture Movement proposes the story of the New Culturalists as being astute entrepreneurs and self-promoters, who took over the intellectual landscape around 1919, and not entirely due to their intellectual merit.[1] By the same token, the motivations behind the

---

[1] Elisabeth Forster, *1919 – The Year that Changed China: A New History of the New Culture Movement* (Berlin: De Gruyter, 2018).

May Fourth intellectual project, which aimed to manipulate discourse and appropriate the notion of modernity by downplaying the views of conservatives, have also received much scholarly attention.²

With the rise of this new historiographical trend, the study of the history of Chinese conservatism in Republican China has gained a new urgency. It is necessary to reveal the intellectual dynamism and political standpoints of Chinese conservatives, which have long been buried beneath modern Chinese historiography. This includes conservatives' personal motivations, cultural and political viewpoints, intellectual networks, political engagement, and the political institutions and systems that they advocated. Nevertheless, mapping modern China's intellectual landscape, particularly in the arena of political thought, is notably difficult. As Wang Fan-sen has recently pointed out, from the late Qing period onwards, major political events and intellectual developments have occurred in a "confused period." This was characterized by shifting ideological and political allegiances and competing visions of Chinese society and polity.³ Indeed, some of the most perplexing problems arise from the vagueness of the term "conservatism." What did the term really mean in late Qing and Republican China and who in fact were the conservatives?

In the West, many scholars have made lists of the key principles of conservatism and there is wide agreement on what these are. Russell Kirk, in his classic 1952 work, developed six canons of conservatism: a belief in a transcendent order, an affection for the variety and mystery of human existence, a conviction in hierarchy and the social orders, a persuasion that freedom and property are closely linked, faith in customs and conventions, and a recognition that change must be tied to social preservation and tradition.⁴ It is obvious that many of these characteristic features are not contextually relevant to China. One of the problems with the term conservatism is that it evokes wildly different things in different cultures. Samuel Huntington made this point in his 1957 essay *Con-*

---

2 See, for example, Milena Doleželová-Velingerová et al., *The Appropriation of Cultural Capital: China's May Fourth Project* (Cambridge, Mass.: Cambridge University Press, 2001); Ya-Pei Kuo, "Polarities and the May Fourth Polemical Culture: Provenance of the 'Conservative' Category," *Twentieth-Century China* 2 (2019): 174–189.
3 Wang Fan-sen, "Tansuo Wusi lishi de liangtiao xiansuo" (Two Paths to Explore the May Fourth History), *Twenty-First Century* 4 (2019): 18–31.
4 Russell Kirk, *The Conservative Mind: From Burke to Eliot* (Washington, D.C.: Regnery Publishing, Inc., 2001), 8–9.

servatism as an Ideology, maintaining that "conservatism itself stresses the particular nature of truth and warns of the danger of overarching principles."[5]

In opposing this type of rationalist explanation, Michael Oakeshott argues that conservatism is a psychological disposition. If to be conservative is "to prefer the familiar to the unknown" and "the tried to the untried", then referring to Oakeshott's words,[6] conservatism reveals an attitude that expresses the instinctive human fear of sudden change. In this regard, Fritz Valjavec remarks that conservatism forms a part of what it means to be human.[7] It follows that conservatism cannot be defined solely in a qualitative manner, but also in a quantitative way. As such, conservatism endorses some form of relativism. By exposure to the views of others, one becomes, even unwittingly, conservative. Antoine Compagnon's "*on est toujours l'antimoderne de quelqu'un*"[8] can be "translated" to "one's conservatism is always somebody else's." If conservatism is understood in a relativist manner, the symbolic figure of the New Culture Movement, Hu Shi (1891–1962), becomes a conservative through the eyes of the more radical generation of the post-May Fourth period.[9]

As a matter of fact, it is widely held that conservativism defies definition and the study of conservatism as an ideology is beset with difficulties. Jan-Werner Müller observes that "the overall result [of defining conservatism] has often been a somewhat desperate resort to nominalism ('conservative is who calls themselves conservative'), or historicism ('conservatism is changing all the time'), or what one might call 'conceptual changism' ('there is a concept but it's changing in crucial periods, like a *Sattelzeit*')."[10] I will argue, instead, that both "absolutism" and "relativism" are relevant in approaching the idea of conservatism. Conservatives in Republican China held some beliefs in common, but their "becoming conservative" also needs to be analyzed in the context of wider disciplinary debates, intellectual competitions, socio-political transformations, as well as psychological developments. Michael Gibbs Hill's 2013 monograph

---

[5] Samuel Huntington, "Conservatism as an Ideology," *The American Political Science Review* 2 (1957), 457.
[6] Michael Oakeshott, "On Being Conservative," in *Twentieth Century Political Theory: A Reader*, ed. Stephan Eric Bonner (New York: Routledge, 2006), 78.
[7] Pekka Suvanto, *Conservatism from the French Revolution to the 1990s*, trans. Roderick Fletcher (New York: St. Martin's Press, 1997), 4.
[8] Antoine Compagnon, *Les Antimodernes. De Joseph de Maistre à Roland Barthes* (Paris: Gallimard, 2005), 442. Unless otherwise indicated, translations are those of the author.
[9] Chiang Yung-chen, *Shewo qishe: Hu Shi* (Who Else but Me: Hu Shi), Vol. 3 (Taipei: Linking, 2018), 247.
[10] Jan-Werner Müller, "Comprehending Conservatism: A New Framework for Analysis," *Journal of Political Ideologies* 3 (2006): 359.

on Lin Shu (1852–1924) and Du Chunmei's 2019 work on Gu Hongming (1857–1928) have provided eloquent examples of this approach.[11]

This book will provide a long-term historical perspective on the making of modern Chinese conservatism. The historical period spans over half a century (ca. 1890–1940). I argue that during the last few years of the Empire, adapting traditional teachings to new Western political ideas was seen as the path to political change. However, progressives and radicals in Republican China condemned traditional culture as incompatible with their modern political values and systems. Conservatism in Republican China can be defined by its insistence on maintaining the unity of the political and cultural orders. It can be divided into several groups with different political agendas according to the traditional cultural elements that conservatives appropriated for political purposes. As such, this book will offer a timely reassessment of modern Chinese conservatism and an entirely new conceptual framework for investigating the multiple layers of conservative thought in Republican China.

The Birth of Modern Conservatism in China

The Chinese term for conservatism is a neologism that entered the Chinese political vocabulary during the early 1900s. As in English, "conservatism" is composed of two parts in Chinese: "conserve" (*baoshou*) and "-ism" (*zhuyi*). Both words have Chinese origins. *Zhuyi* is a rare word in traditional sources.[12] It is in Japan that *zhuyi* (in Japanese: *shugi*) was first used as the standard translation of the nominal suffix "-ism." The Imperial Chinese Embassy's Counsellor in Japan, Huang Zunxian (1848–1905), first used *zhuyi* in its modern sense in Chinese in the book *Riben guozhi* (*Treaties on Japan*).[13] In contrast, *baoshou*, in the sense of "conserve," "defend" or "maintain," is a common word in ancient texts. Nevertheless, the meaning of the word in "conservatism" conveys an essential reaction against progressivism. In other words, to fully appreciate the modern

---

11 Michael Gibbs Hill, *Lin Shu, Inc.: Translation and the Making of Modern Chinese Culture* (Oxford: Oxford University Press, 2013); Du Chunmei, *Gu Hongming's Eccentric Chinese Odyssey* (Philadelphia: University of Pennsylvania Press, 2019).
12 See Wang Fan-sen, "'Zhuyi shidai' de lailin – Zhongguo jindai sixiangshi de yige guanjian fazhan" (The Advent of "-ism" – A Key Development in Modern Chinese Intellectual History), in Wang Fan-sen, *Sikao shi shenghuo de yizhong fangshi: Zhongguo jindai sixiangshi de zaisikao* (Thinking as a Lifestyle: Reconsiderations on Modern Chinese Intellectual History) (Taipei: Linking, 2017), 169.
13 Ibid., 171.

meaning of the word, it cannot be considered without reference to the concept of progress.

Available sources do not indicate the exact moment in which the word *baoshou* acquired this sense. The term was used in the sense of the opposite of progress as late as 1902 in Liang Qichao's (1873–1929) famous article *Xinmin shuo* (*On the New People*), in which he remarked: "All the phonemes of the world can be considered to be structured around two '-isms' (*zhuyi*), the first is to conserve (*baoshou*) and the second is to make progress (*jinqu*)."[14] This new meaning of the word was widely adopted by progressive journals and magazines during the 1900s. In 1903, Jiang Fangzhen (1882–1938), the Chinese army general and military writer, who was sponsored by the government to study in Japan at the Imperial Japanese Army Academy, was probably the first person who used the term *baoshou zhuyi* in Chinese. Jiang associated *baoshou zhuyi* with the mental inertia and the "evil customs" of the Chinese population that impeded the nation's progress.[15] But Chinese intellectuals of the 1900s did not all share these views on conservatism. Du Yaquan (1873–1933), who had spent some time in Japan in 1906, wrote several months before the Wuchang Uprising (*Wuchang qiyi*) that conservatism and progressivism should be the guiding principles of any two major parties aiming to win control of a Western-style parliament in China, once the Republic was established:

> Political parties of our nation in the future should be created in accordance with the general rules [of the West] and be divided into a conservative one and a progressive one. The progressive party is devoted to political reforms … while the conservative party is prudent, committed to tradition and aims to reform politics and strengthen the nation in a moderate way… I think this two-party system resembles the wheels of a car or the wings of a bird; the two parties balance each other and cannot function alone.[16]

Du's interpretation of the relation between conservatism and progressivism resonates with that of Fukuzawa Yukichi (1835–1901). Fukuzawa, one of the intellectual founders of modern Japan, left an influential legacy to Chinese intellectuals

---

14 Liang Qichao, "Xinmin shuo" (On the New People), in *Xinhai geming qianshinianjian shilun xuanji* (Anthology of Political Comments in the Decade before the 1911 Revolution), ed. Zhang Zhan and Wang Renzhi, Vol. 1 (Beijing: Sanlian shudian, 1960), 123.
15 Jiang Fangzhen, "Guohun pian" (On the National Soul), in *Jindai Zhongguo dui xifang ji lieqiang renshi ziliao huibian* (Collected Materials on Knowledge of the West and the Western Powers in China), ed. Academia Sinica, Vol. 5 (Taipei: Academia Sinica, Modern History Research Institute, 1990), 454.
16 Du Yaquan, "Zhengdang lun" (On Political Parties), *Dongfang zazhi* 1 (1911), 14.

in Japan at the time.[17] In fact, similar to *zhuyi, baoshou* (in Japanese: *hoshu*) also acquired its modern meaning in Japan. The interrelation between conservatism and progressivism is one of the key topics of Fukuzawa's 1879 book *Minjō isshin* (*Transition of People's Way of Thinking*):

> English political parties are divided into two branches: a conservative one and a progressive one... Conservatives are not always stubborn and narrow-minded, while progressives are not always brutal and violent... When the popularity of the ruling party diminishes, and the popularity of the opposition is on the rise..., the members of the governing party should leave the government and become only members of parliament... In this way, the change of power goes smoothly and flexibly. [As a result], the British government... with two political parties in opposition, was able to bear the stress of progress of civilization in the 19th century.[18]

Despite his appraisal of the English governmental system, Fukuzawa warned Japan against conservatism. To him, conservatism designates "a conservation of what exists, a preservation of ancient things and a wish for a carefree and stable present," while progressivism evokes "an orientation towards innovation, a search for what is rare and a wish for future prosperity."[19] Likewise, for the author of *Datsu-A Ron* (*Shedding Asia*), conservatism had no place in a Meiji Japan that had to align itself with the West to achieve civilization (*bunmei*).[20] Fukuzawa's negative portrayal of conservatism was a consensus shared by Chinese intellectuals from the late Qing period onwards.

Around 1861, the Manchu government launched the Self-Strengthening Movement after its failure to win the two Opium Wars (1839–1842 and 1856–1860) against the British and French. This period of military, technological, and institutional reforms, lasting over three decades, failed to prevent China from losing her tributary state of Vietnam to France in the 1880s. China's humiliating defeat in the First Sino-Japanese War (1894–1895) demonstrated the failure of the country's first attempts at modernization when compared with Japan's successful Meiji Restoration. The Qing government initiated two more reforms during the following two decades: The Hundred Days' Reform led by the Guangxu Emperor (1871–1908) in 1898 and the New Policies launched by the Empress

---

17 See, for example, Xiao Liang, "Fukuzawa Yukichi to Chūgoku no keimou shisō: Liang Qichao to no shisouteki kanren o chūshin ni" (Fukuzawa Yukichi and Chinese enlightenment: A Study on Liang Qichao's Enlightenment Ideas), *Nagoya Daigaku Kyōikugakubu Kiyō* 40, no. 1 (1993): 63–77.
18 Fukuzawa Yukichi, *Minjō Isshin* (Transition of People's Way of Thinking) (Tokyo: Joshazohan, 1879), 108–116.
19 Ibid., 19.
20 Ibid., 29.

Dowager Cixi (1835–1908) in 1901. Both reforms failed to produce the desired objectives due to the Imperial Court's traditionalist wing and their reluctance to resolutely embrace political reform. The Revolution of 1911 that followed nevertheless failed to bring about China's long-awaited national stability.

What then was there left to preserve, when the political order of the past and the culture associated with it were so corrupt that they had created such a crisis? With the introduction and dissemination of social evolutionism by the turn of the twentieth century, the qualities of "new" and "old" implied a value judgement.[21] In this ideological context, *baoshou* was a tainted custom with negative connotations and became associated with traditionalism (*shoujiu*) and reactionaryism (*fandong*). It was an idea that seemed to protect the obsolete remnants of a traditionalist politico-cultural order against a modern China, that or embracing a conspiratorial immobilism that was regarded as national decadence. Therefore, none of those who were considered to be conservative accepted this label voluntarily and all of them claimed to be active in China's pursuit of progress. To them, this was to be achieved through updated traditional values, which, ironically, is a characteristic feature of conservative thinking.

Since conservatism is a political neologism of the early 1900s, could it be said to have existed in ancient China, even though the concept itself did not exist? Ji Xiao-bin's study on Sima Guang (1019–1086) observes that Sima's opposition to Wang Anshi's (1021–1086) New Policies (*Xining bianfa*) from 1069 to 1079 reflected a conservative temperament, which can best be described as "a reverence for the cumulative wisdom from the past and a concern for preserving the fruits of past achievements."[22] Obviously, it is expected that any reform, once put forward, will attract opposition. Regarding conservatism, Ji made it clear that he had no intention to "dispute the aptness of this term" nor to create a universally applicable definition of the concept. He adopted Abraham Lincoln's (1809–1965) definition of conservatism as "adherence to the old and tired, against the new and untried."[23] This definition is founded upon a human psychological trait of an inherent preference for habit and custom and is closely aligned with Oakeshott and Valjavec's views. With this in mind, Madsen Pirie calls our attention to the distinction between conservative temperament and conservative political tra-

---

[21] Xong Yuezhi, *Xixue dongjian yu wanqing shehui* (The Dissemination of Western Learning and the Late Qing society) (Shanghai: Shanghai renmin chubanshe, 1995), 730.
[22] Ji Xiao-bin, *Politics and Conservatism in Northern Song China: The Career and Thought of Sima Guang (A.D. 1019–1086)* (Hong Kong: The Chinese University Press, 2005), 1–2.
[23] Ibid.

dition.²⁴ As the conservative temperament corresponds to an attitude of mind, this mental disposition requires an opposing dramatic historical moment to be "crystallized into definitive traditions of thought and practice."²⁵ Therefore, the origin of Western conservatism is often associated with the Industrial Revolution, the Enlightenment, and the French Revolution, which overturned the traditional socio-political order and religious authority.²⁶

Although ancient China and its dynasties attempted numerous reforms and faced many rebellions, it barely experienced a dramatic cataclysm that overturned its traditional political and cultural orders. This is thanks to what Jin Guantao and Liu Qingfeng call the "ultra-stable structure" that permeated ancient Chinese society. This structure was upheld and reproduced by the cohesive integration of its moral ideology with its social, political, and economic system.²⁷ Confucianism served to give a firm practical and ethical foundation with a social structure based on kinship, community, and moral values. In this system, the needs of the family superseded those of the individual. The patriarchal relationship between father and son was reflected in the hierarchy of sovereign and subject in ancient China. Confucianism also served as the political rationale that legitimized each new dynasty in its claim to be the new inheritor and guardian of the Mandate of Heaven (*tianming*), while shaping the character of each dynasty as defined by the Five Relationships (*wulun*).²⁸ The highly educated undertook the Imperial examination (*keju*) that recruited officials based on Confucian doctrines. This allowed them to seek appointments as literati who would perform political duties for the government. In this way, the political and cultural order became closely intertwined.

The political dispute between Wang Anshi and Sima Guang did not, however, lead to the dismantlement of this politico-cultural integration. While Sima opposed Wang's political proposals out of his concern that the cumulative wisdom of the past would be bypassed, Wang had no intention to initiate wholesale reforms, since his point of reference was classical Confucianism.²⁹ The connection

---

24 Madsen Pirie, "Why F.A. Hayek Is a Conservative," in *Hayek – On the Fabrik of Human Society*, ed. Eammon Butler and Madsen Pirie (London: Adam Smith Institute, 1987), 147–167.
25 John Gray, *Liberalism* (Milton Keynes: Open University Press, 1995), 78.
26 Suvanto, *Conservatism from the French Revolution to the 1990s*, 3–5; Roger Scruton, *How to Be a Conservative* (London: Bloomsbury Publishing, 2014), 79.
27 Jin Guantao and Liu Qingfeng, *Xingsheng yu weiji: lun Zhongguo shehui chaowending jiegou* (The Cycle of Growth and Decline: On the Ultra Stable Structure of the Chinese Society) (Hong Kong: The Chinese University Press, 1992).
28 Five Confucian Relationships designates the reciprocity and responsibility that bond ruler to ruled, father to son, husband to wife, elder brother to younger brother, and friend to friend.
29 Ji, *Politics and Conservatism in Northern Song China*, 12.

between the political order and the cultural order remained intact. As Timothy Cheek points out, the persistence of this Chinese Confucian pattern was so engrained that even foreign military invasion and major cultural challenges, notably Buddhism, failed to initiate any alternative political and cultural models.[30] In fact, even during the last few decades of the Qing dynasty, when Chinese officials began to be alarmed by foreign imperialists and accepted the need for major changes to invigorate Chinese statecraft, traditional teachings were still the dominant guiding force behind the reforms.[31]

Conservative temperament was progressively transformed into conservative political practice around the 1890s, when the imperial system was on the verge of collapse. The troika of ideology, politics, and economics that bonded together traditional Chinese society and its political system was unable to reunite the country against the threat of the Western powers and the rise of the revolutionary forces within. The abolition of the Imperial examination in 1905 is the symbolic moment when Confucianism was invalidated as the hegemonic and guiding political philosophy. China's revival required the application of specialized and modern knowledge from the West. Added to that was the rise of the revolutionary force who opposed the Manchu Dynasty and called for a Chinese republic. Therefore, conservatism as a cultural and political movement that set out to preserve the politico-cultural order on the verge of collapse really emerged only during the late nineteenth century.

"What will you Conserve?"

As Huntington rightfully points out, conservatism as a political and philosophical movement is highly skeptical of any universal panacea. What might work in one specific regional and historical context cannot necessarily be replicated in another, since every community is different from one another. In the West, this conservative skepticism rests on the mistrust of the optimism on human perfectibility and the idea of political rationalism that came out of the Enlightenment, which led to attempts to recreate society from abstract ideals. However,

---

30 Timothy Cheek, *The Intellectual in Modern Chinese History* (Cambridge: Cambridge University Press, 2015), 3. It should be nevertheless underlined that foreign intruders also voluntarily appropriated this system as a useful political and ideological tool to legitimize and consolidate their reign over China, which, as I shall show later, led many late Qing revolutionaries to denounce Confucianism as an accomplice to the Manchu government's rule.
31 Philip A. Kuhn, "Ideas Behind China's Modern State," *Harvard Journal of Asiatic Studies* 55, no. 2 (1995): 295–337.

humans are imperfect and so are incapable of comprehending society through a purely rationalist and ideological perspective.[32] Edmund Burke (1729–1797) pointed out that the latent wisdom or "prejudice," which drew on tradition rather than a human-invented rationale constituted the cornerstone of conservative virtue.[33] Kirk explains that prejudice is not bigotry or superstition; rather, it is "pre-judgement, the answer with which intuition and ancestral consensus of opinion supply a man when he lacks either time or knowledge to arrive at a decision predicted upon pure reason."[34]

Yet this conservative particularism points to a more fundamental question which Benjamin Disraeli (1804–1881) posed to his own conservative party in 1844: "What will you conserve?"[35] Indeed, if giving a categorical definition to conservatism is beset with difficulties, it is necessary to understand what conservatives aim to conserve. With regard to Chinese conservatism, the answer to this question seems to have already been provided. In *The Limits of Change* – the first English monograph on modern Chinese conservatism published in 1976 – Benjamin Schwartz proposed a thesis of "cultural conservatism." Schwartz argued that, unlike Burkean conservatism which required deference to the static and inherited socio-political status quo, modern Chinese conservatives were well aware of the urgency of political and social reforms. However, the Westernization that accompanied China's pursuit of modernization posed an ongoing threat to her traditional culture and Chinese conservatism was mainly manifested in the preservation of this cultural tradition.[36] This view greatly influenced subsequent studies on Chinese conservatism for more than three decades. Guy S. Alitto's *The Anti-Modernisation Tide in the World: On Cultural Conservatism*, published originally in Chinese in 1991, further confirmed this thesis. He argued that Chinese conservatism was characterized by the insistence upon the value of traditional culture in the country's pursuit of Westernization and translated "cultural conservatism" as *wenhua shoucheng zhuyi*.[37] The thesis of cultural conservatism

---

[32] John Kekes, *A Case for Conservatism* (Ithaca and London: Cornell University Press, 1998), 44.
[33] Bruce Frohnen, *Virtue and the Promise of Conservatism: The Legacy of Burke and Tocqueville* (Lawrence: University Press of Kansas, 1993), 38.
[34] Kirk, *A Case for Conservatism*, 38.
[35] Benjamin Disraeli, *Coningsby, or The New Generation* (Leipzig: Bernh. Tauchnitz Jun., 1844), 91.
[36] Benjamin Schwartz, "Notes on Conservatism in General and in China in Particular," in *The Limits of Change: Essays on Conservative Alternatives in Republican China*, ed. Charlotte Furth (Cambridge, Mass.: Harvard University Press, 1976), 17.
[37] See Chen Lai, *Tradition and Modernity: A Humanist View* (Leiden: Brill, 2009), 3.

has also become enduringly popular among Chinese scholars; Chinese research on this subject during the last two decades still tends to use this term.[38]

However, this widespread thesis suffers from several conceptual and empirical problems. First and foremost, it conceals the profoundly political nature of Chinese conservatism. Although this theory still has defenders today, more recent studies have provided new insights into the social and political conventions of traditional culture and revealed some far-reaching political implications of "cultural conservatism." Lin Chih-hung's study on Qing loyalists shows that traditional culture was used as a pretext for the restoration of the Manchu Court in the 1910s. To them, democracy and republicanism could not be contemplated in China because they ran counter to traditional political culture.[39] Edmund Fung argues that conservatives "sought a moral regeneration of the nation by attempting to transform the valued heritage that could be used to promote social, political, and economic reforms, thereby enabling China to survive and compete in a world dominated by the West."[40] Hon Tze-ki's monograph on the late Qing National Essence School (*Guocui pai*) showcases how Chinese culture found its place in the anti-Manchu revolution and institutional reforms. He further points out that the preservation of cultural traditions became a means for conservatives to retain their elite social status, in an era where the discursive authority of the educated elites was no longer taken for granted.[41]

The political nature of Chinese conservatism has been highlighted in significant studies on the KMT's rule during the Nanjing decade (1927–1937) and the

---

38 See Chen Lai, *Tradition and Modernity*, 3–4; He Xiaoming, *Fanben yu kaixin – jindai Zhongguo wenhua baoshou zhuyi xinlun* (Reconstruction and Renovation – New Arguments on Chinese Modern Cultural Conservatism) (Beijing: Commercial Press, 2006); Hu Fengxiang, *Shehui bianqe yu wenhua chuantong: Zhongguo jindai wenhua baoshouzhuyi sichao yanjiu* (Social Evolution and Cultural Tradition: Study on Chinese Modern Cultural Conservatism) (Shanghai: Shanghai renmin chubanshe, 2000); Li Xizhu, *Wanqing baoshou sixiang de yuanxing: Woren yanjiu* (The Archetype of the Late Qing Conservatism: A Research on Woren) (Beijing: Shehui kexue wenxian chubanshe, 2000).
39 Lin Chih-hung, *Minguo nai diguo ye: zhengzhi wenhua zhuanxing xia de Qing yimin* (Republic of China is the Enemy State: Qing loyalists under the Transformation of Politics and Culture) (Taipei: Linking, 2009), 215–16.
40 Fung Siu Kee, "Zhongguo minzuzhuyi, baoshouzhuyi yu xiandaixing" (Nationalism, Conservatism and Modernity in China), in *Zhongguo jindaishang de minzuzhuyi* (Nationalism in Chinese Modern History), ed. Zheng Dahua and Zou Xiaozhan (Beijing: Shehui kexue wenxian chubanshe, 2007), 49; Edmund S.K. Fung, *The Intellectual Foundations of Chinese Modernity: Cultural and Political Thought in the Republican Era* (New York: Columbia University Press, 2010), 119.
41 Hon Tze-ki, *Revolution as Restoration: Guocui xuebao and China's Path to Modernity, 1905–1911* (Leiden: Brill, 2013).

wartime New Life Movement (*Xinshenghuo yundong*, 1934–1949). In both of these cases, Confucian values were distorted into a justification for an authoritarian and fascist-like rule. Earlier works on the ideological nature of the Nationalist regime that preceded the Communist victory in 1949 were published in the 1970s and 1980s. These were by scholars generally unsympathetic to the KMT regime. Lloyd Eastman's and Tien Hung-mao's monographs, for example, concentrated mainly on the overall institutional failure and bureaucratic corruption of the Nationalist government.[42] In 2017 and 2018, Maggie Clinton and Brian Tsui offered a bold revisionism and exploration of the political nature of the KMT's policies.[43] Clinton emphasizes the "CC" (Central Club) and Blue Shirt cliques (*Lanyishe*), focusing on the violence engendered by the KMT's rule. Tsui, on the other hand, studies the framework of the party's ideology and its political formation, at both the domestic and international levels. Both monographs contribute to a better understanding of the basis of the KMT regime's conservative but radical ideology, its motivation as well as its performance.

Another problem that arises from Schwarz's thesis is that "cultural conservatism" implies that all attempts to preserve cultural heritage are either conservative or nationalist-conservative. On the one hand, such a view minimizes the role played by cultural tradition in China's pursuit of reforms. Under this label, literati and intellectuals of different political allegiances and persuasions are indiscriminately regrouped. For instance, the royalist Qing official Woren (1804–1871) and the revolutionary intellectuals of the National Essence School are all labelled as cultural conservatives.[44] On the other hand, the view reinforces to a certain extent Joseph R. Leveson's famous but debatable argument in *Confucian China and Its Modern Fate* that the Confucian tradition has suffered "museumification" in the modern era. To Levenson, Confucianism, which had been the cornerstone that sustained the Chinese universalist conception of the world, was transfigured into the exclusive identity of the Chinese nation. This occurred at a time when the universal values represented by Confucianism were an obsolete relic, incapable of embracing the complexity and the totality of modern political notions and institutions. Consequently, conservatives' attachment to

---

42 Lloyd Eastman, *The Abortive Revolution: China under Nationalist Rule, 1927–1937* (Cambridge, MA: Harvard University Press, 1974); idem, *Seeds of Destruction: Nationalist China in War and Revolution, 1937–1949* (Stanford, CA: Stanford University Press, 1984); Tien Hung-mao, *Government and Politics in Kuomintang China, 1927–1937* (Stanford: Stanford University Press, 1972).
43 Maggie Clinton, *Revolutionary Nativism: Fascism and Culture in China, 1925–1937* (Durham: Duke University Press, 2017); Brian Tsui, *China's Conservative Revolution: The Quest for a New Order, 1927–1949* (Cambridge: Cambridge University Press, 2018).
44 Li, *Wanqing baoshou sixiang de yuanxing*, 268.

Chinese traditional culture was regarded as purely nostalgia accompanied by an inferiority complex towards the superiority of the West.[45] According to this view, Chinese conservative thought holds the tension which exists between traditional culture and modern politics as well as that between nationalism and universalism.

Although the articles published in *The Limits of Change* do emphasize these two conflicting scenarios, the authors specialized in the late Qing history tend to stress the opposition between culture and politics rather than discussing the intellectual convictions which linked them. Laurence A. Schneider's article on three conservative intellectual associations – the National Essence School, the Southern Society (*Nanshe*), and the Critical Review Group (*Xueheng pai*) – highlights the abolition of the Imperial examination as the symbolic moment wherein the political order and cultural order disintegrated.[46] The abolition of the Imperial examination in 1905 removed the educated elites from the path to a position in the state bureaucracy. As a consequence, cultural elites began to form a modern intelligentsia and appropriated "the discovery of culture" as a tool to declare their independence vis-à-vis the state. To Charlotte Furth, this separation of the political order from the cultural order allowed late Qing revolutionary intellectuals, including scholars of the National Essence School and the Southern Society, to perceive the political sphere and the cultural sphere as two separate realms. This enabled them to turn politically radical and be involved in the anti-Manchu revolution, while being culturally conservative in their defense of the enduring values of traditional culture. Furth further remarks that it was only during the Republican era, or more precisely during the KMT's rule, that culture was once again to be politically relevant.[47]

While I do agree that culture was mobilized and exploited by the KMT to rationalize its rule and politics, Furth's argument fails to capture the transformation that traditional culture underwent to respond to new spiritual, social, and political needs during the late Qing and the early Republican periods. It seems problematic to affirm that "the cultural conservatives… all accepted the necessity of political change and had to consider cultural-moral questions apart from the political progress."[48] As Yves Chevrier points out: "Chinese culture was consid-

---

[45] Joseph R. Levenson, *Confucian China and Its Modern Fate: A Trilogy* (Berkeley and Los Angeles: University of California Press, 1968).
[46] Laurence Schneider, "National Essence and the New Intelligentsia," in *The Limits of Change*, 58.
[47] Charlotte Furth, "Culture and Politics in Modern Chinese Conservatism,' in *The Limits of Change*, 25–50.
[48] Ibid., 30.

ered, [during the 1890s and 1900s]... rich in endogenous resources to put China on an equal footing with the modern world, provided the complementary processes of political change and intellectual criticism were carried out."[49] Peter Zarrow also clearly states that the traditional culture promoted by revolutionaries such as Zhang Taiyan (1869–1936) in the 1900s was politically radical and aimed to overthrow the monarchy and reform Chinese society, while taking into account Western political theories and Chinese cultural traditions.[50]

Schwartz considered nationalism a feature of conservatism and the above-mentioned scholarly work has provided an insightful understanding of the interrelationship between these two political notions. Nevertheless, as Theodore Huters states, Chinese culture constituted "practically a necessary condition for attaining status within the community of nation," with Chinese intellectuals at the time well aware of the importance of the cultural imperative in their pursuit of nation-building.[51] In this regard, Chevrier's findings point to an important question concerning the relationship between nationalism and conservatism in the late Qing period – should the efforts of late Qing intellectuals to preserve the nation's traditional culture be viewed as conservative or nationalist?

Certainly, conservatism seems closely intertwined with nationalism in the current global political context. Nevertheless, while nationalism is a characteristic of the modern age,[52] Western conservatism is initially a philosophical, cultural, and political movement that arose as a reaction against modernity. This explains why many Western conservatives favored the family and the church, as well as local communities, over the nation-state and the idea of national sovereignty. Furth considers that anti-Manchu intellectuals perceived culture and politics as two distinct constructs when the monarchy – the symbol of the political-cultural unity – began to collapse, and this view allowed them to be politically radical and culturally conservative. In contrast, Chevrier argues that their preservation of traditional culture, which was in fact a bold remodeling, did not rest upon a conservative disposition, but rather upon a political choice which encouraged the growth of their nationalism. Chevrier's thesis underlines an important yet frequently neglected link between nationalism and conservatism – that

---

49 Yves Chevrier, "Antitradition et démocratie dans la Chine du premier XXe siècle: la culture moderne et la crise de l'État-nation," in *La démocratie et la Chine*, ed. Mireille Delmas-Marty and Pierre-Étienne Will (Paris: Fayard, 2007), 392.
50 Peter Zarrow, *China in War and Revolution, 1895–1949* (London: Routledge, 2005), 70.
51 Theodore Huters, *Bringing the World Home: Appropriating the West in Late Qing and Early Republican China* (Honolulu: University of Hawai'i Press, 2005), 62.
52 Stuart Hall, "Introduction," in *Formations of Modernity*, ed. Stuart Hall and Bram Gieben (Cambridge: Polity Press, 1992), 6.

it is not the preservation of traditional culture per se that makes an intellectual conservative. A clear distinction should therefore be made between the preservation of traditional culture and the defense of a political system through the preservation of specific cultural elements.

In other words, the differentiation between nationalism and conservatism lies in the way in which the constituent cultural elements of the nation-state translate into concrete political will and action. To draw a comparison, cultural elements within the German nation, from the Romantic age to the Second World War, were mobilized to thwart liberalism and democracy.[53] In this case, nationalism and conservatism overlapped. However, this was not always the case in late Qing China. The traditional culture that shaped late Qing nationalism encompassed a broad political spectrum, including constitutional monarchism, revolutionary activism, democratic republicanism, and even anarchic socialism. Hence, the key question here concerns the preservation of certain cultural elements and which political order they bring about.

The neglect of the vast political possibilities which lie behind the politics of cultural preservation is the major problem relating to the concept of cultural conservatism and one that has prevailed for more than three decades. Although the above-mentioned recent studies have largely uncovered the blind spot with which this thesis is concerned and have also contributed to our understanding of Chinese conservatism as a movement imbued with strong political implication convictions, a variety of intellectual wellsprings, as well as different personal motivations, it still remains unclear what Chinese conservatives of the Republican era wanted to conserve. Is it possible to speak of Chinese conservatism as a political and intellectual movement that shared key values or is Chinese conservatism condemned to be defined by what it opposed, rather than what it stood for? Indeed, without a proper answer to this fundamental question, the concept of conservatism risks being given little thought to its actual meaning in the Chinese historical context. As a result, the vagueness of the term conservatism leads to the many contradictory and ambiguous notions which appear in existing literature. For example, while Tsui terms KMT's wartime National Spiritual Mobilization as a conservative movement, Clinton considers its policies to be highly anticonservative and radical.[54] Meanwhile, Du Chunmei examines Gu Hongming's loyalism towards the Qing dynasty after its collapse in the framework of ultra-

---

53 Shulamit Volkov, *The Rise of Popular Antimodernism in Germany: The Urban Master Artisans, 1873–1896* (Princeton: Princeton University Press, 1978), 321; J.S. McClelland, ed., *The French Right from de Maistre to Maurass* (New York, Evanston, San Francisco and London: Harper Torchbooks, 1970), 213.
54 Clinton, *Revolutionary Nativism*, 3–5.

conservatism, but one might wonder how she differentiates conservatism from reactionaryism.

To answer this question, this book seeks to investigate the Chinese conservatism of the Republican era by taking apart and rebuilding its foundations. During the revolutionary decade of the 1900s, prominent intellectuals supported social and institutional reforms on the basis of a reformulated traditional culture, albeit with different political implications.[55] I will call this insistence on maintaining the unity of the political and cultural orders "culturalist nationalism." Culturalist nationalism here refers to a nationalism based on the homogeneity of both the political nation and the cultural nation, while also implying that political reform should resonate with China's cultural tradition. This culturalist nationalism nourished both Chinese conservatism and radicalism at the end of the Qing dynasty. While conservative reformers remodeled Confucianism to rationalize constitutional monarchy, radical revolutionaries infused traditional culture with modern Western politics to promote the political changes that ultimately overthrew the Qing dynasty. This type of nationalism was the nationalism that prevailed in late Qing China.[56]

Hence, contrary to the customary thesis of cultural conservatism, the insistence on traditional culture as a means to politically establish the nation-state in late Qing China cannot be automatically considered a conservative position, much less culturally conservative. It is only after the foundation of the Republic, and more precisely during the May Fourth and New Culture Movement, that this culturalist nationalism was marginalized and rejected as conservative overall. What conservatives in Republican China stood for was essentially this late Qing ideal of culturalist nationalism that rejected squarely the museumification of traditional culture. Even among political conservatives affiliated with the Nationalist government and who did not hold obvious intrinsic investment in culture and traditions, such as Tao Xisheng (1899–1988) and Chen Lifu (1900–2001), there was a move to "look to a future reintegration of politics and culture as a consequence of the social process of modernization," arguing for the preservation of native traditions for foreign-imported political ideas to take root.[57]

---

[55] See, for example, Hon, *Revolution as Restoration*; Zarrow, *China in War and Revolution*, 53–74; Huters, *Bringing the World Home*, 1–122.

[56] Wang Ke, "Minzu: yige laizi Riben de wuhui" (Nation: A Misunderstanding from Japan), *Twenty-First Century* 77 (2003): 73–83.

[57] Furth, "Culture and Politics in Modern Chinese Conservatism," 47. For a detailed discussion of these two figures, see Eastman, "The Kuomintang in the 1930s," 199; Arif Dirlk, "T'ao Hsi-sheng: The Social Limits of Change," in *The Limits of Change*, 324–5.

Homogenization of the Cultural Nation and the Political Nation

Not unlike conservatism, nationalism comes in many shapes and sizes and is determined by "historically distinct cultural traditions, the creative actions of leaders, and the contingent situations within the international world order."[58] As such, theoretical interpretations of nationalism have been variable or even contradictory. On the one hand, Anthony Smith, for example, emphasizes the existence of a collective identity shaped by a longstanding culture which is passed down from one generation to the next.[59] On the other hand, contemporary debates on nationalism have been greatly influenced by Ernest Gellner or Benedict Anderson's constructivist theory that a nation is brought together by the influence of high culture or print capitalism bringing about the widespread use of vernacular language.[60] As I shall show in Chapter 2, Chinese nationalism in the late Qing and Republican era is a complex phenomenon rather than something single-handedly orchestrated by political parties or intellectuals; the "duty to participate" also playing a significant role in the modeling of national citizenship.[61]

Apart from the duality between primordial and constructivist theories, the dichotomy between ethno-cultural nationalism and political nationalism has also featured in literature on this subject. The origin of political nationalism is usually associated with the French Revolution and the idea of "civilization," from which the notion of *la grande nation* is derived. This form of nationalism is a left-wing ideology that emerged in the aftermath of the French Revolution as a response to imperial sovereignty and royalism.[62] Political nationalism is a voluntarist model of nationalism. It builds on the idea of political togetherness, in which people form a collective based on a shared political identity and objectives, which becomes a source of solidarity within a civil society. In contrast, German models of ethno-cultural nationalism stress blood, language, and historical culture as the criteria for membership of a nationalist grouping. John Hutchinson defines cultural nationalism as "a movement of moral regeneration which seeks

---

[58] Craig Calhoun, *Nationalism* (Minneapolis: University of Minneapolis Press, 1997), 127.
[59] Anthony D. Smith, *The Ethnic Origins of Nations* (Malden: Blackwell Publishing, 1988), 3–4.
[60] Ernest Gellner, *Nation and Nationalism* (Oxford: Basil Blackwell, 1983); Benedict Anderson, *Imagined Communities: Reflections on the Origins and Spread of Nationalism* (London: Verso, 1991).
[61] Steven E. Finkel and Karl-Dieter Opp, "Party Identification and Participation in Collective Political Action," *The Journal of Politics* 53, no. 2 (1991): 369.
[62] François Huguenin, *Le conservatisme impossible: libéralisme et réactionnaires en France depuis 1789* (Paris: Édition de La Table Ronde, 2006), 279.

to reunite the different aspects of the nation... [whose] proponents... are above all historical scholars and artists who form cultural and academic societies, designed to recover this creative force... and project it to the members of the nation."[63] Although these two models of nationalism and the nation are by no means exclusive to each other, the "culture" put forward by the German ethno-cultural model of the nation is often regarded as the conservative resistance to the French model. The French model of a political nation contains a universal idea of "civilization" which requires promoting French political ideals outside their national boundaries. The Napoleonic invasions which attempted to spread French ideas to the rest of Europe triggered a pride in German particularism, derived from an imagined idyllic past and the belief in a common descent.

The characteristic of the Chinese term "nation" is conveyed in its much more discriminating terminology. Like the term conservatism, the Chinese word for nation is also borrowed from Japanese. In both languages there are two words to express the idea of nationhood, *minzu* (in Japanese: *minzoku*) and *guomin* (in Japanese: *kokumin*), which can roughly be said to be the Chinese counterparts of the German and French models of nationhood. In traditional texts, *guomin* is a commonplace word and synonymous with *chenmin* (subject) and *shuren* (commoner). This word acquired its modern sense around 1899. In 1898, The Guangxu Emperor and his reformist-minded supporters, notably Kang Youwei (1858–1927) and Liang Qichao, initiated the Hundred Days' Reform to carry out far-reaching social and institutional changes. The Reform only lasted 103 days before it was suppressed by his aunt Empress Dowager Cixi, who was persuaded by traditionalist officials that the reforms were putting the dynasty on a path to destruction. Launching a *coup d'état*, the Empress Dowager put the Emperor under house arrest and executed several supporters of the Reform, including Tan Sitong (1865–1898). With the help of the Welsh Baptist missionary Timothy Richard (1845–1919) and the secretary of the Japanese embassy, Hayashi Gonsuke (1860–1939), Kang and Liang escaped respectively to Canada and Japan. After several months in Japan, the word *guomin*, in Liang's writings, took on the new meaning of "nation" and "citizen."[64]

The second word *minzu* also has its origins in China, but its usage was quite rare. Originally, it meant a patriarchal clan or was used to differentiate *hua*

---

[63] John Hutchinson, *Dynamics of Cultural Nationalism: The Gaelic Revival and the Creation of the Irish Nation State* (London: Allen & Unwin, 1987), 14.

[64] Shen Sung-chiao, "Jindai Zhongguo de "guomin" guannian, 1895–1911" (The Conception of "Guomin" in Modern China), in *Higashi Ajia ni okeru kindai shogainen no seiritsu* (Formation of Modern Concepts in East Asian), ed. Suzuki Sadami and Liu Jianhui (Kyoto: International Research Center for Japanese Studies, 2012), 191–2.

(China) from *yi* (barbarians, foreigners, outsiders).[65] While Lydia Liu has remarked that the word *minzu* was invented in Japan,[66] *minzoku* entered the Japanese vocabulary only in 1873.[67] However, it is true that the word did not find more popular usage in China before Chinese students and intellectuals, who were sojourning in Japan, introduced the word with its modern meaning to the Chinese public around 1898. This was at a time when cultural nationalism and the notion of "national essence" (*guocui*; in Japanese: *kokusui*) were booming in Japan.

The neologism of the term "national essence" first appeared in the magazine *Nihonjin* (*The Japanese*), the official body of the Society for Political Education (*Seikyōsha*) that the Japanese journalist and geographer Shiga Shigetaka (1863–1927) founded in 1888. Shiga invented this word to translate the English term "nationality." The rise of cultural nationalism, linked to the idea of national essence, was firmly connected to the failure of incorporating political nationalism as a component of Japanese citizenship during the first two decades of the Meiji era (1868–1912). As previously mentioned, integral to the French idea of nation is what Nobert Elias termed the "process of civilization" which embodies a colonialist and expansionist ambition.[68] However, for the 1870s generation of Japanese intellectuals, this idea of process of civilization was extremely attractive. Building a nation in accordance with this French line of nationhood was considered as indispensable if the aim was to climb the universal ladder of progress.[69] This led to the first Japanese expression of nationalism, which was a *kokumin*-nationalism stressing people's political rights in a democratic and constitutional regime.[70]

---

65 Fang Weigui, "Lun jindai sixiangshi shang de 'minzu,' 'nation' yu 'Zhongguo'" (On *Minzu*, Nation and China in Modern Intellectual History), *Twenty-First century* 70 (2002): 33; Hao Shiyuan, "Zhongwen 'minzu' yici yuanliu kaobian" (On the Origin of the Word "Minzu" in the Chinese Language), *Minzu yanjiu* 6 (2004): 62–3. How to translate appropriately the notion of *yi* is beyond the scope of this book; for a detailed discussion on this issue, see Lydia Liu, *The Clash of Empires: The Invention of China in Modern World Making* (Cambridge, MA. and London: Harvard University Press, 2004), 31–69.
66 Lydia Liu, *Translingual Practice: Literature, National Culture, and Translated Modernity – China, 1900–1937* (Stanford: Stanford University Press, 1995), 292.
67 Huang Xingtao, "'Minzu' yici jiujing heshi zai zhongwen li chuxian?" (When Did the Word *Minzu* Appear in the Chinese language?), *Zhejiang xuekan* 1 (2002): 170.
68 Norbert Elias, *La civilization des mœurs*, trans. Pierre Kamnitzer (Paris: Pocket, 2015), 107–8.
69 Kevin Doak, *A History of Nationalism in Modern Japan: Placing the People* (Leiden: Brill, 2007), 170.
70 Ibid.

For a variety of reasons, the 1890s witnessed the advent of *minzoku*-nationalism which progressively overwhelmed *kokumin*-nationalism. One of the most important ones is the collapse of the hopes of the progressive intellectuals for the 1889 Constitution, which gave no legal recognition of popular sovereignty and defined people as subjects.[71] Meanwhile, Japan's policy of westernization also alarmed conservative intellectuals. In an effort to come up with a nationalism independent of the state, but also capable of rejuvenating Japanese cultural tradition, *minzoku* – the alternative Japanese interpretation of nationhood which stresses the traditional cultural heritage and had been submerged by *kokumin* in the early Meiji period – reappeared.[72] Shiga's explicated the notion of *kokusui* appeared in this way:

> Everything shrouding the Japanese archipelago, sky, physiography, environment, climate, temperature, humidity, soil, water and land dispositions, mountains, rivers, animals, plants, natural views, influence of the atmosphere, scientific reactions, habits, incidents, experiences of hundreds of thousands of years... fostered, secretly and unfeelingly, in the Yamato people a special nationality (*kokusui*).[73]

The upsurge of *minzoku*-nationalism does not mean that political ideals in themselves were absent. To Shiga, the Japanese national character could not be easily changed since it was identified with something immutable. Thus, the central issue of this new culturalist nationalism was how to establish a political system in line with Japan's cultural tradition. The founder of the nationalist magazine *Nippon (Japan)* and member of the Society for Political Education Kuga Katsunan (1857-1907) proclaimed: "If a nation wishes to stand among the great powers and preserve its national independence, it must strive always to foster nationalism... If the culture of one country is so influenced by another that it completely loses its own unique character, that country will surely lose its independent footing."[74] In Kuga's original text, nationalism is translated as *kokumin shugi*. Nevertheless, the word *kokumin* here is reminiscent of the persistent argument of the Society for Political Education that cultural self-knowledge was the first step towards political betterment and strength.[75] In this ideological context, *minzoku*

---

71 Ibid., 176.
72 Ibid., 194.
73 Shiga Shigetaka, "Nihonjin ga kaihōsuru tokoro no shigi wo kokuhakusu" (Declaration of the Doctrine Preconized by *The Japanese*), *Nihonjin* 2 (1888), 1.
74 Kenneth B. Pyle, *The New Generation in Meiji Japan: Problems of Cultural Identity, 1885–1895* (Stanford: Stanford University Press, 1969), 75.
75 Ibid.

and *kokumin*, which respectively favored the cultural and political notions of nationhood, became interchangeable.

The influence of this Japanese culturalist nationalism among late Qing Chinese intellectuals cannot be emphasized too much. Chinese students and intellectuals began to seek exile or study in Japan during the last few years of the 1890s. This was a period in which the homogenization of the cultural nation and the political nation prevailed in Japan and penetrated into the Chinese political vocabulary, and it was exactly this idea of the nation as a culturally and politically unified entity that was picked up by Chinese intellectuals in Japan after 1898. Liu Shipei (1884–1919) wrote in 1903 that "*minzu* is the unique character of *guomin*."[76] This culturalist nationalism carried two political imperatives in the 1900s. Chinese culturalist nationalism first emerged as both a doctrine of popular sovereignty and a movement for a Chinese nation-state independent from external oppressors. It follows that the people who formed the Chinese nation-state should be ethnically and culturally identical. Culturalist nationalism also implied that political reforms should be undertaken in the context of preserving selectively updated traditional culture. China could only hope to have a unified and integrated modern nation-state if she borrowed wisely from different political elements within Western civilization.

But the essential point remains that culturalist nationalism reveals a complex exchange between opposing ideas. Conservative reformers and radical revolutionaries thought of culturalist nationalism in the same way. Monarchist reformers in China and in Japan expressed a conservative culturalist nationalism to legitimize the Qing dynasty. They argued that the Manchu had been assimilated into the Chinese nation and began reinterpreting traditional sources to legitimize a constitutional monarchy. In 1898, Liang Qichao, who escaped the Empress Dowager Cixi's purge but continued to press for reform from Japan, even communicated in classical Chinese with Shiga Shigetaka, advisor to the Minister of Foreign Affairs at the time. This was in the hope of obtaining help from the Japanese government to restore the Guangxu Emperor to power.[77] In contrast, revolutionary intellectuals linked with Sun Yat-sen's (1866–1925) Tongmenghui (Revolutionary Alliance), such as those of the National Essence School and the Southern Society, associated national essence with Chinese history. This meant that China's ancient socio-political system and cultural traditions

---

76 Wuwei (Liu Shipei), "Huangdi jinian lun" (On Calculating Years According to the Yellow Emperor), in *Xinhai geming qianshinianjian shilun xuanji*, Vol. 1, 721.
77 Ding Wenjiang and Zhao Fengtian, *Liang Qichao nianpu changbian* (A Chronological Biography of Liang Qichao) (Shanghai: Shanghai renmin chubanshe, 1983), 159–162.

would be seen to comply with radical political reforms inspired by modern Western politics and to rationalize revolution. As Zhang Taiyan remarked, national essence served to "incite ethnic nationalism, promote patriotism and fabricate a history of the Han."[78] Revolutionary culturalist nationalism was forged to fit a republican political construct. To this end, they even borrowed from the French orientalist Albert Étienne Jean-Bapiste Terrien de Lacouperie's (1844–1894) theory that the Yellow Emperor was a native of Mesopotamia.[79] In this way, revolutionary culturalist nationalism could restore ancient Chinese culture while also allowing it to share the same ancestry with Western political culture. The political radicalism of the revolutionary intellectuals was refreshed by a bold reinvention of tradition, which is where the weakness in the theory of cultural conservatism lies. The close connection between political and cultural radicalism clearly distinguishes the revolutionaries' culturalist nationalism from that of the conservative reformers, as well as the early promoters of the idea *zhongti xiyong* (Chinese teaching as substance, Western teaching as application) in which Confucian values would assimilate Western technological advances to strengthen imperial rule. Hence, contrary to what Furth suggests, revolutionaries were radical at both the political and cultural levels.

From Culturalist Nationalism to Nationalist Conservatism

Yet this culturalist nationalism, which informed both late Qing conservatism and radicalism, was universally rejected as conservative by the supporters of the New Culture Movement during the May Fourth period. The New Culturalists blamed late Qing revolutionaries for lumping together the nation, politics, and culture.[80] For them, the efforts of late Qing intellectuals to accommodate traditional culture and modern politics were counterproductive for China's revolution and modernization, while attempts to preserve this culturalist nationalism came to be belittled as conservative. Their aversion to late Qing culturalist nationalism is directly linked to the Republic's institutional failure to endorse political reform based on traditional culture.

Although the Republic of China was proclaimed in 1912 in Nanjing, the Revolution of 1911 failed to deliver China's long-awaited national revival. In the aftermath of the Revolution, the Nanjing government offered the presidency to the

---

78 Zhang Taiyan, "Yanshuo lu" (Speech), *Minbao* 6 (1906).
79 Lin Xiaoqing, "Historicizing Subjective Reality: Rewriting History in Early Republican China," *Modern China* 1 (1999): 30–1.
80 Chevrier, "Antitradition et démocratie dans la Chine du premier XXe siècle," 427.

Prime Minister of the Imperial Cabinet Yuan Shikai (1859–1919). This new role was on the condition that he arranged the abdication of the young Xuantong Emperor, Puyi (1906–1967), who was still holding on to his throne in Beijing. After his inauguration as president, Yuan moved the capital of the Republic to Beijing and ruled China from 1912 to 1916 as a military dictator.[81] He planned to have himself declared emperor of the Chinese Empire in December 1915 but died suddenly in 1916. Although the Empire was never established due to protests throughout China and Yuan's sudden death, multiple attempts were made in the following years to restore the monarchy. Qing loyalist Zhang Xun's (1854–1923) army seized Beijing in July 1917 and restored the Qing Court. Republican troops launched an assault and quickly ended this coup, which lasted only 12 days, but Qing loyalists never ceased to campaign for restoration.[82] Furthermore, despite Yuan's failed attempt of restoration, he turned out to be the last strong charismatic leader capable of holding China together. In the following decades, China lacked a strong central government, with political power passing into the hands of local warlords.

Arguably one of the most important events that shaped the Chinese political and intellectual landscape during the Warlord Era (1916–1928) is the May Fourth Movement. In 1917, the Duan Qirui (1865–1936) government declared war on Germany and the Austro-Hungarian Empire. By participating in the First World War on the side of the Allied Powers, the government hoped to recover the Kiautschou Bay Leased Territory on the Shandong Peninsula, which had been occupied successively by Germany and Japan.[83] However, the Treaty of Versailles, concluded in April 1919, awarded this territory to Japan, rather than returning it to China. In response to the Chinese delegation's weak response to the treaty, students at several universities in Beijing organized a nationwide protest. This patriotic protest led to the New Culture Movement of the mid-1910s to 1920s, which signified a "wide-ranging and radical moment of soul-searching" among the students and intellectuals of the May Fourth period.[84] Anti-imperialist nationalism turned from political protests towards intellectual activities which probed into the failure of the Republic and how it produced such a weak government which was unable to defend the nation's interests. To restore the country, the New Culturalists

---

[81] Klaus Mühlhahn, *Making China Modern: From the Great Qing to Xi Jingping* (Cambridge, Mass.: The Belknap Press of Harvard University, 2019), 228.
[82] Lin, *Minguo nai diguo ye*.
[83] Xu Guoqi, *China and The Great War: China's Pursuit of a New National Identity and Internationalization* (Cambridge: Cambridge University Press, 2005), 248.
[84] Sebastian Veg, "Lu Xun and Zhang Binglin: New Culture, Conservatism and Local Tradition," *Sixiangshi* 6 (2016): 153.

launched the New Culture Movement with a determined effort to introduce Western values and institutions in order to abolish the remnants of the old political system, along with the traditional culture that sustained the traditional order in China.

The "soul-searching" of the May Fourth Movement headed towards the creation of a "New Culture," which was in essence a kind of scientific and democratic enlightenment. While revolutionary intellectuals in the 1900s had also honored science and democracy and denounced the authoritarian elements of Confucianism, the anti-Confucianism of the May Fourth period led to what Lin Yü-sheng calls a "holistic mode of thinking" that rejected wholesale all traditional learning.[85] In this ideological context, national essence – and building on that concept, culturalist nationalism – was denounced as contributing to the stagnation of the nation and the failure of the Republic. In an article published in the *New Youth* (*Xinqingnian*), Qian Xuantong (1887–1939) announced his iconoclasm in these words: "Yuan Shikai's restoration resounds in my mind like a clap of thunder with an awareness that the national essence should be absolutely abandoned."[86] Qian referred here to Yuan's restoration of Confucianism as the state ideology to legitimize his ambition to become sole emperor. In his article, Qian appealed for young Chinese "to be civilized men of the twentieth century and to become new citizens of the Republic of China."[87] Qian's hostility towards the idea of national essence and his hope for enlightened citizens reflect the vision he had in common with his fellow activists – a vision that to save China as a political entity, traditional culture had to be entirely replaced by a New Culture aligned with modern Western civilization. By the same token, Gao Yihan (1885–1968) noted that the Revolution of 1911 "was won through nationalism, and not republicanism, thus, although the emperor renounced the throne, he has not abdicated yet in people's mind."[88] In other words, the foundation of the Republic was an incomplete project in the sense that it failed to uproot traditional culture sustaining the previous political order. Lu Xun (1881–1936) even compared national essence to a "facial abscess" – "although special, it is better to remove it."[89]

---

[85] Lin Yü-sheng, *The Crisis of Chinese Consciousness: Radical Antitraditionalism in the May Fourth Era* (Madison: University of Wisconsin Press, 1979), 29.
[86] Qian Xuantong, "Baohu yanzhu yu huanhui renyan" (Protecting Eyeballs and Regaining Vision), *New Youth* 6 (1918).
[87] Ibid.
[88] Gao Yihan, "Fei junshi zhuyi" (On Not Having the Ruler Also Serve as the Teacher), *New Youth* 6 (1918).
[89] Tang Si (Lu Xun), "Suigan lu" (Random Thoughts), *New Youth* 5 (1918).

The horrors of the First World War also explained the New Culturalists' loathing of culturalist nationalism. Wary of the way nationalism was prone to fall under the sway of folly, Chen Duxiu (1879–1942) showed a distrust of nationalist ideology that he denounced as an idol to be destroyed.[90] Chen and many New Culturalists' hostility towards nationalism does not imply that they did not stand for the independence and self-determination of the Chinese nation-state. Chen urged his fellow citizens to assume political responsibility and devote themselves to national causes in an article published several months after the breakout of the First World War. However, he pointed out that for citizens to sacrifice themselves for their nation, the only justification would be when the nation acted as the guardian of their rights and well-being.[91] This was not the case for the newly-established Republic which failed to deliver on the promise of national regeneration. This, according to Chen, was the direct consequence of the persistence of the "backward" national essence.[92] Like his fellow New Culturalists, Chen blamed the Republic's institutional failure squarely on traditional culture. The political nation and the cultural nation had to be dismantled to redeem China. Any attempts to preserve the culturalist nationalism were belittled by mainstream intellectuals as conservative and counterproductive for the nation. Liu Boming (1887–1923), a member of the conservative Critical Review Group, lamented how the New Culturalists disparaged the cultural nation which they judged to be conservative.[93] Contrary to the conservatives' attempts to insert certain elements of traditional culture into political reforms, the New Culture activists argued that modern political systems could only be strengthened upon the museumification of the traditional culture. This had the meaning of transforming traditional culture as if from a living reality and creative force to a relic in a museum, wherein tradition was not considered for its political uses but for its value as an artifact.[94]

---

90 Chen Duxiu, "Ouxiang pohuai lun" (On Smashing Idols), *New Youth* 2 (1918).
91 Chen Duxiu, "Aiguoxin yu zijuexin" (Patriotism and Self-consciousness), *Jiayin* 4 (1914).
92 See for example Chen Duxiu, "Jiu sixiang yu guoti wenti" (Old Thought and the National Question), *New Youth* 3 (1917).
93 See He Fangyu, *Kexue shidai de renwen zhuyi: Sixiang yu shidai yuekan (1941–1948) yanjiu* (Humanism in the Era of Science: A Research on the Magazine *Thoughts and Epoch* (1941-1948)) (Shanghai: Shanghai shudian, 2008), 35.
94 Luo Zhitian, "Songjin bowuyuan: Qingji minchu quxin renshi cong 'xiandai' li quchu 'gudai' de yingxiang" (Sending the Tradition to the Museum: The Tendency of the Proponents of Novelty to Push the Tradition out of Modernity) in Luo Zhitian, *Liebian zhong de chuancheng – 20 shijiqianqi de Zhongguo wenhua yu xueshu* (Inheritance Within Rupture: Chinese Culture and Scholarship in the Early 20th century) (Beijing: Zhonghua shuju, 2003), 91–130.

Nevertheless, the New Culturalists did not silence the conservatives. However, in opposition to the New Culturalists, conservatives faced the difficult challenge of differentiating themselves from royalists and, later, being accomplices to the KMT's authoritarian rule. After Sun Yat-sen's death in 1925, his successor Chiang Kai-shek (1887–1975) launched the Northern Expedition (*beifa*) that subjugated all warlords and unified the nation by 1927. The capital of the Republic was relocated to Nanjing. As the Nanjing decade began, the state building and national reconstruction continued until the outbreak of the Second Sino-Japanese War in 1937. While the Nationalist government gained a considerable number of solid achievements for the economy, infrastructure, and industry,[95] the Nanjing decade was also a period of "political tutelage," in which the expansion of the state prevented the voicing of political opposition.[96] The repressive nature of the Nationalist government was made clear during the New Life Movement. This was orchestrated by Chiang in 1934 so as to militarize China and permeate society with Confucian morality following the steady campaign of Japanese aggression and the growing influence of the CCP. The Nationalist government's manipulation of Confucianism for political ends further damaged the reputation of the conservatives.

Nevertheless, although some did collaborate with the Nationalist government in the late 1930s, conservatism should be understood pluralistically. The interpretation of the multifaceted phenomenon of traditional culture differed from one intellectual to the next. Since the conservatives' political opinions and beliefs were often expressed through complex cultural discourse, the policies that they wanted to implement cannot be easily reduced to simply imperial restoration or authoritarianism. In fact, the diversification of conservative thought had occurred immediately after the foundation of the Republic. While late Qing reformers and revolutionaries all regarded traditional cultural ideas in the light of modern Western political institutions, albeit with different interpretations, the outbreak of the First World War led many to question the merits of Western civilization and to search for alternatives. Even though conservatives continued to defend the politics of culturalist nationalism, the re-appropriation of traditional culture to evoke cultural elements compatible with modern Western political institutions was no longer unproblematical. By differentiating different types of conservative thought, this book will capture – for the first time –

---

[95] See, for example, Paul K.T. Sih (ed.), *The Strenuous Decade: China's Nation-Building Efforts, 1927–1937* (Jamaica, NY: St. John's University Press, 1970).

[96] See, for example, Parks M. Coble, Jr., *The Shanghai Capitalists and the Nationalist Government, 1927–1937* (Cambridge, Mass.: Council on East Asian Studies, Harvard University Press, 1986).

how Chinese conservatism was in constant evolution, while also showing how its emblematic figures reacted differently to historical circumstances. The challenge now is to strike a balance between excessive homogenization, like that of the "cultural conservative," and an undue heterogenization, which makes each intellectual an isolated case.

Chapter Overview

This book investigates the conservative movement in Republican China by grappling with the expression and evolution of its ideological roots. It will draw on an abundance of Chinese and Japanese textual sources from the 1890s to the 1940s, including newspapers, journals, magazines, and periodicals published by intellectual associations. Making use also of official documents as well as personal diaries of major players, this book develops a practice-based approach to intellectual history by looking into the dynamic intellectual, social, and political interactions that shaped the birth and evolution of conservative thought. Changes in patterns of thought, personal motives, and the semantics which emerged out of intellectual debates, individual political trajectories, and evolving socio-political contexts will all be analyzed. In this way, the book views the Chinese conservative movement as a series of discourses and organizational practices, rather than mere cultural thoughts without practical meaning. Structured thematically and chronologically, the following chapters trace the origins of conservatism in Republican China and how it developed and diversified. Together they reveal the complex strategy by which conservatives attempted to establish stable alternatives to the New Culturalists' visions of liberalism and socialism.

Chapter 1 aims to clarify existing notions of traditionalism, loyalism, and reactionaryism, in order to better capture the meaning of conservatism. This chapter adopts a unifying approach to capture the relative meanings of these political concepts to each other by examining their significances in different historical settings and intellectual backgrounds. It argues that conservatism sets out to preserve the present by holding on to the past, although not uncritically. Understood in this way, conservatism now appears to be like a web of communication that develops theoretically and strategically from specific historical circumstances and should not be confused with obscurantism, immobilism, or traditionalism. It contends that before the revolutionary forces emerged, prompting reformers to embrace monarchism and seeking to re-establish the legitimacy of the imperial order in the late 1890s, opposition to the late Qing reforms, from the Self-Strengthening Movement of the mid-nineteenth century to the New Policies of the early 1900s, was not rooted in conservatism, but in traditionalism. The hostility towards Westernization and denial of the country's backwardness in relation to the outside world sprang from the conviction that China was the civilization, which had nothing to learn from the West. Thus, it reveals a psychological traditionalism, rather than the type of conservatism that accepts reality and reacts accordingly.[97] As to reactionaryism, the book refers to it as a political attitude or course of action that favors a return to a previous political state. Different from conservatives who "want to preserve the present even against the past," reactionaries "want to break with the present and return to a version of the past that omits all the aspects they dislike."[98] As such, the political system cherished by reactionaries is always fatally obsolete. Although reactionaries and conservatives shared the disappointment of the abortive republican revolution of 1911, conservatives squarely rejected imperial restoration to ensure order. Also, political references to radicalism in the late Qing ideological context require a more nuanced discussion. While Kang Youwei and Liang Qichao were conservative reformers during the 1900s, their radicalism was reflected in their bold remodeling of Confucianism, turning Confucianism into a state religion and reinterpreting it so as to justify constitutional monarchy during the Hundred Days' Reform and the New Policies. While they shared a similar cultural radicalism to the revolutionary intellectuals, politically they advocated the preservation of the Manchu dynasty. Through this problematization of the notion of "radicalism," the chapter underlines the argument that it is not the preservation of cultural heritage that determines one to be a conservative; the key question is what type of political order would be produced in this process. This issue will be discussed in depth in the next chapter.

Chapter 2 surveys the role played by culturalist nationalism in the institutionalization of the Chinese nation-state during the anti-Manchu revolution from the late 1890s to the 1900s. In the aftermath of the *coup d'état* initiated by the Empress Dowager Cixi, monarchist reformers, notably Kang Youwei and Liang Qichao, fled overseas to pursue their campaign for reform. In Japan, reformers continued to defend the cause of constitutional monarchy and the legitimacy of the Qing dynasty against anti-Manchu revolutionaries associated with the Tongmenghui. Both parties, albeit with opposing political views, developed the notion of national essence as a means to establish the nation-state. With the help of written tracts, the revolutionaries created a Han-centric national identity

---

[97] For the differences between traditionalism and conservatism, see Karl Mannheim, *La pensée conservatrice*, trans. Jean-Luc Evard (Paris: Édition de la revue Conférence, 2009), 34–48.
[98] Eugen Weber, "Ambiguous Victories," *Journal of Contemporary History* 13 (1978): 823.

to rationalize a revolution against the Manchu dynasty. Traditional teachings, other than that of Confucianism, were rekindled and merged with modern Western political themes such as democracy, liberty, equality, constitutionalism, federalism, shared power, and parliamentarianism. Confucianism was denounced by many as a remnant of the past that did not represent China's true national essence. Its oppressive ethical code and practices, such as the "three principles and five virtues" (*sangang wuchang*), would have to be abandoned in order for China to become socially enlightened. This was considered indispensable for an anti-Manchu republican revolution. Monarchist reformers in China and Japan used the same notion of national essence as the revolutionaries in order to assert a conservative culturalist nationalism and defend the Qing dynasty by implementing reforms. This chapter explores the confrontation between conservative and radical culturalist nationalisms against the backdrop of evolving political events during the 1890s and 1900s.

Chapter 3 follows the historical development through which culturalist nationalism, which sustained both conservatism and radicalism at the end of the Qing, was indiscriminately denounced as conservative in Republican China and, more specifically, during the May Fourth and New Culture Movement. While culturalist nationalists believed in the flexibility of tradition, which was necessary for educating the masses and guiding the nation's quest for political reform, the New Culturalists ruled out political beliefs which were based on the reformulation of traditional culture. The ongoing efforts of culturalist nationalists to accommodate traditional culture and political reforms came to be rejected as conservative and inconsequential. The New Culturalists' aversion to traditional culture is certainly related to Yuan's regime, under which attempts to restore the cult of Confucius took place to justify his restoration in 1915. However, my account of this struggle between traditional culture and new culture does not draw only upon the New Culturalists' progressive political and cultural views, but also Pierre Bourdieu's theories of the cultural field to show how this struggle is inseparable from the development of "professional intellectuals," when educated elites could no longer take their authority for granted. Neither the New Culture nor traditional culture brought practical solutions for socio-political change; they merely represented the New Culturalists and the conservatives' respective and exclusive politico-cultural visions. The struggle between old and new was thus driven by the ambition to dominate the intellectual field, in order to gain legitimacy and to disseminate the exclusivity of their ideas as being in the general interest. Forster's monograph presents the story of the New Culturalists as astute entrepreneurs and self-promoters whose dominance of the intellectual landscape was not totally due to intellectual merit.[99] It follows that once they fell into disgrace, conservatives lost the battle of making culturalist nationalism a universal idea acting in the general interest. However, the New Culturalists' successful domination of the intellectual field did not exclude conservatives from the realm of political discourse. Instead, in response to the continuing political crisis and the First World War, a considerable number of conservatives were led to question the merits of the West they had so admired during the revolutionary decade. This in turn led to the meanings, attributed to traditional culture and shaping culturalist nationalism, being diversified. In this regard, conservatism in Republican China should be understood in pluralistic terms. I divide the Chinese conservatism of the Republican period into four categories, which I will describe in the next two chapters.

Chapter 4 analyses two types of conservatism which existed from the 1910s to the 1930s: liberal conservatism and anti-modern conservatism. I refer to liberalism in the Chinese context from both the institutional and social perspectives. At the institutional level, liberal conservatives were liberal in their outspoken opposition to despotism and their efforts to implement constitutional parliamentarianism. They were at the same time conservative in their beliefs that separation from the nation's cultural traditions would make a Chinese liberal regime impossible. Despite their different outlooks in terms of electoral democracy, presidential power, and centralization, all liberal conservatives refused to blame the nation's cultural tradition for China's inability to adopt a less authoritarian regime. On a social level, liberal conservatives argued that freedom was not simply a personal right, but also closely interwoven with social responsibility. In contrast to the New Culturalists' denunciation of traditional morality as a cannibalistic machine, liberal conservatives believed that if the Republic was to achieve stability, traditional values should be modernized selectively to allow for individual freedom, which would be fully integrated into the overall struggle for national liberation. Contrary to liberal conservatives, antimodern conservatives urged China to reject the Western political system and return to the social and political arrangements consistent with the nation's agrarian cultural tradition. During the May Fourth period, modernity was associated with the ideals of the Enlightenment and its ideas of linear progress and reason. New Culturalists established Chinese traditional culture as being in opposition to modernity and viewed critically the nation's past through the lens of Western modern values. For antimodern conservatives, the First World War signaled the bankruptcy of the concepts of linear evolution and progress, as well as industrialization, capitalism, and political in-

---

99 Forster, *1919 – The Year that Changed China.*

stitutions associated with modernity. Antimodern conservatives believed that Chinese agrarian and communitarian political culture was more than capable of offering a solution to the flaws they perceived in the Western social and political model.

Chapter 5 narrates the story of the conservative movement from the 1920s to the 1940s. By the 1920s, the New Culturalists had established their charismatic domination in the intellectual sphere and New Culture had become a "buzzword" with which political and cultural programs had to be identified if they were to be considered legitimate.[100] Nevertheless, the New Culturalists did not silence the conservatives and Chinese conservatism of the 1920s remained politically active. However, the conservatives' politico-cultural discourses during this period were increasingly limited to a defensive position. It is in this context that philosophical conservatism emerged. Its cultural ideals remained purely at the philosophical level. Unlike the liberal and antimodern conservatives, philosophical conservatives never conceived a political project or institution to transform their cultural vision into reality. Instead, confronted with the New Culturalists' rhetorical hegemony, philosophical conservatives devoted much of their efforts to contesting the New Culture and its related social, political, and cultural reformist ideas, such as the revolution of the family, populist democracy, and replacing the classical language with the vernacular as the national language and literature, the socialist inclinations of a considerable number of the New Culturalists by the early 1920s, as well as the Anti-Christian Movement, which led to a general attack on all sorts of religions or popular beliefs, including Confucianism and Buddhism from 1922 to 1927. Philosophical conservatism's culturalist nationalism was structured around the Confucian values of elitism, individual morality, social hierarchy, and the doctrine of the moderation. They believed that these elements must be reintegrated into society to correct the "excessive" individual liberty, corrupt politics, capitalism, industrialization, and the emerging socialism of the 1920s. The most representative philosophical conservatives were those of the Critical Review Group, an intellectual association whose ideas will be closely analyzed in this chapter. While philosophical conservatives never translated their values into a clear political program, such values could be easily taken to extremes during wartime, when bold leadership was called for to handle the imminent threats that confronted the nation. From the 1930s onwards, Confucian values held dear by philosophical conservatives were refashioned to accommodate authoritarian conservatism – the only state-endorsed conservatism in Republican China. The culturalist nationalism of authoritarian

---

100 Ibid., 195.

conservatives was influenced by the most repressive elements of Confucianism, such as the supremacy of the leader, absolute obedience to one's superiors, and political tutelage. These Confucian values were propagated by the KMT government and fully integrated during its fascistic rule.

I will conclude with a reflection on the importance of conservative thoughts in the wider social and political conditions from the late Qing period to the end of the KMT's rule in mainland China. A final thought will be given to the latest manifestations of Chinese conservatism after China split into mainland China and Taiwan in 1949. I will first survey the so-called New Confucianism that first developed in Hong Kong and Taiwan as a "philosophy of mind" designed to harmonize Confucianism with industrialization and democratization. The relationship between this Confucian conservatism and Chiang's rule in Taiwan will then be discussed. Finally, I will look into modern Chinese conservative thought, notably New Authoritarianism and the Chinese New Confucianism that appeared respectively in the 1980s and 1990s, and its connection to the CCP government.

1 Conservatism: An Idea Rooted in the Present

If understood simply as a mental disposition inherent in customary and contingent human nature, rather than a political ideology, conservatism risks becoming a sour language of fear towards changes, while traditions dear to conservatism risk being regarded as obsolete. As a consequence, the term conservatism suffers from identity issues that require some scrutiny. Indeed, conservatism is often mistaken for other political notions, movements or doctrines that are radically different from it, yet considered to be its synonyms, namely traditionalism, reactionaryism, royalism, and loyalism.

In different spatial and historical contexts, these seemingly related political labels have all been self-proclaimed or attributed, with fundamentally different political connotations. "Loyalism," for instance, is a telling case in point. In 1902, Zhang Taiyan organized the *"Zhina wangguo erbai sishi'er nian jinianhui"* (Commemoration ceremony for the two hundred and forty-second anniversary of the destruction of China) in Tokyo. He initially wished to hold the ceremony on the same day the last emperor of the Ming dynasty, Chongzhen (1611–1644), hung himself on Coal Hill to avoid capture, when Li Zicheng (1606–1645) and his army seized the Forbidden City.[1] Li was a rebel leader who only ruled over Northern China briefly, before the Manchu forces captured Beijing in 1644 and proclaimed to have received the Mandate of Heaven to rule China. The death of Chongzhen marked the end of the last Han Chinese regime. When the Japanese police arrived at the scene to interrupt the event and asked Zhang which province of the Qing he came from, he responded that he was not a subject of the Qing dynasty but was Chinese (*Zhina ren*) and a loyalist (*yimin*) – one of the "left over persons" of the Ming period. Loyalism here refers to the revolutionaries' allegiance towards the overthrown Chinese government and harbors a defiant attitude towards the Manchu dynasty. After the foundation of the Republic in 1912, loyalism became a synonym for reactionaryism and royalism, the adherents of which refused – at least for a relatively long time – to cut their Manchu queue and continued to press for Puyi's restoration. During the New Culture Movement, loyalism, with this pejorative connotation, was foisted upon conservatives by some New Culturalists to silence their intellectual opponents.[2]

---

[1] Tang Zhijun, *Zhang Taiyan nianpu changbian* (A Chronological Biography of Zhang Taiyan) (Beijing: Zhonghua shuju, 1979), 134.

[2] See for example Liu Yazi, "Wo he Nanshe de guanxi" (My Relationship with the Southern Society), in *Nanshe Jilue* (History of the Southern Society), ed. Liu Wu-chi (Shanghai: Shanghai renmin chubanshe, 1983), 102.

The erroneous equivocations that bear upon the notion of conservatism do not constitute a phenomenon peculiar to China. In France, the heritage of the French Revolution is so inscribed in the political process that French conservatism is usually considered to assimilate the reaction rooted in the counterrevolution, reactionaryism, immobilism, and traditionalism of which Joseph de Maistre (1753–1821) and Louis de Bonald (1754–1840) are often regarded as architypes.[3] The history of modern Chinese conservatism is also somewhat unique in the sense that conservatism, as a neologism at the turn of the twentieth century, was rarely evoked as a political ideology in late Qing and Republican China. This does not mean that conservatism as a political movement did not exist at that period. Rather, the term itself was stigmatized by accusatory abuse in an era that venerated newness and progress; even conservatives did not seek to ascribe to this political label and strived to be involved in the mainstream intellectual sphere that sullied the use of this term. Hence, if conservatism is not de-demonized and theoretically and empirically differentiated from its seemingly related but in fact distinct concepts, conservatism will be condemned to remain an intellectual current and political movement, the existence of which is commonly believed to be obvious, while also being associated with confusion.

Another problem that arises from the term conservatism is its relationship with its most directly opposed political view. In every political contest, conservatism is usually believed to denote an irreconcilably opposed worldview of radicalism. The former emerges in terms of oppositional politics that attempts to preserve deeply rooted historical connections under threat from modernity and only salutes incremental change, while the latter seeks an overarching transformation blueprint for social and political lives. The political languages of conservatism and radicalism in interwar Europe and Latin America are, however, far from being incommensurable. Francoist Spain is an interesting example. In Brian Tsui's study of the KMT's "conservative revolution," from the victory of the Northern Expedition to the CCP's seizing of mainland China, he highlights various contradictions in the KMT's project: participating in the United States-led international order, while claiming to confront Western colonialism; holding on to the goal of universal harmony yet insisting upon the maintenance of traditional hierarchical relations, or relying on spontaneous, individual self-discipline as a means of reinvigorating society but featuring mass organizations in

---

[3] See for example Jean-Phillipe Vincent, *Qu'est-ce que le conservatisme? Histoire intellectuelle d'une idée politique* (Paris: Les belles lettres, 2016), 22–7.

its revolutionary strategy.⁴ To the author, conservative revolutionaries were certainly not models of consistency.⁵

Indeed, conservatism has long had two faces – its ideological foundations and its practice. This is not to say that conservatives are hypocritical or oblivious to the contradiction. When conservatives became radicals during the interwar period, historical complexity attempted to explain the strange entente between conservatism and radicalism. However, it should be emphasized that without establishing ideas in the political and cultural spectrum, "conservatism" and "radicalism" are both too vague and elusive to enable a rigorous analysis when applied to empirical cases. The American Socialist Daniel Bell once self-identified as "a socialist in economics, a liberal in politics and a conservative in culture."⁶ While I disagree with Charlotte Furth's view that late Qing revolutionaries were politically radical and culturally conservative for the reasons stated in the introductory chapter, her words hint that intellectuals of late Qing and Republican China can be described in a similar way using a variety of combinations. Without relating conservatives to other intellectual and political currents or movements to which they responded or in which they positioned themselves, building up a history of conservatism risks being ahistorical. It is therefore necessary to explore how conservative and radical political-cultural lenses shaped views about important issues and how they shaped the views of one another.

Traditionalism

Traditionalism is probably the most "meaningful" utterance of the conservative standpoint, since conservatism, after all, is about retaining what is valuable and useful in tradition. However, when the preservation of tradition becomes a stiff doctrine of traditionalism, it departs from the conservative political value of belief during the piecemeal change through the time-honored wisdom of the past. Jean-Philippe Vincent distinguishes traditionalism from conservatism with these words:

> Traditionalism is a nostalgy, more or less structured, by the past. Conservatism does not constitute in any case a nostalgy, because it is deeply rooted in the present and one of its intellectual fortes is to perpetuate the best of the tradition by imposing reforms. Besides, conservatism attaches an essential importance to the History, but not to a past idealized or

---

4 Tsui, *China's Conservative Revolution*, 246–7.
5 Ibid., 247.
6 Malcolm Waters, *Daniel Bell* (New York: Routledge, 1996), 19.

painted in a holy way. Especially, it is the relation to tradition that separates conservatism from traditionalism. For the latter, tradition is usually rigid: it is stopped, or fixed in whichever moment of the History according to traditionalists' preferences. For conservatives, tradition is a living thing: it constitutes a way to live in the present, which contributes to the enrichment of tradition in constant formation.[7]

While it may not be totally plausible that conservatism does not encapsulate a nostalgic feeling towards tradition at all, conservatism does not embody an overall rejection of change, neither does it mean re-invention so much as rediscovery: a return to the sources may be necessary, with fresh consideration as to how they might be applied to current problems. "Present" is indeed the keyword to understanding conservative political and moral thinking.[8] Vincent's argument echoes Karl Mannheim's classic studies on conservatism. Mannheim found in traditionalism a universal psychological attitude in most people for most of the time, which translates into an instinctive reliance on the past and a revulsion towards change that constitute a purely reactive act.[9] In this regard, Mannheim's traditionalism resembles the conservative mindset theorized by Michael Oakeshott. Conservatism, by contrast, is an articulation of the past that provides meaningful orientation in a specific situation and designates "a continuity that... appears in a very precise sociological or historical situation, [which] develops with what is historically vibrant in an immediate solidary way."[10] While traditionalists may not be self-conscious of their reactive mentality, conservatism is a mental structure endowed with an objective existence, as compared with "the lived experience *hic et nunc* of a particular individual."[11] Although this mental structure cannot be regarded as independent of the individuals who initiated it, since its development is entirely dependent on the fate and the spontaneity of the latter, this structure can never be produced by a single and isolated individual.[12] It follows that conservative politics admits and accepts diverse objective realities and is constantly transforming itself to suit the world. To this end, tradition and historical experiences are subjected to scrutiny, reflection, and selection.[13]

Conservatism is context-specific. It is not likely to be replicated in its exact form under other political, social, and temporal circumstances. In contrast, the

---

7 Vincent, *Qu'est-ce que le conservatisme?*, 23–4.
8 Weber, "Ambiguous Victories," 823.
9 Mannheim, *La pensée conservatrice*, 38.
10 Ibid., 39.
11 Ibid., 35–36.
12 Ibid., 36–7.
13 Vincent, *Qu'est-ce que le conservatism?*, 24.

traditionalist reaction to the new is perfectly predictable.[14] Mannheim gave an interesting example that can be fully related to Chinese traditionalism during the last few decades of the Qing dynasty: the traditionalist reaction to the introduction of the railway. While scientific writing and technological innovation were not absent in ancient China, when the Qing dynasty was forced to initiate a series of open policies to foreign countries from the second half of the nineteenth century onwards, Western technologies and especially machineries introduced to China terrified Chinese civilians and provoked significant opposition amongst traditionalist officials. In 1863 – three years after the establishment of the First Convention of Peking (*Beijing tiaoyue*) was concluded between China and Britain, France, and Russia – the consulates-general of the United States, France, and Britain in Shanghai requested the permission of the Beijing government to build a railway between Shanghai and Suzhou in order to facilitate transportation. The request was refused by the Qing high official, Li Hongzhang (1823–1901). Western countries made several subsequent requests during the following year without success, because the Qing government suspected that the railway network across China would allow foreigners to penetrate areas other than the treaty ports.[15] In 1865, an English businessman constructed a railway of approximately half a kilometer to exhibit in Beijing. The train was pushed forward by manpower, which, however, resulted in a great panic.[16] According to the records of Qing official Li Yuerui (1862–1927), "The train ran at the railway like flying. People in Beijing had never been frightened by such a thing that they were convinced... had been made by a demon. The whole country was in such a panic that the scene went almost out of control."[17] The Infantry Commander ordered an immediate destruction of the railway to appease the population, but this dramatic event was only a prelude to the traditionalist resistance to Western "material" civilization that China would experience in the following decades.

Contrary to Li Hongzhang, whose opposition to the foreigners' railway project was rooted in his concern of Western nations' encroachment on China, other Qing officials' loathing of railway construction revealed a visceral traditionalist hostility to Western civilization that they perceived not only to be the root cause of all societal ills but also intrinsically inferior to Chinese civilization. Li Hongzhang, after all, turned out to be a prominent promoter of the Self-Strength-

---

14 Mannheim, *La pensée conservatrice*, 35.
15 Lee Kuo-chih, *Zhongguo zaoqi de tielu jingying* (Early Railway Management in China) (Taipei: Institute of Modern History, Academia Sinica, 2015), 12–4.
16 Ibid., 14.
17 Li Yuerui, *Chunbingshi yecheng* (Li Yuerui's Anecdotes), Vol. 3, in *Guanzhong congshu* (Series of the Central Shaanxi plain), Vol. 8, ed. Song Liankui (Shanxi Tongzhiguan edition, 1936), 65.

ening Movement. The Movement was initiated in 1861 following China's disastrous military defeat in the Opium Wars and aimed to empower China through an imitation of Western feats of arms and technology. Despite the initial suspicion of the railway project, Li realized that the country's backwardness placed China at the mercy of Western powers and the lack of modern machinery could only further retard progress. In 1880, under Li's instruction, Liu Mingchuan (1836–1896) submitted a memorial to the throne concerning the construction of a national railway network, in order to strengthen military power and facilitate commerce and mining.[18] The reformers' suggestion quickly triggered a backlash from traditionalist officials.[19] While reformers tried to convince the traditionalists of the benefits that railway transport would bestow on other industries such as mining, Yu Yue (1821–1907) – a prominent official-scholar in philology and teacher of Zhang Taiyan – warned that the extraction of carbon would lead to a complete depletion of natural resources on which human survival depended.[20] Perhaps an environmentalist far ahead of his time? This, however, was not really the case, because Yu Yue aspired to maintain the status quo, with little regard for China's backwardness in technology and its consequent vulnerability vis-à-vis the West. From the traditionalists' perspective, railway construction would allow women to travel freely and lose moral integrity, ruin ancient customs, and enrage the mountain gods and river gods who would punish the country with drought and flood and destroy China's *fengshui* and "dragon veins" (*longmai*) that blessed the dynasty – even though the fate of the dynasty was already in question.[21]

The controversy that followed the railway construction project is only one of the many examples that illustrated the late Qing traditionalists' limited understanding of the outside world. This was a defensive interpretive method, and their anti-modernization mentality was deeply rooted in the worldview that China was the civilization, which was far superior to the West in the spiritual domain. While supporters of the Self-Strengthening Movement embraced Western technology as a panacea for extant problems, traditionalists despised it as a ves-

---

18 Tang Wenquan, "Qingting jie yangzhai de zhengzheng" (Political Struggle Concerning the Qing Courts' Foreign Debts), in *Jingji lunli yu jinxiandai Zhongguo shehui* (Economic Ethos and modern Chinese society), ed. Liu Xiaofeng and Lin Liwei (Hong Kong: The Chinese University Press, 1998), 99.
19 Lee, *Zhongguo zaoqi de tielu jingyi*, 46–7.
20 Sun Kuang-Teh, *Wanqing chuantong yu xihua de zhengduan* (Disputes between Tradition and Westernization at the End of the Qing) (Taipei: Taiwan Commercial Press, 1995), 36.
21 Li Zhiting, *Qingshi* (A History of the Qing Dynasty) (Shanghai: Shanghai renmin chubanshe, 2002), 1704; Lee, *Zhongguo zaoqi de tielu jingyi*, 18.

tige of the early development of Chinese civilization. Zeng Guofan's (1811–1872) son Zeng Jize (1839–1890) once said:

> I think Europe was inhabited by barbarians. European writings, learnings, political and administrative systems all came from China. European customs and culture are particularly close to that of ancient China... Everything in Europe could be found in ancient China and has nothing new... Some people say that machineries, such as steamers and cars, did not exist in ancient China. In fact, these people forgot that the level of development of machineries is proportional to the quantity of natural resources. When there are not enough resources, tools become primitive and it will not be possible to fabricate these machineries. As a matter of fact, in ancient China, there were a lot of machines. With the decline of natural resources, people have become lazier and lazier. As a consequence, they no longer know how to fabricate these machines. By looking at Europe today, one understands how China was during ancient times. By looking at China today, one understands what Europe will become in the future. These ingenious machines will surely be replaced by primitive tools... because the natural resources that we have are limited and cannot sustain the continual use and exploitation by all countries.[22]

Although this reasoning may not seem to be very convincing, it reflected a consensus among traditionalists, for whom Western civilization was nothing but a reproduction of some abandoned and obsolete elements of Chinese civilization, that it was not necessary to learn from the West. Some even claimed that not only did Western technology originate in China, the fact that the fabrication of machines was still disregarded as an inferior skill, which only thrived in the West, demonstrated that China was a country of empathy and morality, while the materialistic West had renounced all spirituality.[23]

Traditionalism can also serve to mobilize political activity and becomes radical in the sense of political extremism in order to re-establish past conditions. As stipulated in the treaty settlement of the Second Opium War, Western missionaries were granted the right to travel, buy land for the construction of churches, and proselytize in China, with Chinese civilians being able to convert to Christianity without punishment. The missionaries who arrived in the nineteenth century aimed to convert villagers and townspeople to Christianity and were intolerant of traditional rituals that were practiced in rural China. Despite the missionaries' belief in the superiority of their creed, they combined their evan-

---

[22] Wang Erh-min, "Zhongxixue yuanliushuo suo fanying zhi wenhua xinli quxiang" (Psychological Tendencies Reflected in the Discussion of the Origins of Western and Chinese Learnings) in *Zhongguo jindaixiandaishi lunji* (Collection of Articles on Modern and Contemporary Chinese History), ed. Zhongguo wenhua fuxing weiyuanhui (Committee for the Restoration of Chinese Culture), Vol. 18 (Taipei: Taiwan Commercial Press, 1985), 247–8.
[23] Sun, *Wanqing chuantong yu xihua de zhengduan*, 20.

gelical activities with programs aimed at transforming the intellectual and physical life of their Chinese converts, including health, hygiene, education, and dissemination and translation of scientific knowledge, albeit with a theological twist.[24] However, the missionaries' aversion to Chinese traditional values and practices inevitably provoked the antipathy of local gentries, who saw the presence of missionaries as a threat that upset traditional society. Some of the behaviors most unacceptable to local gentries included the missionaries' protection of the converted who were involved in lawsuits and the converts' refusal to worship Confucius, ancestors, and spirits.[25] Natural disasters that affected millions of people in several provinces were interpreted by superstitious victims, officials, and gentries as punishment from Heaven for the damage of the dragon veins and *fengshui* when foreigners constructed the railways.[26] Added to that were numerous rumors instigated to depict missionaries as immoral perverts in order to stir up anti-Christian activities, claiming that foreigners molested children and extracted their organs to make medicine.

From the late 1880s onwards, riots and violence towards the missionaries, churches, and houses of foreigners became commonplace. By the middle of the 1890s, things began to change decidedly for the worst. The militia organizations, whose slogan was "Support the Qing dynasty and destroy the foreign" (*fuqing mieyang*), formed in areas from Shandong to Shanxi. Men associated with these societies were commonly called Boxers, or *yihetuan* – the Fists of Righteous Harmony. Boxers were mostly poor young men, whose primary rituals were a type of spiritual possession which involved the "whirling and twirling of swords, violent prostrations and incantations to Taoist and Buddhist saints" as well as "monk-like austerity."[27] In 1899, the Boxers revolted in Northwest Shandong and reached the border of Zhili, where local officials were unable to control this anti-foreign revolt.[28] Although the Qing Court usually repressed secret soci-

---

24 Benjamin A. Elman, *On Their Own Terms: Science in China 1550–1900* (Cambridge, Mass.: Harvard University Press, 2005), 320–52; Sonya Grypma, *Healing Henan: Canadian Nurses at the North China Mission, 1888–1947* (Vancouver and Toronto: UBC Press, 2008); Chen Kaiyi, "Missionaries and the Early Development of Nursing in China," *Nursing History Review* 4 (1996): 129–49; Geoffrey Pen, *The Fist of Righteous Harmony: A History of the Boxer Uprising in China in the Year 1900* (London: Leo Cooper, 1991), 19.
25 Immanuel C.Y. Hsu, *The Rise of Modern China* (New York: Oxford University Press, 1970), 388–9.
26 Ibid., 390.
27 Larry Clinton Thompson, *William Scott Ament and the Boxer Rebellion: Heroism, Hubris and the Ideal Missionary* (Jefferson, NC: McFarland, 2009), 7.
28 Joseph W. Esherick, *The Origins of the Boxer Uprising* (Berkeley: University of California Press, 1987), 241.

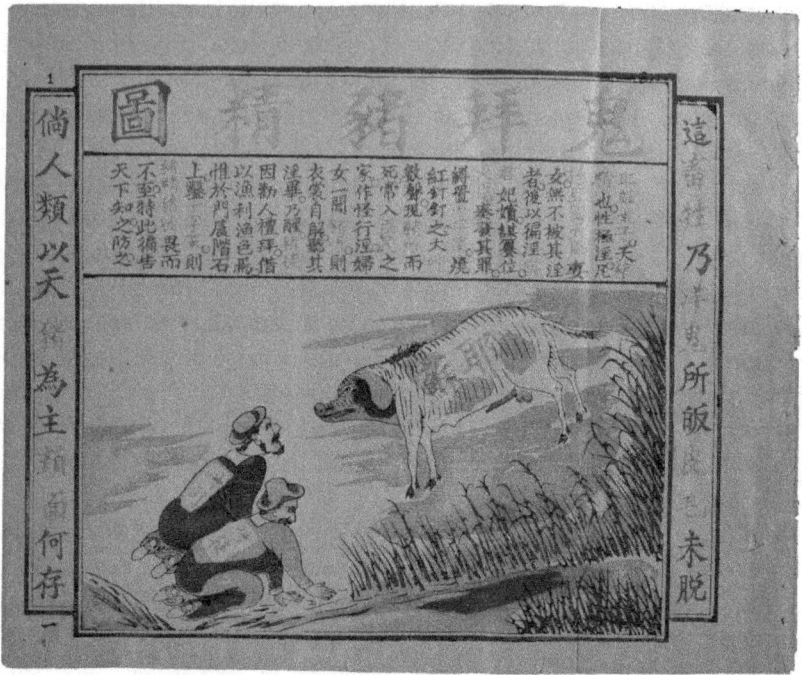

Figure 1.1: Devils [i.e. foreigners] worshipping the Hog [i.e. Jesus]. **Text (center top):** Jesus was a hog with a mythical goblin spirit. Extremely promiscuous, he raped all the women in the Kingdom of Judah. Later he put his hand on the King's concubines and aspired to the throne. Ministers revealed his crime and crucified him with nails. He screamed, transformed into a hog, and died. The hog often does mischief and violates women in civilians' homes. Once women hear the hog's oink, they automatically take off their dress, waking up only when the hog has violated them. The hog's followers persuade people to attend Sunday service so that they could reap profits and commit adultery. But if a cross is put at the gate, the hog's followers will be terrified and do not dare to enter. Please be advised so that you can protect yourselves. **Text (left):** It is a disgrace for humankind to worship a hog as a god. **Text (right):** This animal is transformed from a "gweilo," therefore he still has fur. From *The Cause of the Riots in the Yangtse Valley: A "Complete Picture Gallery"* published in 1891 in Hankou by the missionary Griffith John, which is a reproduction of a Chinese anti-Christian pamphlet.

eties of this kind out of fear that the ethnic Han Chinese revolt would threaten their rule, the government, which was dominated by a xenophobic group, granted the Boxers the right of association in January 1900.[29]

---

29 Ibid., 273–4.

The consequences were disastrous, even worse – murderous. From Shandong to Beijing, Boxers burnt churches and massacred hundreds of missionaries and thousands of Chinese converts.[30] In May 1900, diplomats requested that foreign soldiers intervene in Beijing to protect their legation and citizens. Hundreds of foreign civilians and soldiers and around 3,000 Chinese converts took refuge in the Legation Quarter.[31] Following the foreign armies' attack on the Dagu forts on June 17, the Empress Dowager Cixi ordered all foreigners to leave the capital within 24 hours. She took the side of the Boxers and sent the army to assist in the Boxers' attack on foreigners in Beijing and Tianjin. One of the most outspoken supporters of the Boxers was the Grand Secretary of the Tiren Cabinet (*Tirenge daxueshi*) Xu Tong (1819–1900), a senior first rank official. His traditionalism was so rigid that *Qingshigao* – the official history of the Qing dynasty – describes him as an "old fogy (*shoujiu*), who abhorred Western learnings as deadly foes."[32] When Cixi declared war on the allied armies, Xu submitted a memorial, advising the Empress Dowager to issue an imperial edict authorizing Chinese civilians to kill any foreigners in any province.[33]

The Eight-Nation Alliance was dispatched in June and ended the Boxer Rebellion in August 1900. The Emperor and the Empress Dowager fled to safety to Xi'an and returned to the Forbidden City only after the conclusion of the Boxer Protocol (*Xinchou tiaoyue*), which was signed on September 7, 1901. The Alliance demanded the death penalty for 11 high-ranking officials, as well as the posthumous degradation of several traditionalist officials responsible for the hostilities, including Xu Tong, whose family committed mass suicide when the Alliance took over the capital.[34]

Late Qing traditionalism was rigorous but was ill-situated to provide the prompt and decisive action that the crisis of the Qing government demanded and needed. In contrast to conservatism that I will discuss in the following pages and throughout this book, traditionalists idolized the Confucian moral and political tradition as the primordial and perennial truth that endured over time and stood in stark and superior opposition to the West. Traditionalism is

---

[30] Stacey Bieler, *"Patriots or Traitors"? A History of American-Educated Chinese Students* (London and New York: Routledge, 2015), 19.
[31] Thompson, *William Scott Ament and the Boxer Rebellion*, 84–5.
[32] Zhao Erxun et al., "Liezhuan erbaiwushier" (Biography No. 252), in *Qingshigao* (Draft History of Qing), Vol. 465 (Taipei: Dingwen, 1981), 12750.
[33] Tong Te-Kong, *Wanqing qishinian* (The Last 70 Years of the Qing Dynasty), Vol. 4 (Taipei: Yuanliu, 1998), 101.
[34] Ray Huang, *China: A Macro History* (Armonk and London: M. E. Sharpe, Inc., 1997), 249–50; Tong, *Wanqing qishinian*, 136–7.

an unthinking attachment to the status quo. This is a kind of immobilism that is characterized by a thorough absence of the sense of new realities, believing that people's "hearts and minds" could defend China against Western arms and ammunition.

Monarchism and Cultural Radicalism

Following the suppression of the Boxer Rebellion and the conclusion of the Boxer Protocol, the Empress Dowager Cixi, writing in the name of the Guangxu Emperor, launched the New Policies in January 1901, which were the Qing dynasty's last-ditch effort to restore its vigor. The threat that confronted the Qing dynasty at the time did not only come from outside; the government had already lost control and credibility in some parts of China. When the Empress Dowager declared war against foreign powers in June 1900, the provincial governor-general and officials, notably Li Hongzhang of Zhejiang, Zhang Zhidong (1837–1909) of Hubei and Hunan, Liu Kunyi (1830–1902) of Jiangsu and Jiangxi, Yuan Shikai of Shandong, and Yu Lianruan (1844–1901) of Shanghai, all refused to be involved in the war and reached the agreement of the "Mutual Protection of Southeast China" (*Dongnan hubao*) with the aim of maintaining peace in their own provinces.[35] It was even rumored that Li had come into contact with Sun Yat-sen and was prepared to establish himself as King or President if the Emperor died.[36] Although the Emperor did not die, the defeat of China in the war greatly undermined the central government's authority and credibility.[37]

The New Policies of 1901 were based on the experiences of the Self-Strengthening Movement and the Hundred Days' Reform. While the Self-Strengthening movement aimed primarily at promoting economic, military, and infrastructural developments, the New Policies demanded more than the abortive Hundred Days' Reform and pushed for a profound educational, political, and institutional reform. The famous slogan "Chinese learnings as substance, Western learning for application" during the Self-Strengthening Movement was no longer sufficiently in tune with the new political reality. There is no relationship between *ti* (sub-

---

[35] Chan Hok-yin, "Gexing qishi: 'Dongnan hubao' yu 'Liangguang duli' zhong zhi gefang zhengzhi choumou" (Dynamics of Different Policies: From Southern Mutual Protection to Guangdong and Guangxi Independence), *Historical Inquiry* 49 (2012): 71.
[36] Ibid., 92.
[37] Luo Zhitian, "Yiduan de zhengtonghua: gengzi yihetuan shijian biaoxianchu de lishizhuanzhe" (Turning Heterodoxy into Orthodoxy: Historical Transitions Manifested in the Boxer Rebellion of 1900), in *Liebian zhong de chuancheng*, 19.

stance) and *yong* (application), because they are different aspects of one thing.³⁸ By borrowing Western technologies, reformers had to abandon the insistence upon the integral preservation of Neo-Confucian principles and find a way to adapt foreign political and institutional systems in China. Zhang Zhidong, who had popularized this slogan in his 1898 *Quanxuepian* (*Exhortation to Learning*), emphasized in 1901:

> Every country around the world is willing to reform and people of determination are willing to listen to talk about reform. Reform means changing old Chinese political methods to Western political methods rather than superficially revising and reorganizing them... The situation of the world today is such that a country will perish if isolated and will survive if it stands together with others. Thus, to save China from its dismal state, the only choice is to adopt Western political methods... When China adopts Western political methods, China will longer hate the Westerners and then all countries will stop antagonizing China and the Court. Only by adopting Western political methods will religious cases come to an end, business and diplomatic affairs be put on a fair basis, so that inland foreigners will stop flaunting superiority and making troubles.³⁹

This political vision overturned his earlier defense of the Confucian orthodoxy (*zhengxue*), which he put forward to criticize Kang Youwei and his followers during the Hundred Days' Reform in 1898. This short-lived reform constituted a genuine introspection of the Self-Strengthening Movement that ended in China's defeat in the First Sino-Japanese War (1894–1895). During the same year that the War ended, Kang Youwei obtained the highest degree *jinshi* in the Imperial examination. In 1898, he organized his students, notably Liang Qichao, to petition the Emperor to introduce a number of proposals designed to reform the Chinese governmental and institutional system. Reform came to be seen not only as a means of gaining technical and military expertise but, more crucially, as political change.

A native of Guangdong, Kang Youwei, as was the case of many of his generation, was sent by his family to study the Confucian classics in order to pass the Imperial examination at an early age. In 1879 and 1882, Kang visited Hong Kong and Shanghai. Deeply impressed by the sheer modernity of the two cities under foreign administration, he bought numerous Western works in Shanghai to study.⁴⁰ In 1891, he opened a school called Wanmu Caotang (Myriad Trees Acad-

---

38 Peter Zarrow, *After the Empire: The Conceptual Transformation of the Chinese State, 1885–1924* (Stanford: Stanford University Press, 2012), 119.
39 Luo, "Yiduan de zhengtonghua: gengzi yihetuan shijian biaoxianchu de lishizhuanzhe," 18.
40 Tang Zhijun, *Kang Youweizhuan* (Biography of Kang Youwei) (Taipei: Taiwan Commercial Press, 1997), 8.

emy) in Guangzhou to teach Western learnings and offer his own interpretations of Confucianism, known as the "New Text" (*jinwen*) scholarship. Declaring that the Old Text (*guwen*) was fabricated by Liu Xin (46–23 BC) and resulted in the sclerosis of Chinese intellectual tradition, Kang drew on foreign political systems to inform his reformist ideas that he rationalized in the framework of the New Text.[41]

The New Text school harks back to an ancient debate and refers to the Confucian classics, re-compiled during the early Han dynasty by Confucian scholars who had survived "the burning of books and burying of scholars" (*fenshu kengru*) ordered by Qin Shi Huang (259–210 BC) in 213 BC. This was alleged to have destroyed Confucian texts and buried hundreds of Confucian scholars. During the second century, Prince Liu Yu (?–128 BC) discovered new versions of Confucian classics in the walls of Confucius' old residence in Qufu. Since these classics were written in an archaic pre-Qin script, the newly found editions were referred to as "Old Text" and became incrementally predominant. One of the major differences between the two versions of Confucian classics lies in the role of Confucius. In the Old Text scholarship, Confucius was a transmitter and not a maker of the classics (*shu er bu zuo*). In contrast, the New Text scholars were interested in the esoteric readings of the classics and cosmological speculation. To them, Confucius was an institutional reformer and a holy prophet who had received the Mandate of Heaven to save China from the chaos of two thousand generations.[42] Their voluntarist interpretation of Confucianism aimed at applying the classics to the practical use of administration (*jingshi zhiyong*).

The *Chunqiu* (*Spring and Autumn Annals*), which to Old School scholars only traced the history of Confucius' state of origin Lu, and the *Gongyang* commentary of these historical annals, heightened Kang's particular interest. The Eastern Han New Text scholar He Xiu (129–182) found the evolutionary scheme of the "Three Ages" (*sanshi*) – chaos (*juluanshi*), rising peace (*shengpingshi*) and great peace (*taipingshi*) – in the Confucian texts.[43] By interpreting the *Gongyang* commentary, He Xiu argued that Confucius divided the chronicle of the State of Lu into three time periods that corresponded to the Three Ages:

> From the transmitted era (*suo chuanwen shi*), Confucius saw that government emerged from chaos. He used his mind in a still unrefined way. Therefore, he included his own state and excluded the other Zhou states. He made detailed records about the inside first, and only then dealt with the outside… From the heard era (*suowenshi*), he saw the government of the

---

41 Anne Cheng, *Histoire de la pensée chinoise* (Paris: Seuil, 1997), 621–3.
42 Ibid., 617.
43 Ibid., 609–617.

rising peace. He included all the Zhou states and excluded the barbarians. He wrote down the departures and meetings outside [his own state]. From the seen era (*suojianshi*), he recognized the government of Great Peace. Barbarians came to Court and attained noble ranks. All-under-Heaven, no matter how far or close, he treated as one.⁴⁴

Nevertheless, Kang pointed out that Great Peace had not been achieved in "the seen era" of the State of Lu; rather, Chinese history was a process of civilization decadence. Kang considered this contradiction to be the proof that Confucius did not elaborate on the Three Ages to resume the history of Lu; instead, what the *Chunqiu* revealed was a universal historical evolutionary scheme that predicted the political developments for the centuries to come.⁴⁵ As strange as it may sound, Kang associated the Three Ages with three types of political systems. The age of chaos was ruled by the absolute monarchy. The age of rising peace came when the absolute monarchy gave way to the constitutional monarchy. Constitutional democracy and republicanism would be achieved in the age of great peace.⁴⁶ At the final stage of human evolution, all nations would be dissolved into a Grand Unity (*datong*), in which "there will be no longer any nations, no more radical distinctions, and customs will be everywhere the same."⁴⁷

Without explaining why, he maintained that the age of chaos was coming to an end in China. Hence, he hoped the Emperor would implement reforms to transform the Qing dynasty into a constitutional monarchy, in which the Emperor would not only be the head of the state but also an "enlightener" who would instruct the politically indifferent and submissive people in state affairs.⁴⁸ From 1888 to 1898, Kang sent a total of seven memorials to the Emperor expressing his ideas for the urgent reform of government, including the promotion of modern education, the development of industries, the modernization of the army, the relocation of the capital and the establishment of a political mechanism, allowing low-rank officials to communicate directly with the Emperor and local assemblies, the lifting of bans on newspapers and social associations, as well as the

---

44 Zhao Lu, *In Pursuit of the Great Peace: Han Dynasty Classicism and the Making of Early Medieval Literati Culture* (Albany: SUNY, 2019), 154–5.
45 Kang Youwei, "Chunqiu Dongshixue" (Dong Zhongshu's Teaching of the *Spring and Autumn Annals*), in *Kang Youwei quanji* (Complete Works of Kang Youwei), Vol. 2 (Shanghai: Shanghai guji chubanshe, 1990), 671.
46 Tseng Chun-hai, *Zhongguo jindangdai zhexueshi* (History of Modern and Contemporary Chinese Philosophy) (Taipei: Wunan, 2018), 40–1.
47 Theodore de Bary and Richard Lufrano, *Sources of Chinese Tradition*, Vol. 2 (New York: Columbia University Press, 2000), 731–2.
48 Kang Youwei, "Mengzi wei" (On Mencius), in *Kang Youwei quanji*, Vol. 5 (Beijing: Zhongguo renmin daxue chubanshe, 1998), 421–2.

destruction of illicit shrines (*yinsi*) in order to erect community schools.⁴⁹ The last two suggestions were particularly embedded in his cultural radicalism.

In an 1898 memorial, Kang advised the government to uphold Confucianism as the state religion, establish shrines to worship Confucius, adopt the Confucian calendar, and prohibit religious activities in illicit temples.⁵⁰ Although the memorial was not submitted to the throne, it reinforced Kang's repeated requests to seize the landed property of improper cults to build schools for village children.⁵¹ Kang did not specify who these improper cults were but he manifested a strong sensibility to the Western perspective on Chinese religious practices. Westerners worshiped God and religious institutions were dedicated solely to him; religious worship, held in a solemn and civil manner, was a peculiar kind of homage, which was due only to God. In contrast, the Chinese religion was a superstitious practice that made Westerners despise the Chinese as barbarians.⁵² Hence, he urged the throne to purge all the popular cults and wished to banish the dualism between official and non-official cults by establishing a state religion – the Confucian religion (*kongjiao*) – common to both the throne and the people, and capable of resisting the dissemination of Christianity in China.⁵³ Liang Qichao echoed his professor's argument in 1897 by saying "no one in the world can be governed without religion, thus no country can be well founded without religion."⁵⁴ In other words, the importance of religion precedes that of the country, which, in turn, should obey religious doctrines. Liang even planned to establish the Association for the Protection of Religion (*Baojiao gonghui*) to promote Confucianism as the state religion.⁵⁵ In the following year, the German

---

49 Tang Zhijun (ed.), *Kang Youwei zhenglunji* (Political Comments of Kang Youwei) (Beijing: Zhonghua shuju, 1981), 52–61, 114–36, 139–162, 201–221.
50 Kang Youwei, "Qing zun Kongsheng wei guojiao lijiaobujiaohui yi kongzijinian er fei yinsiqi zhe" (Memorial Concerning the Establishment of Confucianism as the State Religion, Ministry of Religion and Shrines, the Adoption of Confucian Calendar and the Ban on Worships in Illicit Temples), in *Kang Youwei zhenglunji*, 279–283.
51 Vincent Goossaert, "1898: The Beginning of the End for Chinese Religion?" *Journal of Asian Studies* 65, no. 2 (2006): 313.
52 Kang Youwei, "Qing zun Kongsheng wei guojiao lijiaobujiaohui yi kongzijinian er fei yinsiqi zhe," 279–80.
53 Ibid., 282; Peng Chunling, *Rujia zhuanxing yu wenhua xinming: yi Kang Youwei, Zhang Taiyan wei zhongxin (1898–1927)* (The Transformation of Confucianism and the New Culture Movement: On Kang Youwei and Zhang Taiyan (1898–1927)) (Beijing: Beijing daxue chubanshe, 2014), 171.
54 Liang Qichao, "Fu youren baojiao shu" (Letter to a Friend Concerning the Protection of Religion), in *Liang Qichao quanji* (Complete Works of Liang Qichao), ed. Zhang Pinxing et al. (Beijing: Beijing chubanshe, 1999), 150.
55 Ge Zhaoguang, "Kongjiao, fojiao yihuo yejiao? – 1900 nian qianhou Zhongguo de xinliweiji yu zongjiao xingqu" (Confucianism, Buddhism or Christianity? – Psychological Crisis and Reli-

Empire forced the Qing Court to accept a leasehold of the Kiautschou Bay (*Jiaozhou wan*) in Shandong for 99 years. As a consequence of this event, Kang Youwei and his followers founded the Association for the Protection of China (*Baoguo hui*) in Beijing, with the aim of protecting China's sovereignty (*baoguo*), Chinese autonomy (*baozhong*), and the perpetuity of the Confucius religion (*baojiao*).[56]

The lifting of bans on newspapers and social associations was another significant request that Kang made to the Emperor in order that the reform become established. Since communication between the people and the government was proven to be non-existent, reformers used newspapers as a vehicle to communicate government policies to the people and political suggestions were printed.[57] In 1896, the *Shiwubao* (*Chinese Progress*), a reformist newspaper founded by Liang Qichao, attempted to concretize what Tan Sitong later called the "communication between the top and the bottom" (*shangxiatong*) and the "communication between China and outside" (*zhongwaitong*), in the sense that the newspaper should publish foreign reformist ideas, decisions made by the Court, and should give the people a voice.[58] While modern newspapers might be considered an updated tradition of the official bulletin *Peking Gazette* (*Jing bao*) that contained information on Court decisions and official memorials, the aforementioned act of association was firmly prohibited by the Qing Court. *Huidang* (gang) or *pengdang* (political faction) were usually condemned as being provocative, dissident but also perilous, as they were capable of undermining central political power. To rationalize this act, Kang stated that Confucius himself opened the way for one form of alignment among gentries and cultural elites, that is, in the form of *qun* (grouping). Confucius maintained: "the gentlemen are noble and do not strive to surpass others, thus they gather together to form a community (*qun*) but not a faction in service of their personal ambitions (*dang*)."[59] Kang took the politics of *qun* as the ideological weapon to defend the formation of the group. In 1895, he founded the Education Society for Strength (*Qiangxuehui*), which he claimed was the embryo of the modern political party in

---

gious Interests around 1900 in China), in *Zhongguo jindai sixiangshi de zhuanxing shidai* (The Age of Transformation of Modern Chinese Intellectual History), ed. Wang Fan-sen (Taipei: Linking, 2007), 203.

56 "Baoguohui zhangcheng" (Memorandum of the Association of the Protection of China), *Guowenbao* 185 (1898).

57 Liang Qichao, "Lun baoguan youyi guoshi" (Discussion on the Benefits of Newspapers for Public Affairs), *Liang Qichao quanji*, 66–8.

58 Tan Sitong, "Renxue" (Study of Benevolence), in *Tan Sitong quanji* (Complete Works of Tan Sitong), Vol. 1 (Beijing: Sanlian shudian, 1954), 6.

59 The original text reads: 君子矜而不爭，群而不黨。

China.⁶⁰ If the Qing Court was to be transformed into a constitutional monarchy during the age of rising peace, the government should grant people popular power (*minquan*), the lack of which led inevitably to the weakness of the state.⁶¹ As such, China's revival could not count on the government's initiatives alone; it required the efforts and actions of all individuals. This political vision was legitimized in the framework of the New Text scholarship, which attempted to revive the political activism of the Western Han dynasty.⁶²

Unsurprisingly, Kang's reform proposals that built upon his cultural radicalism attracted a storm of criticism from the Qing officials adherent to the Confucian orthodoxy, including Zhang Zhidong. Zhang's *Quanxuepian* can be read as a riposte to Kang's reformist ideas and the New Text scholarship. With regard to religious matters, not unlike Kang, Zhang supported the confiscation of a large percentage of Buddhist and Taoist monasteries and their landed property to establish new schools and believed that a degree of "Confucianization" would invigorate these two declining religions.⁶³ Nevertheless, he exhibited no tolerance to Kang's prioritizing of the Confucian religion over the strengthening of the country. He made it clear in the first chapter of *Quanxuepian* that the force of a religion was proportional to the strength of a country. Zhang stated: "Given the political situation today, the only way [to save China] is to instill loyalty [of the population], search for wealth and strength, respect the Court and defend the state (*sheji*)… If the country is lost, how would it be possible to protect Confucianism and Chinese?"⁶⁴ To him, Kang and his fellow reformers mistakenly inverted the political pyramid by downplaying the urgency of protecting the Court and fortifying the country. Worried that the progressive political comments and foreign ideas might inspire and inflame the readers' resentment towards the Court, Zhang acted to censor the eventual antigovernment content in the *Shiwu-*

---

60 Kang Youwei remarked that he had planned to set up with his colleagues *Qiangxuehui* as the first Chinese political party; see Tang, *Kang Youwei zhenglunji*, 163. Liang Qichao had also spoken of *Qiangxuehui* as a combination of education society and political party; see Tang, *Kang Youwei Zhuan*, 140.
61 Liang Qichao, "Xixue shumubiao houxu" (Epilogue of the Bibliography of Western Knowledge), in *Liang Qichao quanji*, 86.
62 Anne Cheng, "Nationalism, Citizenship, and the Old Text/New Text Controversy in Late Nineteenth Century China," in *Imagining the People: Chinese Intellectuals and the Concept of Citizenship*, ed. Joshua A. Fogel and Peter G. Zarrow (Armonk: M.E. Sharpe, 1997), 61–81.
63 Goosaert, "1898: The Beginning of the End for Chinese Religion?," 312.
64 Zhang Zhidong, *Quanxuepian* (Exhortation to Learning) (Zhengzhou: Zhongzhou guji shubanshe, 1998), 51.

bao, which led to Liang's resignation.⁶⁵ In sharp contrast to Kang and his leading disciple, Zhang argued that the notion of popular power itself not only had no specific role in boosting China, it also risked damaging the political structure based upon the Five Relationships of Confucianism, since the popular power would indubitably challenge the souverain's power (*junquan*):

> The idea of popular power will cause nothing but damage... Although China today is not a great power, people can still live and work in peace and contentment, thanks to the Court's policies. Once put forward, the notion of popular power will only please the imbecile, incite the rebels to revolt, erode the moral principles and turn the whole country upside down. In this scenario, can the proponents of popular power expect to be able to live in peace?... Because of the despotism of the French King, indignation that surrounded the whole country provoked people to revolt against the government. France then became a democratic regime. But our Court is gracious and righteous. Never once has it imposed any tyrannical politics. What is the point of disseminating this idea that shatters the social and political orders and endangers everyone? This is what I mean by saying that popular power will cause nothing but damage.⁶⁶

Zhang maintained further that the notions of popular power and the right to autonomy (*zizhu zhi quan*) stemmed from a misinterpretation of Western parliamentarianism, which, according to him, allowed peoples' voices to be heard through intermediaries, namely officials, rather than transferring power directly to the people. Hence, the right to autonomy was simply nonsense. Furthermore, China did not need parliament, since the country already had its own political mechanism for hearing the complaints and grievances of civilians and allowing officials to provide suggestions directly to the Court. Thanks to the clear-sightedness and graciousness of the Court, Zhang did not believe that petitions would be void. Furthermore, liberty and autonomy would only endanger the solidarity required to help the country redress from foreign intrusions.⁶⁷

Zhang compared the French *Ancien régime* with the "gracious and righteous" Qing dynasty to refute the constitutional monarchy extolled by Kang and Liang during the Hundred Days' Reform. Several years later, the latter used much of the same rhetoric to argue against the abolition of the monarchy and the revolutionaries.⁶⁸ Despite the Empress Dowager's bloody suppression of the 1898 reform, Kang and Liang, who fled into exile, continued to call for a con-

---

65 Ezra F. Vogel, *China and Japan: Facing History* (Cambridge, Mass.: The Belknap Press of Harvard University Press, 2019), 136.
66 Ibid., 86.
67 Ibid., 86–7.
68 See, for example, Mingyi (Kang Youwei), "Faguo gemingshi lun" (On the History of the French Revolution), in *Xinhai geming qianshinianjian shilun xuanji*, Vol. 2, 304.

stitutional monarchy and pinned their hopes on the restoration of the Guangxu Emperor. Kang established the Chinese Empire Reform Association in July 1899 in Canada, which led a series of cross-national activities in the name of "protecting the Emperor" (*baohuang*). Liang Qichao resurfaced as editor and writer of several magazines in Japan. The *Qingyibao* (*The China Discussion*), published in Yokohama from 1898 to 1901, was the first official publication of the Chinese Empire Reform Association. The *Xinmin congbao* (*Journal of New People*) that replaced *Qingyibao* initiated a heated debate concerning monarchism and republicanism with the revolutionary *Minbao* (*People's Tribune*).

Overall, Kang adopted the fundamental political reformist ideas from the West but made sense of them through the heterodoxic Confucianism. Liang Qichao once praised his teacher as "the Martin Luther of the Confucian religion" for his uncovering of the authentic meaning of Confucianism and saving China by means of her own learnings and tradition.[69] Kang would probably not have said less of himself, but did his cultural radicalism lead to a political radicalism? After all, a constitutional monarchy was only one stage in Kang's political scheme; he aspired for democracy and even the utopian Great Unity, albeit not immediately but over a longer period of time. However, as Peter Zarrow remarks, despite reformers' sometimes ambiguous political views, their goal for the immediate future was to instigate fundamental institutional reforms to maintain the monarchy.[70] The traditional cultural order and political order remained integral.

Kang Youwei and Liang Qichao remained monarchists during the revolutionary decade. Liang made it clear that popular power and the souverain's power were not at variance to one another and should be sustained conjointly as important vessels for the state's power.[71] However, it must be stressed that the challenge of the Confucian orthodoxy in the framework of Confucianism actually deconfucianized the tradition and paved the way for a much more direct and large-scale attack on Confucianism in the following decades.[72] Kang's advocation of the adoption of the Confucian calendar along with the reign of Emperor Guangxu and the worship of Confucius as the "uncrowned king" (*suwang*), who was

---

[69] Liang Qichao, "Nanhai Kang xiansheng zhuan" (Biography of Kang Youwei), in *Liang Qichao quanji*, 486–7.
[70] Peter Zarrow, "The Reform Movement, the Monarchy, and Political Modernity," in *Rethinking the 1898 Reform Period: Political and Cultural Change in Late Qing China*, ed. Rebecca E. Karl and Peter Zarrow (Cambridge, Mass.: Harvard University Press, 2002), 22.
[71] Liang Qichao, "Gu yiyuan kao" (On Ancient Parliament), *Liang Qichao quanji*, 61–2.
[72] Cheng, *Histoire de la pensée chinoise*, 623; Wong Young-tsu, *KangZhang helun* (On Kang Youwei and Zhang Taiyan) (Taipei: Linking, 1988), 71–3.

the bearer and transmitter of the divine inspiration, embodied, in reality, a revolutionary seed that would later take root in the early 1900s.[73] In other words, despite Kang and Liang's monarchism, their cultural radicalism indicated that China and the Qing Court were no longer the same entity. For this reason, they were accused by the Qing Court of "saving China to the detriment of the Qing dynasty," even though they remained monarchist in the sense of advocating the Manchu government's legitimacy and condemned revolutionary upheavals.[74] Nevertheless, this revolutionary hint in their cultural radicalism attracted many revolutionaries, who first sympathized with Kang and Liang and supported the Qing Court's New Policies, before calling for the downfall of the Manchu dynasty in favor of a Chinese republic.

Political and Cultural Radicalism

After the brutal suppression of the Hundred Days' Reform, the overseas reformers' greatest enemy was no longer the Qing Court under the Empress Dowager. She now belatedly supported the New Policies, despite the fact that Kang and Liang were never granted amnesty by the Empress Dowager when she launched the New Policies, nor by Zaifeng, Prince Chun (1883–1951) who served as the Prince-Regent during the reign of his son Puyi, following the death of the Empress Dowager Cixi and the Guangxu Emperor in 1908.[75] Their utmost rivals were now anti-Manchu revolutionary activists and intellectuals, against whom they mobilized intellectually, socially, and politically to defend the cause of the constitutional monarchy and the legitimacy of the Manchu regime.

Several months after the outbreak of the First Sino-Japanese War, Sun Yat-sen founded the Xingzhonghui (Society for Regenerating China) in November 1894 in Honolulu, under the slogan, "expel the barbarians (i.e., the Manchu), revive China and establish a federal government."[76] During the following years, multiple uprisings organized by the Society ended in failure. In August 1905, in Tokyo, the Xingzhonghui was merged with Huang Xing's (1874–1916) Huaxinghui (China Revival Society, 1904) and Cai Yuanpei's (1868–1940) Guangfuhui (Restoration Society, 1904) to form the Tongmenghui, with Sun elected as leader.

---

73 Peng, *Rujia zhuanxing yu wenhua xinming*, 76.
74 Ibid.
75 Edward J.M. Rhoads, *Manchus and Han: Ethnic Relations and Political Power in Late Qing and Early Republican China, 1861–1928* (Seattle and London: University of Washington Press, 2000), 134.
76 The original text reads: 驅除韃虜, 恢復中華, 創立合眾政府。

Common to all three societies was a strong anti-Qing sentiment, born out of Han-centric nationalism. Since the Manchu was not considered to be part of the Han Chinese, the Qing government was denounced as an illegitimate foreign regime which had to be brought down, so that the Chinese could reclaim the sovereignty lost at the end of the Ming dynasty and establish a republic. As the Tongmenghui's wordsmith Zhang Taiyan stated in 1903, "Political reform within the same nation is called revolution (*geming*), while expulsion of foreign rule is called restoration (*guangfu*). Since China has been destroyed by barbarians, what we should do now, is nothing but restoration."[77]

Tongmenghui waged its underground resistance movement against the Qing dynasty by means of assassination (*ansha*), revolt (*jubin*), and propaganda (*xuanchuan*).[78] As the political leaders of the association were mainly devoted to raising funds and recruiting activists in order to assassinate Qing officials and agitate and coordinate revolts in China, revolutionary intellectuals stirred up anti-Manchu sentiment and theorized the political foundations of the revolution and the Chinese Republic through the platform of print media. Major intellectuals who associated with the Tongmenghui were those of the Southern Society and the Society for the Preservation of National Learnings (*Guoxue baocunhui*), which was commonly known as the National Essence School.

As Laurence Schneider remarks, the abolition of the Imperial exams in 1905 generated a new intelligentsia in China.[79] However, it was not easy for the educated elites who were deprived of access to the state bureaucracy by traditional exams to compete with other social groups to secure jobs.[80] The exclusion of the intellectuals from the political realm did not prevent them from renouncing political activism. This was to be expected, considering that China's politico-cultural tradition requires educated elites to assume the responsibility for promoting the betterment of society and state. The emergence of intellectual professionals in China resembled those seen in France and in Russia, where intellectuals engaged collectively to reflect on politics vis-à-vis the state, by upholding values such as "morality," "truth," "justice", and "the freedom of consciousness and speech," as in the Dreyfus Affair or the case of the Decembrists, who inspired

---

[77] Zhang Taiyan, "Gemingjun xu" (Preface of *Revolutionary Army*), in *Meng huitou: Chen Tianhua Zou Rong ji* (Fierce Back: Works of Chen Tianhua and Zou Rong), ed. Zhi Zhi (Shenyang: Liaoning renmin chubanshe, 1994), 2.
[78] Chen Gonglu, *Zhongguo jindaishi* (Chinese Modern History) (Hong Kong: Open Page, 2017), 470.
[79] Laurence Schneider, "National Essence and the New Intelligentsia," 58.
[80] Hon, *Revolution as Restoration*, 8; 115.

the first generation of the Russian intelligentsia.⁸¹ However, the marginalization of the educated elites in the political sphere resulted in the radicalization of the intellectual sphere.⁸² In the nationalist context of the 1900s, more and more intellectuals acquired a sense of revolution. Through associations, they carried out collective actions as the most effective way of establishing discursive power and influencing society.⁸³ The Society for the Preservation of National Learnings and the Southern Society were two of the most prominent revolutionary associations established in this political and ideological context. What Habermas termed as the venue of the public sphere in eighteenth-century Europe had already aroused the attention of some Chinese intellectuals as an effective means of political engagement. Chen Qubing (1874–1933), founder of the Southern Society, wrote in 1907 that the literary associations in China during that period were founded in the same spirit as Western teahouses.⁸⁴

Although the Society for the Preservation of National Essence and the Southern Society were founded in 1904 in Shanghai and in 1909 in Suzhou respectively, the members' involvements in social and political engagements dated back to the late 1890s. Chen Qubing, who obtained the lowest degree in the Imperial examination *shengyuan*, founded the Society to Avenge the Insult (*Xuechi xuehui*) in Tongli in 1897, following the defeat of China in the First Sino-Japanese War.⁸⁵ Initially a supporter of the New Policies, he soon realized that reformism could not save China. He changed his birth name from Qinglin to Qubing, which was inspired by the military general of the Western Han dynasty, Huo Qubing (140–117 BC), who led several campaigns to defeat the Xiongnu. Chen compared the Xiongnu with the Manchu – to liberate China, the latter must be expulsed.⁸⁶ Invited by Cai Yuanpei, he went to Shanghai in 1902 with a view of forming the Chinese Association for Education (*Zhongguo jiaoyuhui*), a nationalist educational institution.⁸⁷ He left for Japan in 1903, where he became chief editor of the rev-

---

81 Michael Confino, "On Intellectuals and Intellectual Traditions in Eighteenth- and Nineteenth- Century Russia," *Daedalus* 2 (1972), 117–149; Christophe Charles, *Naissance des intellectuels:1880–1900* (Paris: Les Éditions de Minuit, 1990).
82 Yü Ying-shih, "Zhongguo zhishifenzi de bianyunhua" (The Marginalization of Chinese Intellectuals) *Twenty-First Century* 15 (2003): 15–25.
83 Michel Hockx, "Playing the Field: Aspects of Chinese Literary Life in the 1920s," in *The Literary Field of Twentieth-Century China* (Honolulu: University of Hawai'i Press, 1999), 72–6.
84 Zhang Yi, *Chen Qubing nianpu* (Chronological Biography of Chen Qubing) (Shanghai: Shanghai guji chubanshe, 2009), 87.
85 Zhang, *Chen Qubing nianpu*, 18.
86 Liu Wu-chi and Yin Anru, *Nanshe renwuzhuan* (Biographies of the Members of the Southern Society) (Beijing: Shehui kexue chubanshe, 2002), 314.
87 Zhang, *Chen Qubing nianpu*, 30.

olutionary magazine *Jiangsu* and was actively involved in the Association for National Military Education – the latter was formed in the context of the Anti-Russia Movement (*Ju'e yundong*), which erupted in April 1903 as a result of the refusal of the Russian Empire to retreat from Northeast China, as stipulated in the treaty signed with the Qing government in 1902. Immediately after his return to Shanghai, Chen Qubing joined the editorial board of the *Jingzhong ribao* (*Alarm Bell Daily*) and published China's first theatre journal *Ershi shiji dawutai* (*Great Stage of the 20th Century*). The magazine served as a platform to publish the plays of the "improved Peking opera" (*gailiang xinju*), promoting the anti-Manchu Revolution.[88]

To commemorate his friend Qiu Jin (1875–1907) who was tragically executed by the government in June, Chen Qubing founded the Society of Spiritual Union (*Shenjiao she*) in Shanghai. It aimed to bring together intellectuals of the same mindset to restore the associated tradition of the Ming literati, in order to resist the ideological control of the Qing Court.[89] The political and intellectual orientation of this society foreshadowed that of the Southern Society. Indeed, the Southern Society saw itself as the inheritor of the late-Ming literati resistance group known as the Society of Restoration (*Fushe*) and the Society of Potency (*Jishe*).[90] Both associations appeared during the reign of the Emperor Chongzhen. After the collapse of the Ming dynasty, many of the members sacrificed their lives during the massacres committed by the Manchu in their conquest of China. In a different historical era, the Southern Society bridged the gap of time and "communicated" with the literati of nearly 300 years previous, since they fought against the same enemy – the Manchu dynasty.

Chen's initiative soon attracted Liu Yazi (1887–1958) and Gao Xu (1877–1925). As Chen, Liu's idols at the beginning of the 1900s were overseas reformers, notably Liang Qichao.[91] It was through *Xinmin congbao* that he learnt about Rousseau's (1712–1778) ideas, inspired by which he adopted the art name (*hao*) Yalu – the Asian Rousseau. Liu's father Liu Nianzeng (1865–1912) was Chen Qubing's classmate and follower of Kang Youwei. After obtaining the degree *shengyuan* in 1902, Liu met Chen in a small town near Suzhou. Recommended by Chen,

---

[88] Kwok-kan Tam, *Chinese Ibsenism: Reinventions of Women, Class and Nation* (Singapore: Springer, 2019), 16.
[89] Zhang, *Chen Qubing nianpu*, 86–7.
[90] Yang Tianshi and Wang Xuezhuang, *Nansheshi changbian* (Chronological History of the Southern Society) (Beijing: Zhongguo renmin daxue chubanshe, 1995), 133.
[91] Guo Changhai, "Gao Xu nianpu" (Chronological Biography of Gao Xu), in *Gao Xu ji* (Works of Gao Xu), ed. Guo Changhai and Jin Juzhen (Beijing: Shehui kexue wenxian chubanshe, 2003), 87; Liu, *Nanshe jilue*, 9.

he abandoned the traditional official career and became a student at the Chinese Association for Education that Chen helped to organize. His reformist ideas completely changed in this nationalist school, where his teacher was Zhang Taiyan. Zhang once required the students to write an autobiography. In his studies, Liu reflected on his political journey from reformer to revolutionary. Considerably impressed, Zhang wrote him a letter:

> When I was 14 or 15 years old... I already had the determination to expulse the Manchu. I later took a post at the agency of *Shiwubao* in 1897, a period during which I heard, for the first time, of Sun Yat-sen. Despite the fact that I did not detach myself from reformers' absurd words, Sun reassured me, saying that I was not the only one with such an ambition. Some texts in *Qiushu*, such as the *Kedilun*, are proof of my past errors. I give you this book today. I hope that by reading this, you will understand that the initial intelligence of every person is more or less the same. [Hence], at the beginning of evolution [of mentality], nobody can surpass [the reformist ideas, such as] the establishment of Confucianism as the state religion and the defense of the emperor.[92]

This was an encouraging reply that noted several key moments in Zhang's own political trajectory. In 1890, Zhang entered the Gujing Academy (*Gujing jingshe*) to study under Yu Yue, a great Old Text scholarship master, whose research accentuated the importance of the philological methods in Confucian studies and history.[93] The impact of this view of philological exegesis (*kaozheng*) on Zhang Taiyan was particularly acute. Six years later, *Shiwubao*'s editors Liang Qichao and Wang Kangnian (1860–1911) approached Zhang. As previously explained, Liang had established a direct causal link between the press and the state's power and appealed to the educated elites to serve as mediators between the state and the people.[94] This, however, went against Yu Yue's conviction that the study of Confucian classics was an end in itself.[95] Nevertheless, for Zhang, it was time to leave the ivory tower.

Against the will of his professor, he joined the editorial board of the *Shiwubao* in 1897, despite different intellectual convictions in terms of the New and Old Text scholarships. The intellectual discordance was tempered by the promotion of reforms.[96] Cixi's bloody suppression of the Hundred Days' Reform in the fol-

---

92 Liu Yazi, *Wushiqinian* (Memoir Written at the Age of 57), in *Zizhuan • Nianpu • Riji* (Autobiography, Chronology and Diaries), ed. Liu Wu-chi and Liu Wufei (Shanghai: Shanghai renmin chubanshe, 1986), 152.
93 Wong Young-tsu, *Search for Modern Nationalism: Zhang Binglin and Revolutionary China 1860–1936* (Hong Kong: Oxford University Press, 1989), 5.
94 Liang, "Lun baoguan youyiyu guoshi."
95 Wong, *Search for Modern Nationalism*, 5.
96 Tang, *Zhang Taiyan nianpu changbian*, 36.

lowing year did not dishearten him; he still believed in the possibility of institutional reforms in the framework of the political structure at the time. To escape persecution, he fled to Taiwan, where he found a job at the agency of *Taiwan nichinichi shinbun* (*Taiwai Daily*).[97] He produced a number of articles, praising Kang Youwei, denouncing Kang's political enemies such as Zhang Zhidong, defending the Guangxu Emperor, and warning of the imminent danger that revolution would bring to China.[98] One of these articles was the famous *Kedilun* (*The Guest Emperor*) that first appeared in the *Taiwan nichinichi shinbun* in 1899 and was later embodied in the first edition of *Qiushu* (*Urgency*). In this text, Zhang insisted that the reform of the constitutional monarchy could be realized by the Guangxu Emperor. The only issue was that the Qing Court was not Han Chinese, therefore, the Manchu emperor could only be the "guest emperor" of China. To maintain the stability and integrity of Chinese society, it was deemed necessary to establish a symbolic monarchy alongside the constitutional monarchy, with the Duke Successor of Confucius (*yansheng gong*) as the symbolic monarch (*gongzhu*) who upheld the royal line (*wangtong*).[99]

Zhang became hostile towards the Court because of its incapacity to defend China during the Boxer Rebellion.[100] In July 1900, Tang Caichang (1867–1900) organized the political association called Chinese Parliament (*Zhongguo guohui*) in Shanghai and convened a meeting to address the Boxer Rebellion in the context of the "Mutual Protection of Southeast China." Zhang was one of the participants. Two contradictory principles were put forward during the meeting. While the hosts declared that the Manchu government no longer had political legitimacy, the Chinese Parliament would help to reestablish the Guangxu Emperor. Zhang found this ambiguous attitude towards the Qing Court to be an outrage. He left the meeting and immediately cut off his Manchu queue, proclaiming his formal separation from the Qing dynasty.[101]

During the following decade, Zhang was closely involved in the revolutionary network in Shanghai. This enraged his teacher Yu Yue considerably. Upon his visit to Yu's house in 1901, Yu reproached him for "leaving for Taiwan and aban-

---

97 Wong, *Search for Modern Nationalism*, 8–9.
98 Peng, *Rujia zhuanxing yu wenhua xinming*, 47–55.
99 Zhang Taiyan, "Kedilun" (The Guest Emperor), in *Zhang Taiyan zhenglun xuanji* (Political Comments of Zhang Taiyan), Vol. 1, ed. Tang Zhijun (Beijing: Zhonghua shuju, 1977), 84–9. This article was initially published in Taiwan under the administration of Japan. It conveyed also a sentiment of resistance against the colonization of the Japanese Empire; see Peng, *Rujia zhuanxing yu wenhua xinming*, 68–77.
100 Wong, *Search for Modern Nationalism*, 8–9.
101 Tang, *Zhang Taiyan nianpu changbian*, 109–10.

doning the parents' tombs" and "disseminating anti-Manchu ideas to harm China and reprimanding the Emperor." Unfilial and disloyal, Zhang was "not even a human" in his teacher's eyes.[102] Due to his anti-government statements, he was forced to flee to Japan in 1902 and developed a friendship with Sun Yat-sen. It was also during this time of exile that he organized the aforementioned Commemoration ceremony for the two hundred and forty-second anniversary of the destruction of China. Upon his return in 1903, he was sentenced to three years in prison for insulting the Emperor in the radical newspaper *Subao* (*Journal of Jiangsu*).[103] After his release, he joined the Tongmenghui in Japan and became the chief editor of its official body, *Minbao*. In the ninth issue, he published a text, announcing his formal disassociation from his traditionalist teacher.[104]

Zhang left a legacy which influenced a number of future members of the Society for the Preservation and the Southern Society. One of them was the third founder of the Southern Society, Gao Xu. His brother Gao Zeng (1881–1943), his uncle Gao Xie (1877–1958), whose grandson Charles K. Kao (1933–2018) was the laureate of the 2009 Nobel Prize in Physics, and his cousin Yao Guang (1891–1945) were all members of the Southern Society. Under the influence of Zhang, Gao Xu organized the Society to Awaken the People (*Juemin she*) in 1903 in Shanghai, which published the anti-Manchu magazine *Juemin* (*Awakening the People*). In 1904, he enrolled in the Hosei University in Tokyo. In 1905, he published the magazine *Xingshi* (*Awakening the Lion Nation*) and adhered to the Tongmenghui, only to be forced to go back to China during the same year because of the "Rule on the Control of Qing Students" (*Shinkoku ryūgakusei torishimari kisoku*) put forward by the Japanese government, which many Chinese radical students denounced as a measure destructive to their revolutionary activities.[105]

In 1909, Chen Qubing, Liu Yazi, and Gao Xu formally declared the foundation of the Southern Society in Suzhou. The members vowed to "be devoted to a literary revolution (*wenzi geming*) to achieve a non-literal objective," that is, to create a new literature which would serve as a vehicle for the anti-Manchu na-

---

102 Ibid., 121–2.
103 Wong, *Search for Modern Nationalism*, 39–43.
104 Tang, *Zhang Taiyan nianpu changbian*, 122.
105 The official name of this regulation was *Shinkokujin wo nyūgaku seshimuru kōshiritsugakkō ni kansuru kitei* 清国人ヲ入学セシムル公私立学校ニ関スル規程 (Regulation Concerning the Admission of Qing Students to Private and Public Schools). Nagai Kazumi, "Iwayuru Shinkoku ryūgakusei torishimari kisoku jiken no seikaku" (The Nature of the Rule on the Control of Qing Students), *Shinshū daigaku kiyō* 2 (1952): 11–33.

tionalism and the expression of revolutionary sentiment and ideas.[106] Literature needs to be understood in the broad sense here. It included not only traditional poetry, essays or novels but also news reports, political comments, and editorials, because before the mid-1920s, journalism was not a specialized profession and journalists were called *wenren* (man of letters) as literary writers.[107] Members of the Southern Society included traditional men of letters, newspaper journalists, and political activists who would later occupy significant roles in the future political landscape, such as Wang Jingwei (1883–1944), Song Jiaoren (1882–1913), Ye Chucang (1887–1946), Dai Jitao (1891–1949), and Chen Qimei (1878–1916).

Figure 1.2: Photograph of the First Assembly of the Southern Society. Liu Yazi is in the front row, third from the left; Chen Qubing, second row, second from the left. Gao Xu was absent that day. From Zheng Yimei, *Nanshe congtan* (History of the Southern Society). Shanghai: Shanghai renmin chubanshe, 1981.

Apart from individual newspapers and journals, members of the Southern Society collectively published the periodical *Nanshe congke* (*Literary Collection of the Southern Society*). When it was first published, the reformist newspaper *Shibao* (*Eastern Times*) mocked the writers of the Southern Society as nothing but old

---

106 Zhang Yi, *Chen Qubing nianpu*, 105.
107 Xu Xiaoqun, *Chinese Professionals and the Republican State: The Rise of Professional Associations in Shanghai, 1927–1937* (Cambridge: Cambridge University Press, 2004), 161.

loyalists of the Ming dynasty.[108] However, it is highly likely that the Southern Society took it as a compliment rather than an insult. Chen Qubing had compiled two volumes of Ming loyalists' biographies to commemorate these martyrs' deeds in resisting the Manchu invasion. Indeed, the "Southern Society" encapsulated two ideologies. On the one hand, the Southern Society was first and foremost the antithesis of the Northern Court (*beiting*). On the other hand, its denomination conveyed the sense of "speaking the Wu language and not forgetting our origin."[109] Indeed, most of the members came from Southern China and spoke the Wu language as their mother tongue. The Wu language was also a metaphor for authentic Chinese culture, which was considered to have been preserved intact in Southern China, while the culture of the North was periodically stained by "barbarian" nomadic people.

The two Cantonese founders of the Society for the Preservation of National Essence, Huang Jie (1873–1935) and Deng Shi (1877–1951), had both taken the Imperial examination several times, albeit without success. Deng Shi was born and lived in Shanghai until the age of 19 with his father, who worked at the Jiangnan Arsenal (*Jiangnan zhizaoju*), with which the Shanghai School for the Diffusion of Languages (*Shanghai guangfangyan guan*) was affiliated. The School was founded in 1863 as part of the educational program of the Self-Strengthening Movement to teach foreign languages and instruct interpreters. Thanks to a great variety of foreign books translated by the School, Deng was quite familiar with Western culture at an early age.[110] After returning to Guangdong, he became Huang Jie's classmate at the School of Jian'an (*Jian'an caotang*). The school was founded by Jian Chaoliang (1851–1933), who was a fellow student of Kang Youwei under Zhu Ciqi (1807–1881). Huang had spent some time in Japan between 1900 and 1901, where he witnessed and was impressed by Meiji Japan's successful modernization. This experience convinced him that the new learnings of the West were much more important and pressing for China than success in the Imperial examination.[111] In 1902, ordered by his mother, he tried for the third and final time to pass the Imperial examination. He failed, but this failure did not discourage him at all. Obtaining a scholarly honor (*gongming*) could not save China, but publication could. In the same year, he arrived in Shanghai,

---

[108] See Luan Meijian, *Minjian de wenren yaji: Nanshe yanjiu* (Elegant Assembly of the Men of Letters: A Study on the Southern Society) (Shanghai: Dongfang chubanshe zhongxin, 2016), 83.
[109] The original text reads: 操南音, 不忘舊也。 Yang and Wang, *Nansheshi changbian*, 13.
[110] Wu Zhongliang, "Deng Shi shixue sixiang lunxi" (Interpretation of Deng Shi's Historical Thought), *Dongfang luntan* 2 (2003): 111.
[111] Sun Yingying, *Cultivation Through Classical Poetry: The Poetry and Poetic Studies of Huang Jie (1973–1935)*, unpublished PhD dissertation, University of Hong Kong, 2013, 18–20.

where he finally had the opportunity to familiarize himself with Western knowledge and created the *Zhengyi tongbao* (*Journal of Politics and Technology*) with Deng Shi. At this time, both men were convinced reformers. The politics section of the *Zhengyi tongbao* echoed that of *Xinmin congbao*. Deng Shi's statement against revolution in 1903 resonated with the intellectual tradition of Western conservatism's suspicion that society could be remodeled through ideology:

> The proponents of Western learnings have learnt about civil rights, liberty and the bloody revolution preached by Western politicians. Impressed by the progress of Western society, they are convinced that the same revolution should be reproduced in China. Nevertheless, they ignore the fact that Chinese people's mentality has yet to be enlightened... If they are provoked to destroy the current regime, I fear that they will all die before the end of the drama. Is it the intention of these gentlemen to encourage innumerable people to die for their own ideal vision?[112]

This was also the case for Ma Xulun (1885–1970). When he learnt through the Shanghainese newspapers that the Empress Dowager and the Emperor had fled to Xi'an to escape the Alliance of Eight Nations, he suffered a mental breakdown as if the whole world had ended.[113] He became a revolutionary under the influence of his professor at the School of Yangzheng (*Yangzheng shushu*) in Hangzhou, who encouraged him to read books forbidden by the Qing Court, namely the Ming loyalists' work and the historical records of the massacre of the Han Chinese by the Manchu, when the latter invaded China.[114] After being expelled by the school and consequently missing the opportunity to study in a military academy in Japan because of his involvement in a revolt against the school bureaucracy, he went to Shanghai in 1912 and began his collaboration with Huang and Deng in order to support his family financially.

Ma once recalled that Zhang Taiyan's public speech at Zhang Garden (*Zhangyuan*) in the Shanghai International Settlement was extremely popular, with numerous attendees listening enthusiastically to his revolutionary appeals. Zhang's allure was indeed potent and many previous reformers, such as Deng and Huang, changed their political stances immediately after they started collaborat-

---

112 Deng Shi, "Lun zhengzhi sixiang" (On Political Thought), in *Jindai Zhongguo shiliao congkan xubian di nianba ji* (The 28th Volume of the Second Edition of Sources on Modern Chinese history), ed. Shen Yunlong (Taipei: Wenhai chubanshe, 1976), 93.
113 Ma Xulun, *Wo zai liushisui yiqian* (My Life before the Age of 60) (Chongqing: Shenghuo, 1947), 11.
114 Ibid., 12 and 18.

ing with him around 1904. Huang joined the Tongmenghui in 1909.[115] More importantly, Zhang brought to the Society for the Preservation of National Learnings an active and gifted intellectual, Liu Shipei, whom Hao Chang praised as "the only person among the revolutionary intelligentsia to rival Zhang's intellectual prestige."[116]

Liu's outstanding intellectual accomplishments greatly endeared him to political figures at the time. He obtained the degree *shengyuan* at the age of 18 and the degree *juren* one year later.[117] He supported the New Policies and aimed to pass the last level of the Imperial examination. Had he succeeded, it is very likely that he would have become a reformist official. His failure to obtain the final degree obliged him to find a job outside the political realm. He later made his acquaintance with Zhang Taiyan and Cai Yuanpei in Shanghai, with Cai offering him a teaching position in the Chinese Association for Education.[118] Following his conversion to the revolutionary camp, he decided to change his name to Guanghan – restore the Han Chinese.[119] A prolific writer, he published his political comments widely in anti-Manchu newspapers and his academic essays in *Guocui xuebao* (*Journal of National Essence*) – the official body of the Society for the Preservation of National Learnings that he cofounded with Deng Shi and Huang Jie in 1904. He later became arguably the most radical member of the National Essence School. In 1907, the Qing government issued a wanted notice for him. He fled to Japan and joined the Tongmenghui.

Only three months after his arrival, he founded the Society for the Study of Socialism (*Shehui zhuyi jiangxihui*) with Zhang Ji (1882–1947). Together with his feminist wife, He Zhen (1886–?), he was actively involved in the Japanese anarchist network, notably Kōtoku Shūsui (1871–1911).[120] The Society published the anarcho-socialist journal *Tianyi* (*Natural Justice*), which later became *Hengbao* (*The Chinese Anarchist News. EQUITY*). Members affiliated with the Society formed *Dongjing pai* (The Tokyoite Party) as part of the Chinese anarchism. Although they shared many of their ideas with the Chinse anarchists in Paris (cf.

---

115 Liu Jun, "Shiren Huang Jie de sixiang yu fengge" (Huang Jie's Thought and Literary Style), in *Huang Jie shixuan* (Selected Poems of Huang Jie), ed. Liu Sifen (Guangzhou: Guangdong renmin chubanshe, 1993), 3.
116 Hao Chang, *Chinese Intellectuals in Crisis* (Berkeley: University of California Press, 1987), 146.
117 Chen Qi, *Liu Shipei nianpu changbian* (Chronological Biography of Liu Shipei) (Guiyang: Guizhou renmin chubanshe, 2007), 33.
118 Ibid., 44.
119 Ibid., 41.
120 "Preface," in *Tianyi · Equity* (Tianyi & Equity), ed. Wan Shiguo and Liu He (Beijing: Zhongguo renmindaxue chubanshe), 3.

Chapter 2), they were antimodern in the sense that they sought to construct a universal agrarian lifestyle and political system.[121] A Japanese newspaper observed that the Chinese adherents of this Society were no longer satisfied with liberal democracy or the anti-Manchu cause.[122] This observation was confirmed by the Society itself, for which the most fundamental revolution was socialist revolution, while the victory of nationalism would trigger more violence since nationalism was built upon racial and cultural differentiation.[123]

Arif Dirlik professed that "anarchists imagined a society where individual freedom could be fulfilled only through social responsibility, but without being sacrificed to it."[124] This observation is partially inconsistent with the reality. Although the Society abhorred liberation without a sense of responsibility and denounced cities like Shanghai as cluttered with thieves and prostitutes,[125] it is clear that the aspiration for equality overwhelmed liberty. As Peter Zarrow explains, Liu Shipei "postulated a complex, technologically advanced society wherein individuals performed only one economic task at a time... Jobs could not be assigned according to some system based on the individual's strength or talent... Nor could individuals simply be allowed to choose their own jobs as their inclinations tended."[126] Thus, Liu's anarchism was characterized by a collectivism that suffocated individuals' autonomy for independence.[127] This tendency was also reflected in the name of the Society. Inspired by Kōtoku's Society of People (*Heimin sha*), the Society for the Study of Socialism was renamed the Society for the Equality of People (*Qimin she*).[128] Unlike Chinese anarchists in

---

121 Eugene Chiu, *Qimeng, lixing yu xiandaixing: jindai Zhongguo qimeng yundong, 1895–1925* (Enlightenment, Reason and Modernity: Chinese Modern Movement of Enlightenment, 1895–1925) (Taipei: National Taiwan University Press, 2019), 154.
122 Tomita Noburu, "Shakai shugi kōshūkai oyobi Ashū washin kai kanren shiryō" (Sources of the Society for the Study of Socialism and the Asiatic Humanitarian Brotherhood), in *Chūgoku bunka to sono shūhen* (Chinese culture and its peripheries), ed. Hosoya Yoshio (Sendai: Tōhoku gakuin daigaku chūgokugaku kenkyūkai, 1992), 114.
123 "Shehui zhuyi jiangxihui guanggao" (Advertisement of the Society for the Study of Socialism), in *Tianyi · Equity*, 294.
124 Arif Dirlim, *Anarchism in the Chinese Revolution* (Berkeley, Los Angeles and London: University of California Press, 1991), 4.
125 Zhida, "Nandao nüchang zhi Shanghai" (Shanghai, Where Men Are Thieves and Women Are Prostitutes), in *Tianyi·Hengbao*, 283–5.
126 Peter Zarrow, *Anarchism and Chinese Political Culture* (New York: Columbia University Press, 1990), 85.
127 Cho Se-Hyun, *Qingmo minchu wuzhengfupai de wenhua sixiang* (Cultural Thoughts of Anarchism at the End of the Qing and the Early Republic) (Beijing: Kexue wenxian chubanshe, 2003), 12.
128 "Shehui zhuyi jiangxihui guanggao", 612.

Paris, Tokyoite anarchists understood that their utopian agrarian society could be not achieved immediately but would begin to be concretized when weak nations declared their independence vis-à-vis the strong nations. This was the reason why members of the Society co-founded the Asiatic Humanitarian Brotherhood (*Yazhou heqinhui*) that federated revolutionaries from colonized Asian countries to resist Western and Japanese imperialism.

The National Essence School and the Southern Society that acted as the mouthpiece of the Tongmenghui formed a broad network of associations that linked the foreign concessions in Shanghai and the revolutionary congregations of Chinese dissidents in Tokyo, within which political radicalism was transmitted through radical cultural ideas in numerous journals and magazines issued by the two associations. If Kang Youwei's cultural radicalism created a revolutionary seed that he was not willing to plant,[129] political radicals, many of whom had once been his followers, sowed the revolutionary seed and watered it. Since the late 1890s, China had been submerged in the rising tide of nationalism. Although Kang stated that the uncrowned king, Confucius, had provided universally valid political lessons and models for thousands of generations, he had to nationalize the Confucian religion in order to fit it into a modern national society.[130] In 1902, he reconfirmed that Confucius' so-called Three Ages deemed it imperative that China followed the path to a constitutional monarchy.[131] Zhang Taiyan responded in *Subao* by accusing him of manipulating Confucianism to fawn on the Qing government that precipitated the decline of China for his own interests to the detriment of those of the nation.[132] It was also in this text that Zhang referred to the Guangxu Emperor as a clown, who could not tell beans from wheat – a lèse-majesté that costed him three years in prison.

At a cultural level, many revolutionary intellectuals accused Confucianism of being the spiritual culprit responsible for China's backwardness and an accomplice in the dictatorship of the absolute monarchy. Refuting Kang's advocacy of the adoption of the Confucian calendar, Liu Shipei argued in favor of calculat-

---

129 Although Kang Youwei opposed the revolution to overthrow the Qing dynasty, he did not deny the righteousness of revolution if the circumstances required it, which, however, was not the case in the 1900s. See, Peng, *Rujia zhuanxing yu wenhua xinming*, 60–1.
130 Ibid., 19.
131 Kang Youwei, "Da Nanbeimeizhou zhu huashang lun Zhongguo zhi kexing lixian bukexing gemingshu" (Responding to Chinese Businessmen in North and South America, Arguing that Only Constitutionalism, not Revolution, Is Feasible in China), in *Kang Youwei zhenglunji*, 474–94.
132 Zhang Taiyan, "Bo Kang Youwei lun gemingshu" (Essay in Refutation of Kang Youwei and on Revolution), in *Zhang Taiyan zhenglun xuanji*, 194–208.

ing years according to the Yellow Emperor.¹³³ The revolutionaries' veneration of the Yellow Emperor as the common ancestor of the Han Chinese not only aimed at excluding the Manchu from the Chinese nation and delegitimizing their rule over China but was also motivated by the desire to find an authoritative alternative to Confucian learning. Hence, they accepted Terrien de Lacouperie's theory, on which I will elaborate more fully in Chapter 2. According to the French orientalist, the authentic beginnings of the Chinese civilization could date back to the time of the Yellow Emperor – a Caucasian who had moved with his tribe to China from Mesopotamia. Terrien de Lacouperie's somewhat fanatical theory was extremely attractive to revolutionaries, in that "he pointed the way to demonstrating inherent virtue in Chinese culture despite its present decline... [which] flowed from the intervention of Confucian learning towards the end of the Zhou dynasty, when the Yellow Emperor's social order was being swept away."¹³⁴ As such, the authentic Chinese culture uncovered ancient philosophies other than Confucianism that shared the same civilizational origins with the modern Western political system. According to this logic, the political radicalism that was aimed at revolutionary praxis and Chinese republic building was rationalized in a cultural radicalism built upon an alternative Chinese tradition.¹³⁵

In 1906, the library of the Society for the Preservation of National Learnings – the first private library in China – opened its door in Shanghai. Deng Shi welcomed the fact that more and more foreigners were interested in ancient Chinese objects and books. To him, it showed that Chinese civilization had an important role to play in the progress of humanity. However, he was deeply concerned by the hemorrhaging of Chinese antiquities and books to overseas collectors, museums, and researchers.¹³⁶ Thus, the library assumed the mission of preserving Chinese material civilization in China and had a collection of 60,000 ancient books, most of which were donated. The library issued a statement declaring: "To promote nationalism, the preservation of ancient learnings and the promotion of national radiation now bears upon the shoulders of everyone."¹³⁷ Book-collecting became a highly political act; many ancient books collected were

---

133 Liu Shipei, "Huangdi jinian lun" (On Calculating Years according to the Yellow Emperor), in *Xinhai geming qianshinianjian shilun xuanji*, Vol. 1, 721.
134 Lin, "Historicizing Subjective Reality," 31.
135 Aymeric Xu, "Mapping Conservatism of the Republican Era: Genesis and Typologies," *Journal of Chinese History* 4 (2020): 140–1.
136 Deng Shi, "Aiguo suibi" (Patriotic Essay), *Guocui xuebao* 10, no. 49 (1908): 5692.
137 "Guoxuebaocunhui baogao dierhao" (Second report of the Society for the Preservation of National Learnings), *Guocui xuebao*, Vol. 6 (1906), 2997.

those of the Ming loyalists.[138] As such, traditional culture was upheld, not to obstruct Westernization but to forge a national identity and a nationalist sentiment against the Manchu government.

Figure 1.3: Photograph of the 14 participants at the opening of the library of the Society for the Preservation of National Learnings on October 12, 1906. From right to left, the names of the guests are not registered: Shen Tingyong, guest, Huang Jie, Deng Shi, guest, Gao Tianmei, Zhu Baokang, Ma He, guest, guest, guest, Wen Yongyu, Lu Shaoming, guest. From *Guocui xuebao*, Vol. 1 (1906), 92.

The debates between reformers and revolutionaries ended with the Wuchang Uprising that preluded the Revolution of 1911. In the late Qing ideological context, the meaning of radicalism cannot be fully appreciated without an awareness of the cultural and political dynamics on which it was built. Reformers aligned with Kang Youwei and Liang Qichao, who were radical in the sense of being anti-Confucian orthodoxy but were not supporters of the revolution. They were culturally radical but their heterodox New Text scholarship, combined with Western political theories, focused on the preservation of the Qing dynasty. Eugen Weber describes conservatism as a doctrine that seeks to preserve the present even

---

138 Wang Fan-sen, "Qingmo de lishijiyi yu guojia jiangou – yi Zhang Taiyan weili" (Historical Memories and the Construction of Nationhood – A Case Study of Zhang Taiyan), in *Zhongguo jindai sixiang yu xueshu de xipu* (A Genealogy of Modern Chinese Thought and Scholarship) (Taipei: Linking, 2003), 97.

against the past.[139] The "present" during the 1900s was the Qing dynasty. At variance with the traditionalists, conservatives reacted accordingly to the present situation. To defend the Manchu's reign, orthodox Confucianism, which constituted the "past," could, and should, be sacrificed, so that the Qing government could continue to claim legitimacy to rule over China. With the rise of the revolutionary force, their monarchism was indubitably conservative. Political conservatism and cultural radicalism worked in tandem to resist revolution. Revolutionary intellectuals, affiliated with the Tongmenghui, were also culturally radical, with their point of reference being the Yellow Emperor and other traditional learnings that were overshadowed by Confucianism. Combining these cultural elements with Western political theories to institutionalize a future Chinese republic, their political radicalism complemented cultural radicalism in developing an anti-Manchu revolution in the 1900s. Indeed, the crucial question here is to which political order does the preservation of a certain cultural element give rise? It is through this lens that the meaning of late Qing radicalism and conservatism can be fully seized.

Reactionaryism

Reactionaryism reveals a desire to reestablish the political systems and institutions of the past. As the Chinese translation of the word suggests, reaction (*fandong*) embodies revulsion towards the reactionaries' perception of a polluted present and a will to return to the past. Therefore, the political institutions dear to reactionaries are fatally obsolete. It is thus important to distinguish reactionaryism from conservatism and traditionalism. On the one hand, the Qing officials, by opposing any reforms, incarnated the traditionalist spirit, but they were not reactionary, since the reforms put forward by reformers operated within the framework of traditional sources, albeit with a heterodox twist. Thus, there was nothing to restore at the political level during the last few years of the Qing dynasty and the attachment of the anti-reformers to the tradition was fueled by a psychological aversion, rather than a concrete movement of reestablishment which remains nevertheless at the core of reactionaryism. On the other hand, unlike conservatives whose focal point is the present, reactionaries are ready to sacrifice the present that they joined against their will for a return to the past.

---

139 Weber, "Ambiguous Victories."

An eloquent example that differentiates conservatism from reactionaryism is shown by the different political paths that Liang Qichao and Kang Youwei followed after 1912. Immediately after the outbreak of the Revolution of 1911, Liang published a text, in which he still argued for a strong central government and constitutional monarchy.[140] However, when the Republic was formally proclaimed in 1912, Liang supported the Republic, provided it would be conservative.[141] He stated that the "form of state" (*guoti*) should be maintained as it was, while the "form of government" (*zhengti*) was the real issue.[142] The form of state referred to the monarchy or democracy, whereas constitutional monarchy, absolute monarchy, and democratic constitutionalism constituted three forms of government.[143] Liang was a monarchist during the 1890s and 1900s because he wanted to maintain the status quo, however, since the Qing dynasty and the imperial system were overthrown in 1912, the status quo was now the Republic. In this regard, Liang was indeed a conservative *par excellence*.

Kang Youwei, on the other hand, still promoted the cause of monarchism. On the eve of Puyi's abdication in mid-February 1912, he still counted on Liang Qichao to join his league and save the Qing dynasty. Liang answered him curtly and firmly that he would prefer to break with his former professor once and for all than denounce the Republic.[144] In 1917, the question about whether to join the Allied Powers in the First World War led to a confrontation between President Li Yuanhong (1864–1928) and the Premier Duan Qirui, with the President and the Parliament opposing Duan's call for a declaration of war on Germany.[145] The confrontation resulted in the dismissal of Duan as Premier, who immediately called on the provincial troops to intervene and rebel against Li. Li demanded that General Zhang Xun enter Beijing and mediate the situation. Taking advantage of the unrest, Zhang seized the capital on July 1 and proclaimed the restoration of Puyi.[146] Kang was one of the key figures in the restoration that only lasted for 12 days before it was crushed by the Republican troops. As a re-

---

140 Liang Qichao, "Xin Zhongguo jianshe wenti" (On the Construction of a New China), in *Liang Qichao quanji*, 2433–43.
141 Chang Peng-yuan, *Liang Qichao yu Minguo zhengzhi* (Liang Qichao and Republican Politics) (Shanghai: Sanlian shudian, 2013), chapter 1.
142 Liang Qichao, "Biren duiyu yanlunjie zhi guoqu ji jianglai" (My Opinions on the Press in the Past and the Future), in *Liang Qichao quanji*, 2509.
143 Liang Qichao, "Li xianfa yi" (On the Establishment of Constitution), in *Liang Qichao quanji*, 405.
144 Chang, *Liang Qichao yu Minguo zhengzhi*.
145 Andrew J. Nathan, *Peking Politics 1918–1923: Factionalism and the Failure of Constitutionalism* (Ann Arbor: University of Michigan Center for Chinese Studies, 1998), 91.
146 Ibid.

action to this event, Liang called Kang a "shameless scholar who knew nothing about politics" and a "sinner of the nation."[147] Although Liang spoke highly of Kang after his death,[148] this event did mark the end of the relationship between a teacher and a student, as in the case of Zhang Taiyan and Yu Yue.

The Provisional Government of the Republic of China (1912–1913) allowed the Imperial family to continue to live in the Forbidden City, with Puyi holding his imperial title and the Republican government providing an annual subvention. Consequently, many royalists remained assured that the Qing Court still existed, and the restoration was only a matter of time and chance.[149] The political turmoil from 1912 onwards reinforced their expectation, since "the tranquility, stability and dignity [in the Forbidden City] promise a ready solution by going back, if to move forward as a republic would be a domed failure."[150] Particularly active loyalists were those of the Royalist Party (Zongshe dang) organized in 1912.

With Yuan Shikai's accession to the presidency, the Royalist Party tried to undermine the Republic and Yuan in any way possible in Northern China, including military rebellions.[151] Yuan himself was not interested in a restoration of the Qing dynasty and began to take action to suppress the Royalist Party.[152] However, Yuan was not averse to starting his own monarchy. Shortly after the Revolution of 1911, Sun Yat-sen and Song Jiaoren merged the Tongmenghui and several pro-revolutionary parties to create the KMT. While the founding of the Republic of China was proclaimed in Nanjing in 1912, the Xuantong Emperor still held on to his throne in Beijing. Without military power, the Nanjing government negotiated with the Prime Minister of the Imperial Cabinet, Yuan Shikai, to arrange the abdication of the Emperor. In exchange, Sun ceded the presidency to Yuan. This negotiation between the two parties was not based on mutual trust. Vigilant towards Yuan, the KMT advocated constitutional parliamentarism to put a check on the power of the President. Led by Song, the KMT secured an over-

---

[147] Liang Qichao, "Fandui fubi dian" (Opposing the Restoration), Dagongbao (July 3, 1917); Jingtang, "Kang Youwei yu Liang Qichao" (Kang Youwei and Liang Qichao), Gujin yuekan 1 (1942).
[148] Liang Qichao, "Gongji Kang Nanhai Xiansheng wen" (Elegiac Address to Kang Youwei), in Yinbingshi wenji (Works of Liang Qichao), Vol. 44 (Taipei: Zhonghua shuju, 1960), 29–31.
[149] Feng Jia, "The Dragon Flag in the Republican Nation: The Dowager Empress Longyu's Death Ritual in 1913 and Contested Political Legitimacy in Early Republican China," in Transnational Histories of the 'Royal Nation', ed. Milinda Banerjee, Charlotte Backerra, and Cathleen Sarti (London: Palgrave Macmillan, 2017), 225.
[150] Ibid.
[151] Phil Billingsley, Bandits in Republican China (Stanford: Stanford University Press, 1988), 56–7.
[152] Ibid., 59.

whelming majority in the first National Assembly election in December 1912 and Song was to become Prime Minister. But before taking office, Song was assassinated in Shanghai in 1913. The KMT accused Yuan of the assassination and launched a Second Revolution (*Erci geming*). Following its failure, the KMT was purged by Yuan and its members fled into exile in Japan.

In August 1915, at Yuan's behest, the American political scientist and constitutional adviser to the Chinese government Frank Johnson Goodnow (1859–1939) suggested that a monarchy suited China better than republicanism, if the population and foreign nations were not against the restoration.[153] Monarchist Yang Du (1875–1931), who had reasoned that the Chinese had long been used to monarchic rule and suggested to Yuan that the Republic was a transitional period during which Yuan could emulate Napoléon Bonaparte (1769–1821), organized the Chouanhui (Peace Planning Society) almost immediately to support Yuan's restoration.[154] One of the members of the Chouanhui was the former revolutionary Liu Shipei.

The political trajectory of Liu Shipei is somewhat mysterious. In 1908, Liu Shipei and his wife returned temporarily to China. In Shanghai, he met Chen Qubing, Liu Yazi, Gao Xu, and other revolutionaries. During the meeting, Chen proposed the foundation of an anti-Manchu association that would later become the Southern Society. The proposition was well received by everyone, except Liu, who neither accepted nor opposed the idea.[155] It is believed that at the time Liu had already been bribed by the Qing high official Duanfang (1861–1911) to provide information about revolutionaries' activities to the Court.[156] Upon his return to Japan, he continued the publication of the anarchist magazine *Hengbao* until it was censured by the Japanese police in August. Subsequently, Liu began to exalt the Qing Court publicly and officially broke with the revolutionary camp.[157] Liu was willing to settle for the monarchy in any form, whether it be Han or Manchu. The overthrow of the Republic, restoration, anti-democracy, and establishment of the Confucian religion were the keywords of Liu's discourses during Yuan's reign. He even attempted to involve his former colleagues of the

---

153 Kuo Ting-yee, *Jindai Zhongguo shigang* (History of modern China) (Hong Kong: The Chinese University Press, 2012), 433.
154 Ibid., 432–3.
155 Zhang, *Chen Qubing nianpu*, 87.
156 Chen, *Liu Shipei nianpu changbian*, 266.
157 For a possible explication of Liu's treason, see Wang Fan-sen, "Fan xihua de xifang zhuyi yu fan chuantong de chuantong zhuyi – Liu Shipei yu 'Shehui zhuyi jiangxihui'" (Anti-Western Westernization and Anti-traditional Traditionalism – Liu Shipei and the Socialism Conference), in *Zhongguo jindai sixiang yu xueshu de puxi*, 235–6.

Society for the Preservation of National Learnings in choosing an era name (*nianhao*) for Yuan's intronization.[158]

Loyalism, in this political context, became a synonym for reactionaryism. Qing loyalists refused to cut off their Manchu queues and many of them became increasingly traditionalist. Luo Zhenyu (1866–1940) is a case in point. Sent by the Qing government to Japan in 1901 to study the modern educational system, he was well known for advocating popular and women's education. However, after the collapse of the dynasty, he sought to restore the traditional educational model from which women were strictly excluded.[159] Like many Qing loyalists, he later followed Puyi and served the Manchukuo, a puppet state in the Empire of Japan in Manchuria.[160] Qing loyalists also undermined the Republic at an intellectual level. Lao Naixuan (1843–1921), for example, argued that the word *gonghe* (republic) meant in reality monarchist government, in ancient sources.[161] To him, the prosperity and the stability of Chinese society was not comparable with that of the imperial period. This argument was widely expressed by former Qing officials instigating a re-intronization of Puyi, such as Song Yuren (1857–1931), Zhang Qin (1861–1949), Zhao Bingjun (1859–1914), and Luo Zhenyu.[162]

Resorting to traditional sources to rationalize political endeavors was a common strategy for reactionaries and conservatives. Nevertheless, while conservatives and reactionaries all valorized the cultural root and shared the disappointment with regard to the political chaos of the new Republic, a return to the past was not feasible from the perspectives of the conservatives, whose selective references to the past, which were by no means idealized as they were in the case of the reactionaries, were deeply anchored in the reality of the present.

Conservatism: Preserving the Present

Although conservatism has historically claimed to conserve the past, conservatism is only interested in conserving particular elements of the past, so that the present can build upon the achievements of the past. Either as a political doctrine or mindset, conservatism differs from traditionalism and reactionaryism

---

158 Ma, *Wo zai liushisui yiqian*, 48.
159 Lin, *Minguo nai diguo ye*, 192.
160 See Chu Wan-li, "Huiyi Fu xiansheng zai Taida de wangshi" (Mr. Fu at the University of Taiwan), in *Huainian Fu Sinian* (In memory of Fu Sinian) by Hu Shi et al. (Taipei: Showwe, 2014), 107.
161 Lin, *Minguo nai diguo ye*, 214–5.
162 Ibid., 68–9; 215–8.

primarily in that it does not express itself as a fear of innovation or a tendency to cling to an idealized past. To preserve the present, conservatives are not afraid to even break with the past. Kang Youwei's cultural radicalism that contested the reliability of orthodox Confucianism aimed to preserve the political present – the Qing dynasty. The conservation of the present implies an inclination towards stability. This means that, without criticism, conservatives will not embrace the present, of which they might have been highly suspicious or the fundamentals of which they rejected, until its final concretization. In this regard, Liang Qichao after the foundation of the Republic could probably be compared with François Guizot (1787–1874), who accepted the consequences of the French Revolution without being adherent to the fundamental historical ruptures that the political ideas of the Revolution advocated.[163] However, on a social and intellectual level, conservatives were often mistakenly labeled as reactionaries in Republican China. This stigma was bestowed on them by their intellectual rivals to delegitimize their voices in the intellectual sphere – an issue I will elaborate on in Chapter 3. In the late Qing era, the relationship between conservatism and radicalism is perhaps the most perplexing one. Both were expressed in a culturalist form, which was applied to entirely different ends of the political spectrum. This is the major difference between a cultural radicalism destined to preserve the present and a cultural radicalism looking towards the future. It is the nuanced culturalist nationalism that fueled both the late Qing political conservatism and radicalism, which I will consider next.

---

163 Huguenin, *Le conservatisme impossible*, 67–79.

2 Culturalist Nationalism from the 1890s to the 1900s

Culturalist nationalism is defined as a nationalism that rested upon the homogenization of the political nation and the cultural nation. The population that formed a nation should share the same cultural origin and the political reforms inspired by the West should resonate with the nation's cultural tradition. Why did this type of nationalism prevail in the late 1890s and 1900s in China? The most obvious reason was Japan's intellectual and political influence on China. But the eagerness of Chinese intellectuals to appropriate a Western modern political system cannot be overemphasized. As Wang Jingwei explained:

> The difference between England, France, and the United States on one side and our nation on the other side is not that their political systems do not exist in China, but that they have better developed these systems. In fact, it is difficult to imitate something that does not exist [in our tradition]. But imitation becomes simpler if we try to ameliorate what exists already.

In other words, attempts to homogenize the cultural nation and the political nation facilitated the nation's political modernization in line with Western experiences. The global political situation forced China to be transformed from an empire into a nation-state. China should join a restructured world in which the traditional conception "All under Heaven" (*tianxia*) that defined China as the civilization was obsolete.[1] This awareness was not developed as easily as might be imagined, since there was no corresponding concept of nation in the Chinese language before the late 1890s. Qing officials were deeply confused by the idea of nation when they negotiated treaties with foreigners in the 1840s.[2] Chen Duxiu only acquired an awareness of nationhood in 1901, when the Alliance of Eight Nations entered the capital:

> Only at this time, I realized that the world population was divided by nations... China was also one of the many nations in the world and I was Chinese. The honor and the humiliation of a nation also bared upon each member of the nation... Only when I was in my 20s

---

[1] Luo Zhitian, "Lixiang yu xianshi – Qingji Minchu shijiezhuyi yu minzuzhuyi de guanlian hudong" (Ideal and Reality – The Interaction between Cosmopolitanism and Nationalism at the end of the Qing and the Early Republican Era), in *Zhongguo jindai sixiangshi de zhuanxin shidai*, 272–3.
[2] Wang Fan-sen, "Wanqing de zhengzhi guannian yu 'xinshixue'" (Political Conceptions of Late Qing China and the "New Historiography"), in *Zhongguo jindai sixiang yu xueshu de xipu*, 197.

did I learn about the concept of nation and understand that nation was the big family of all nationals and everybody should do their part for the betterment of the nation.³

Indeed, reformers of the 1890s explained China's embarrassing backwardness with respect to the outside world as the consequence of the lack of a consciousness of political togetherness among the people. The salvation of China could not be dependent solely on the government's initiatives. Instead, a "nation" should be built as a modern political sphere that transcended traditional familial allegiance. Chen Duxiu was clearly a good student of the late Qing reformers in this regard. It also worth highlighting that he only mentioned the nationalist awareness as the core of national salvation, but before the advent of nationalism it was the concept of *qun* – groups, communities, or society – that endowed reformers with a potent ideological weapon to push for institutional change.

As I have mentioned in the introductory chapter, the concept of nationhood entered the Chinese political vocabulary only after 1898, when reformers who escaped the Empress Dowager Cixi's purge learnt about this idea in Japan. Qun, on the other hand, abounded in reformers' writings at least as early as 1895, when Yan Fu (1854–1921) published the famous *Yuanqiang* (*On the origins of strength*) in *Zhibao* (*Journal of Zhili*). In this article, Yan Fu explained how the lack of a sense of social togetherness affected China's strength by citing Xun Zi (c. 310– c. 235 BC):

> The human force is inferior to that of a bull and a horse runs much faster than a man. Why, then, are men able to hold the bull and the horse in slavery? Because they know how to form a social group (*qun*). Why are they able to do so? Because they assume different social roles (*fen*). Why can they put in place such an institution? Because they share the principle of justice and duty (*yi*). So, thanks to this principle, everyone has a place in this hierarchical structure and lives in harmony with each other. Harmony makes solidarity, solidarity is a force, and force leads to prosperity.⁴

In 1877, Yan Fu was sent to the Royal Naval College to study for two years. In 1896, he translated *Evolution and Ethics* by T.H. Huxley (1825–1895) under the name of *Tianyan Lun* (*On Evolution*). This popularized two concepts in China: natural selection (*wujing tianze*) and survival of the fittest (*shizhe shengcun*). These two ideas became widespread and accepted in China because they helped to explain the humiliation that the West had imposed on a regressive China. In

---

3 Ibid., 198.
4 Chen Xulu, "Wuxu shiqi weixinpai de sheshuiguan – qunxue" (The Concept of Society among the Reformers of the Wuxu Era – *qunxue*), *Jindaishi yanjiu* 2 (1984): 163–4.

the preface of *Tianyan Lun*, Yan Fu commented, "those who are able to form a social group survive and those who cannot perish in the evolution."⁵ This view of society is consistent with that of Kang Youwei, who urged the Qing Court to remove prohibition on social associations to form a viable social congregation in which people could collectively strive for China's rightful standing in the world.

If *qun*, or society, had already acted as a political reference that transcended people's consanguineous bond, the political usage of nationalism requires more scrutiny. After all, society as a political concept is often "nationalized." Liah Greenfeld notes that "modern societies are nations by definition."⁶ Alain Touraine explains that the abstract concept of society is inconceivable from national society, because "society is defined as a network of institutions, controls and education, which refer the matters to a government, or a territory, or a political collectivity. The idea of society was and always is the ideology of nation under formation."⁷ Late Qing intellectuals often identified society with cultural elements, which are used in general to characterize a cultural nation. Deng Shi, for instance, conceived society to be formed by people of the same racial group who shared and developed common customs, characters, language, and enterprises within a given territory.⁸ In this regard, is one or the other political reference redundant?

As Kevin Doak points out, society and nation can be at odds if society is identified with people while nation is regarded as synonymous with the state. However, nationalist expectations from society may also be explored in various "social forms" when society is mobilized against the state.⁹ Thus, reformers and the Qing Court made it clear that bottom-up initiatives should be regulated by the government. Nevertheless, once people realized that they were no longer passive imperial subjects, but nationals entitled to political rights, the political upheavals mobilized in various social forms in the name of a newly-obtained political identity would be out of the reformers' control. The word *min* (people) became consequently highly politicized in this context. Chinese civilians were no longer *chenmin* (subject), but *gongmin* (citizen), *guomin* (national), and *renmin*

---

5 Yan Fu, *Tianyan Lun* (On evolution) (Beijing: Commercial Press, 1981), 32.
6 Liah Greenfeld, "Nationalism and Modernity," *Social Research* 63, no. 1 (1996): 10.
7 Alain Touraine, "Une sociologie sans société," *Revue française de sociologie* 22, no. 1 (1981): 5.
8 Deng Shi, "Yuyan wenzi duli zhier" (The Second Chapter on the Independence of the Language and the Letters), in *Jindai Zhongguo shiliao congkan xupian diershibaji*, 173; idem, "Zhengqun dier" (The Second Chapter on the Politics of *Qun*), in *Jindai Zhongguo shiliao congkan xupian diershibaji*, 109.
9 Doak, *A History of Nationalism in Modern Japan*, 127.

(people, in the sense of citizens). Sun Yat-sen's "Three Principles of the People" (*sanmin zhuyi*) is a telling example. Society, in this sense, acted as a counterweight that forced the government to take action and as a watchdog against state-based intruders into the social realm.

While there was harmonization between society and political nationalism, the nationalization of society at the cultural level remained an artificial process. In contrast to the common view that society is equal to national society, Jean-Claude Milner reasons that "society" as an "organizing point of the political vision of the world" is a regime without limitation, because "the function of modern society includes all the possible variants – human or not human, animated or unanimated... Modern society has the vocation to cover the whole world and to embrace the totality of all beings."[10] In other words, society as a political reference is not subject to any geographical, cultural or racial restrictions. This connotation of the concept of society was known by Chinese intellectuals at the time through the ideology of socialist internationalism. As Doak underlines, "Society – and building on that concept, socialism – is supposed to provide an alternative to the hierarchical, oppressive ideology which they associate with 'nationalism': the more of one, the less of the other."[11] As such, there is an intrinsic contradiction in the concept of society. While delimiting a closed territory in which nationalist expectations are expressed, society is supposed to provide an alternative to nationalism, which to many late Qing intellectuals was the founding ideology of imperialism of which China was a victim.

Chinese anarchists in Paris reasoned that since the nation was rooted in cultural and ethnical particularism and posed a limit on society, it should be abandoned. Dismissing revolutionaries' anti-Manchu actions as mere Han-centric xenophobic sentiment, they saw the nation as nothing but an unsustainable impediment to the implementation of a universal pacific society. It was argued that Chinese culture should be eliminated, since it delineated the characteristics of the Han and differentiated them from others. Although not denying the legitimacy of the anti-Manchu revolution, they maintained that this legitimacy could only be justified in the framework of a socialist movement rather than nationalism.[12]

---

10 Jean-Claude Milner, *Les penchants criminels de l'Europe démocratique* (Lagrasse: Éditions Verdier, 2003), 21–3.
11 Doak, *A History of Nationalism in Modern Japan*, 127.
12 Fan, "Guocui zhi chufen" (Sanction on national essence), in *Xinhai geming qianshinianjian shilun xuanji*, Vol. 3, 191–2; Li Shizeng, 'Shenlun minzu, minquan, shehui sanzhuyi zhi yitong' (On the Similarities and Differences between Nationalism, People's Power and Socialism) in *Xinhai geming qianshinianjian shilun xuanji*. Vol. 2, 1003–8.

Aspirations to reassemble all the nations and attachment to the Chinese nation-state were common intellectual traits of reformers and revolutionaries. Universalism constituted the ultimate purpose of the evolutionary scheme and it was not contradictory to recognize the reality of the nation and to dream of a universal peace at the same time.[13] As Deng Shi once wrote: "In the scheme of evolution of the world, individualism transforms into nationalism, nationalism will become imperialism to finally reach world pacifism, which is socialism."[14] Nevertheless, world pacificism was simply too utopian at that time because the underlying premise of power struggle between nations still constituted a serious threat to China. If nationalism did not constitute an end in itself, it was the only pragmatic political principle capable of guiding China to find its due place in a world of "survival of the fittest."

Since political legitimacy was considered to be found in society, granting of political rights within the state to a part of the population within society had to be "based on the exclusion of foreigners."[15] However, the limitless political reference of society did not provide a meaningful conceptual framework to differentiate Chinese from foreigners. In this context, the most immediate cultural elements that characterized a cultural nation provided a necessary limit on society and stipulated an indispensable distinction between national society and foreign society. The nationalization of society was therefore an artificial process: membership of a nation was acquired, not only by one's implications in various social forms through which nationalist expectations were expressed, but also by an individual's cultural and ethnic roots.

In this way, society, the political nation, and the cultural nation became a coherent, integral, and homogenous entity. Revolutionaries and reformers all theorized Chinese nationalism in this manner. They differed from each other primarily on the relationship between the Manchu and the Han Chinese, as well as the cultural elements mobilized to rationalize opposing political projects. In this way, cultural nationalism based upon the cultural radicalism of the two parties nourished both the late Qing political radicalism and conservatism. The two parties, though resorting to the same ideological framework, came up with entirely different political aims. To understand how this had happened, we must scruti-

---

13 WONG Young-tsu, "The Ideal of Universality," in *Reform in 19th Century China*, ed. Paul A. Cohen and John E. Schrecker (Cambridge, Mass.: East Asian Research Center, Harvard University, 1976), 150.
14 Deng Shi, "Lun Shehui Zhuyi" (On Socialism), in *Jindai Zhongguo shiliao congkan xubian diershibaji*, 98–9.
15 Hu Hanmin, "Paiwai yu guojifa" (Xenophobia and International Law), *Minbao* 4 (1906): 58.

nize the cultural elements, or *guocui*, that informed their culturalist nationalism and the strategies to politicize it.

The Society for Political Education and National Essence in Japan

The Society for Political Education is generally seen as the counterpower of the Westernization of Meiji Japan. As mentioned in the introductory chapter, the Society defended the politics of *kokusui*, which was invented by the founder Shiga Shigetaka to translate the word "nationality" in the Society's official body *Nihonjin* in 1888. The magazine's existence was a long but precarious one. Due to its anti-government position, the magazine was censured multiple times and changed its name twice to *Ajia* (*Asia*) so that the publication could be ensured. The *Nihonjin* ceased its publication in 1906 and was replaced by *Nihon yobini Nihonjin* (*Japan and Japanese*). The Society itself outlived the Meiji era and existed until 1945.

Many of the most active members of the Society had received both a traditional and a Western education. Shiga, for example, received his early education from his maternal grandfather whose family had been Confucian scholars for generations. Shiga himself became a great scholar in Chinese classics.[16] In 1880, he was enrolled in the Sapporo College of Agriculture (*Sapporo nōgakkō*) and obtained a diploma in 1884.[17] Education in this college was very Western. Many of his fellow students were Christians, and English was the lingua franca.[18] Two years after graduation, he traveled to Australia and the Pacific Islands – a trip that would change his political and intellectual trajectory. He was stunned by the social and economic achievements that British colonizers had realized in Australia and New Zealand, but what impressed him the most was the rising Australian nationalism and the miserable living conditions of the aboriginal population. While white Australians were chanting for "Australian for Australians" and Australian independence, the Maori population lost control of their

---

[16] Masako Gavin, *Shiga Shigetaka, 1863–1927: The Forgotten Enlightener* (London: Routledge Curzon, 2001), 47.
[17] Sadahira Motoshiro, "Shiga Shigetaka, hito to shisō" (Shiga Shigetaka, Life and Thought), *Kwanseigakuindaigaku shakaigakubu kiyō* 24 (1972): 33.
[18] Pyle, *The New Generation in Meiji Japan*, 57.

lands, as well as their lives, due to the virus brought by Europeans.[19] Shiga believed that Japan should learn an important lesson from the submission of the Australian and New Zealand aboriginal population to the colonizers:

> Alas! Japan could be another New Zealand. As I look up at the autumn sky of these Southern Seas, I fear the threat to my home country far away. Having witnessed such cultural and radical opposition in New Zealand, I – as a son of the new Japan – must take immediate actions to make my people aware of this possibility happening back home.[20]

However, Japan at that time was not open to the kind of "cultural resistance" Shiga strove to capture. The nation replaced China and became the biggest power in Asia thanks to its successful Westernization. One of the symbols of Japan's Westernization was the Rokumeikan (Banqueting House), a large two-story building in Tokyo, commissioned to house foreign guests. The construction was ordered by the Foreign Minister Inoue Kaoru (1836–1915) and was completed in 1883. Parties and balls were regularly held to impress Westerners and to convince them that Japan was their equal. But Westerners' comments on Rokumeikan were not always favorable. The French marine official Pierre Loti (1850–1923) once wrote that the tasteless imitation of these Japanese had a "close resemblance to a monkey."[21] If Shiga had been invited to one of the balls, he would probably have said no less than this. To him, Japan's Westernization was encapsulated in the "principle of daub" (*tomatsu shigi*) and the "principle of breaking with all Japanese elements" (*Nihon bunshi daha shigi*), which consisted of superficially imitating Western lifestyles and buildings, rather than selectively assimilating the progressive elements of Western civilization.[22]

Miyake Setsurei (1860–1945) was another founder of the Society. He started studying Chinese classics at the age of six. In 1876, he was enrolled in the preparatory school of the University of Tokyo and obtained a bachelor's degree in philosophy in 1883. Although not all the founding members received a literary or philosophical education, most of them held a modern university degree. Sugiura Jūgō (1855–1924) was a chemist trained in England.[23] Kikuchi Kumatarō (1864–1908) and Kon Sotosaburō (1865–1892) were graduates of the Sapporo

---

19 Ogiwara Takashi, "Shiga Shigetaka ni okeru kokusuishugi no kannen" (Shiga Shigetaka's Concept of National essence), *Nagoya gakuindaigaku ronshū* 45, no. 2 (2008): 29; Gavin, *Shiga Shigetaka*, 35–6; 81–3.
20 Gavin, *Shiga Shigetaka*, 84.
21 Pierre Loti, *Japoneries d'automne* (Paris: Calman Lévy, 1889), 88.
22 Shiga, "Nihonjin ga kaihōsuru tokoro no shigi wo kokuhakusu."
23 Haga Noboru, "*Nihonjin* no kaisetsu" (On *Japanese*), in *Nihonjin*, ed. Nihonjin kankōkai, Vol. 34 (Tokyo: Nihon tosho senta, 1984), 39.

College of Agriculture. Miyazaki Michimasa (1853–1916), Inoue Enryō (1858–1919), and Tanahashi Ichirō (1863–1942) were educated at the University of Tokyo.

However, despite the members' shared nationalism, they turned national essence into a polysemic and multidimensional concept, which tied into each member's attitude towards tradition, the West, as well as the imperial house. While Shiga related national essence to Japan's geography and climate, Kikuchi associated it with the imperial family.[24] Miyake attempted to link this concept directly to Japanese people. In 1891, Miyake published the famous *Shin zen bi Nihonjin* (*Truth, Goodness, Beauty and Japanese*). Truth suggested new ideas to the open mind that were selectively assimilated into Japan, goodness highlighted Japan's contribution to justice on the international political scene, and beauty underlined the unique Japanese estheticism.[25] Miyake's cultural nation seems to be detached from the often-deprecating judgement on the *Kulturlnation*. He maintained that the Japanese "must value truth, goodness, as well as beauty, and strive for peace in the world."[26] Therefore, the Japanese had no reason to feel inferior to Westerners, since it was intelligence, not physical advantage, that dominated struggles between nations.[27] The *Shin zen bi Nihonjin* was published at the same time as the *Gi aku shu Nihonjin* (*Hypocrite, Badness, Ugliness and Japanese*), in which Miyake denounced Japan's servile mentality with regard to the West.

Indeed, the question of beauty was a recurrent subject of the Society. National essence often centered around the resurrection of the national and traditional beauty: fine arts, theater, music, and esthetic lifestyle.[28] Shiga's masterpiece, *Nihon fūkeiron* (*Japanese landscape*), was an exaltation of Japan's national essence that was reflected in its natural landscape. Published in 1894, *Nihon fūkeiron* quickly became a bestseller and remained so decades after the first edition. The book is known for its scientific rigor expressed in an exquisite language. Shiga was, after all, a professional geographer and one of the most eloquent

---

24 Pyle, *The New Generation in Meiji Japan*, 69–70.
25 Ibid., 153–155.
26 Gavin, *Shiga Shigetaka*, 183.
27 Pyle, *The New Generation in Meiji Japan*, 152.
28 Nakanome Tōru, *Meiji no seinen to nashonnarizumu: Seikyōsha • Nippon shinbunsha no gunzō* (The Young Generation of the Meiji Era and Nationalism: The Society of Political Education and the Society of Japanese Daily) (Tokyo: Yoshikawa kōbunkan, 2014), 124–44.

poets of Chinese style of his time.²⁹ Typically, the scientific description of a landscape was written in an elegant language, followed by a Chinese poem or *haiku*.³⁰ What makes this book even more interesting is the year of its publication – 1894, the year that marked the start of the First Sino-Japanese War. In fact, the Japanese landscape, under Shiga's pen, constituted an expansive space. To him, the annexation of the Shandong province would add Mount Tai to the Japanese landscape, and the incorporation of Taiwan would complete the Japanese landscape with a tropical zone.³¹ Japan achieved both.

Despite the fact that Shiga opposed Fukuzawa's "Shedding Asia" and appealed for a common struggle of the "yellow race" again the West,³² the Society continued to see China as a threat rather than an ally. Like many Japanese supporters of pan-Asianism, the Society believed that Japan should lead the fight in this "racial struggle" between the Orient and the West. The beauty in Miyazake's work also expressed a wish for Japan to become a nation powerful enough to protect its weak neighbors.³³ An editorial of the *Nihonjin* says: "We do not hesitate to affirm that the fate of our nation depends on the repression of this monster at the east."³⁴ If Japan failed to convince China to cede its leading place in Asia in a peaceful way, then resorting to violence in the name of virtuous objectives was the only way to show China who was the new leader of the region.³⁵ To Shiga, the condition of the armistice of the First Sino-Japanese War should force China to cede to Japan the peninsula of Liaodong in Shandong and Taiwan – an ambition that corresponded perfectly to what he expressed in *Nihon fūkeiron*. Contrary to Shiga, the sinologist Naitō Konan (1866–1934) acknowledged the heavy presence of Chinese culture in the shaping of a Japanese national essence. This was not to deny the cultural originality of the Japanese nation; rather, it served to rationalize Japan's political control over China. To him, Chinese revolution could not win for two reasons: the lack of the consciousness of nation of the population, as well as foreigners' military intervention to protect their interests and citizens. Thus, Naitō suggested a small central government and a re-

---

29 Anzai Shinichi, "Unmediated Nationalism: Science and Art in Shigetaka Shiga's *The Japanese Landscape* (1894)," *Journal of the Faculty of Letters, The University of Tokyo, Aesthetics* 34 (2009): 68.
30 See, for example, ibid., 68–9.
31 Ibid., 106.
32 Sadahira, "Shiga Shigetaka, hito to shisō," 35.
33 Pyle, *The New Generation in Meiji Japan*, 154.
34 Nakanome, *Meiji no seinen to nashonnarizumu*, 114.
35 Li Xiangying, "Seikyōsha no tai shinninshiki: zasshi *Nihonjin* wo chūshin ni" (The Qing Dynasty in the Eyes of the Society for Political Education: On the Magazine *Japanese*), *Nihon kenkyū* 18 (2015): 92–101.

vival of the political system of *xiangtuan*, which was a kind of rural self-government.[36] Due to the weak central government, China would need foreigners' aide to protect its integrity. Japan was the best candidate because, as Naitō argued, the Chinese cultural center had been relocated in Japan. Thus, not like the West, Japan would only promulgate and execute policies suitable to Chinese' mentality and traditional culture.

After the First Sino-Japanese War, the Society became increasingly invested in the politics of "dismemberment of China" (*Shina bunkatsu ron*).[37] Shiga, Miyake, Sugiura, and 14 other members of the Society were involved in the foundation of the Tōhō Kyōkai (Union of Asian Unity) in 1891. Shiga and Miyake were also members of the Tōa Dōbunkai (East Asian Shared Culture Association), which, in the name of education and cultural exchange, sent 5,000 Japanese students to China as secret agents.[38]

In summary, it is impossible to formulate a categorical definition of national essence since the Society for Political Education never chose to do so. Nevertheless, it should be emphasized that, although national essence was invented as an ideological counterpower to Westernization, the Society was not traditionalist and did not fear new ideas. Shiga made this point clearly when he was questioned by his intellectual rival and friend, Tokutomi Sohō (1863–1957), a fervent proponent of Westernization.[39] After the First Sino-Japanese War, the term national essence began to be popularized among intellectuals and politicians who were not members of the Society to defer all reforms, which deviated from the Society's intention in the 1880s.[40] Given the fact that the Society pursued their expansionist project in the framework of national essence, were Chinese intellectuals' interests in national essence illogical?

---

36 Tam Yue-Him, "'On Intellectual's Response to Western Intrusion: Naitô Konan's View of Republican China," in *The Chinese and the Japanese*, ed. Akira Iriye (Princeton: Princeton University Press, 1980), 184–180.
37 Ibid., 106.
38 Sheng Banghe, "Riben yazhou zhuyi yu youyi sichao yuanliu" (The Origin of the Japanese Asianism and the Japanese Ring Wing), *Lishi Yanjiu* 3 (2005): 135; Urs Matthias Zachmann, "The Foundation Manifesto of the Toa Dobunkai (East Asian Common Culture Society), 1898," in *Pan-Asianism: A Documentary History, Volume 1: 1850–1920*, ed. Sven Saaler Sven and W. A. Szpilman (Maryland: Rowman & Littlefield Publishers, 2011), 118.
39 Gavin, *Shiga Shigetaka*, 11; Haga, "*Nihonjin* no kaisetsu," 37.
40 Pyle, *The New Generation in Meiji Japan*, 184.

National Essence in the Late Qing Political and Ideological Context

Chinese intellectuals may not have perceived the expansionism encapsulated in the Japanese politics of national essence. Even if they had, caution with respect to Japanese imperialism was very likely to have been played down. Despite the outbreak of the First Sino-Japanese War, many still sincerely wished for a Sino-Japanese alliance against the West. Zhang Taiyan even suggested to Li Hongzhang that certain Chinese territories should be ceded to Japan, since the military power of Japan and the two nations' shared culture meant that Japan was the only country that China could count on.[41] Zhang believed that Japan's entry into Weihai could contain Russia and Germany.[42] Furthermore, Zhang's suggestions would offer Japan more than what Japan wanted at the time, since he believed that Japan's aspiration to become a great Asian power was established upon the policy of "preserving China" (*Shina honzen*) to resist Western imperialism. This policy was supported by the Japanese government and many decisive officials, notably Itō Hirobumi (1841–1909) and Konoe Atsumaro (1863–1904).[43] Hence, Japan's foreign policy towards China was not consistent at the turn of the twentieth century. While some wished to partition China, others argued for a preservation of China.

In any case, national essence was a neologism whose signification in the Chinese political and intellectual vocabulary had greatly deviated from its original meaning. In the Chinese historical context, it was Liang Qichao who, in a text of 1901, first used the phrase "national essence." He drew on this concept to argue against the adoption of anno Domini in China.[44] He even extended "national essence" to "Asian essence" (*yacui*) in the slogan of the *Qingyibao*: "Discover the Eastern Asian Teaching to Preserve the Asian Essence."[45] The notion had later federated a large number of defenders of Chinese culture, who appropriated this concept to put forward different, or even opposite, political projects. At the state level, Luo Zhenyu had learnt about this concept from the Japanese

---

41 Zhang Taiyan, "Lun Yazhou yi ziwei chunchi" (Asian Countries Should be Interdependent), in *Zhang Taiyan zhenglun xuanji*, Vol. 1, 5–8.
42 Zhang Taiyan, "Shangshu Li Hongzhang" (Letter to Li Hongzhang), in *Zhang Taiyan Nianpu changbian*, 61.
43 Asanuma Chie, "Guanyu Mingzhi moqi riben jiaoxi paiqian beijing de yanjiu" (On the Context in Which Japan Sent Consultants to China at the End of the Meiji Era), in *Zhejiang yu Riben* (Zhejiang and Japan) (Beijing: Zhongguo shehui kexue chubanshe, 2019), 265–284.
44 He, Fanben yu kaixin, 158
45 The original text reads: "發明東亞學術，以保存亞粹。"

senator and educator Isawa Shūji (1851–1917) during his trip to Japan to investigate modern educational systems for the Qing Court. Isawa told him that Western learnings could not be adopted by sacrificing the national essence.⁴⁶ In 1907, when Zhang Zhidong was nominated the Minister of Education, he proposed to preserve national essence by implementing the School for the Preservation of Antiquity (*Cungu xuetang*) in all provinces.⁴⁷ To him, national essence was national literature (*guowen*), which comprised the writing, the language, and the classics.⁴⁸

Although the project might sound at odds with the Qing Court's plan to popularize modern education, Zhang was not hostile towards Western learnings. In fact, in these schools, science, Western history, geography, and mathematics were taught alongside traditional learnings. He also proposed to establish foreign language schools near each School for the Preservation of Antiquity to encourage students to learn Western languages, so that they would be capable of "studying and consulting Western books in the future."⁴⁹ Moreover, the regulation concerning the implementation of the Schools for the Preservation of Antiquity enacted by the Ministry of Education substantially augmented the proportion of courses on Western learnings. These schools were never conceived to challenge modern educational institutions. In fact, the study of classics was limited to some explanatory books on Confucianism and the study of original Confucian classics was not required.⁵⁰

National essence quickly captured the attention of many reformist-minded scholars. Song Shu (1862–1910), a renowned scholar and educator involved in the reformers' network, for example, reasoned that the prosperity of national essence was proportional to the spread of Western learning. To him, different civilizations constituted the multiple manifestations of the universal truth. Thus, if people did not understand the essence of their cultural tradition, it was naturally impossible for them to select and assimilate useful Western resources.⁵¹

---

46 Ibid., 154.
47 Ibid., 153.
48 Ibid.
49 Luo Zhitian, "Qingji baocunguocui de chaoye nuli jiqi guannian yitong" (Efforts to Preserve the National Essence by Government and Society and Differences and Similarities Between the Two), in *Jindaishi yanjiu* 2 (2001): 77.
50 Luo Zhitian, "Dushu yu chuantong: Qingji minchu shiren yixiang chixu guanhuan de yanbian" (Study and Tradition: The Evolution of a Continuous Preoccupation Among Intellectuals at the End of the Qing and the Early Republican Period), in *Liebian zhong de chuancheng*, 136–7.
51 Song Shu, "Shang Dongfu qingzou chuang Cuihuaxuetang yi" (Proposition to the Governor of Shandong Concerning the Foundation of School of National Essence), in *Shenwen yu mingbi-*

Revolutionaries' visions of national essence were as diverse as that of the Society of the Political Education. To Ma Xulun, national essence comprised three dimensions: ancient art, traditional thought, and political systems.[52] Indeed, *Guocui xuebao* was not only an academic journal. In addition to political comments and academic papers, each issue published the portraits of ancient sages and Ming loyalists, photos of fine art, antiquities, animal and plant species that could only be found in China, and reproductions of ancient paintings and calligraphies. If one is to love the nation, one must first know why the nation is lovable. Following this logic, *Guocui xuebao* added in 1907 the sections of "natural science" (*bowu*) and "fine arts" (*meishu*) in order to arouse the affection for the nation that was crucial to the nation-state's survival:

> Our country is vast and rich in abundant natural resources. The country is nourished by the best resources in the world. But, sadly, nobody cares to study these resources. As a consequence, the essence of the country has remained imperceptible for a long time. We begin to edit today the section of natural science, which is subdivided in topics dedicated specifically to animals, plants and minerals. Through the research, study and analyze, we wish to incite love for the landscape... which will lead to the affection for the nation... Oriental civilization had germinated very early and invented, before the West, the technics of epigraphy, printing, tincture, painting, calligraphy and lyrics. They are all exquisite. Our country is particularly gifted in literature and elegant poems. In the section of fine arts, we will interpret and explain the epigraphy, the music, the sculpture, the painting, the calligraphy and lyrics to show the nobility and the elegance of our ancestors... We hope that all the patriots cherish these national radiations.[53]

Although the National Essence School did not reference Shiga Shigetaka, these words resonate with Shiga's national essence in terms of national landscape. The section on geography was added to the *Guocui xuebao* in 1908. Liu Shipei and Chen Qubing were actively involved in the redaction of the "textbook of local history" (*Xiangtu lishi jiaokeshu*). To them, a profound knowledge of local history that instilled a sense of local collectivity would extend to nationalism. As Ching May Bo points out, "love for the nation begins with love for one's native

---

an: *wanqing minguo de "guoxue" lunzheng* (Interrogation and Apprehension: The Question of National Learning at the End of the Qing and the Early Republican Period), ed. Liu Dong and Wen Tao, Vol. 1 (Beijing: Beijing daxue chubanshe, 2012), 149.
52 Luo, "Qingji baocunguocui de chaoye nuli jiqi guannian yitong," 62.
53 "Guoxuebaocunhui baogao diwuhao" (Report of the Society of the Preservation of National Learning, no. 5), *Guocui xuebao* 6 (1906): 3002.

place."⁵⁴ This type of localism also informed revolutionaries' support for local self-government and eventually federalism, as will be shown later. Furthermore, a basic concern for the geography section of the journal was to protect national territory. Authors sought to delineate the frontiers from a historical perspective. One text exhorts the government to take actions to protect the islands far away from the continent, before they would be invaded and occupied by foreigners.⁵⁵

In 1903, while in prison, Zhang Taiyan expressed his belief that Heaven had endowed him with the responsibility of fostering and carrying forward the national essence.⁵⁶ At the Tokyo Chinese students' welcoming event in 1906, Zhang explained that the promotion of national essence set out to preserve the history of the Han race (*hanzhong*), which included Chinese language and literature, traditional political laws, and institutions and biographies of patriotic Chinese. He specifically mentioned that national essence had nothing to do with the Confucian religion.⁵⁷ Revolutionaries rarely alluded to the Society of Political Education when conceiving the Chinese national essence. Huang Jie was an exception:

> The thought, the ancient things and the laws different from others constitute the essence of our nation. But the essence of the nation does not systematically form our national essence. National essence is the fundamental nature (*yuanzhi*), incorporated in the form of state, particular to a nation and embedded in people's mind. National essence is an independent thought... and a refined... powerful and open spirit... that push forward the society... Thus, the elements rooted in our culture and useful to us constitute the national essence, but foreign learning that can serve us and can be adapted in China also constitute our national essence. Inoue Kaoru acknowledged only the indigenous things as national essence and ignored useful foreign experiences.⁵⁸

But what if foreign political elements could not be adopted in China? Was national essence an updated version of the "Chinese teaching as substance, Western teaching as application?" In fact, ancient political culture, institutions, and

---

54 Ching May Bo, "You ai xiang er ai guo: Qing mo Guangdong xiangtu jiaocai de guojia huayu" (From Loving One's Native Place to Loving the Nation: The National Discourse on Native-Place Textbooks of Guangdong in Late Qing), *Lishi yanjiu* 4 (2003): 68–84.
55 Tian Beihu, "Yadandao ji" (On the Island of Yadan), *Guocui xuebao* 10, no. 39 (1908): 5166–7.
56 Zhang Taiyan, "Zhang Taiyan guimao (yu) zhong manbi" (Zhang Taiyan's Essay in (the Prison)), *Guocui xuebao* 4, no. 8 (1905): 1390.
57 Zhang Taiyan, "Yanshuo lu" (Speech), *Minbao* 6 (1906): 9.
58 Huang Chunxi (Huang Jie), "Guocui baocunzhuyi" (Doctrine of the Preservation of National Essence), in *Jindai Zhongguo shiliao congkan xubian di nianqi ji* (The 27th Volume of the Second Edition of Sources on Modern Chinese history), ed. Shen Yunlong (Taipei: Wenhai chubanshe, 1976), 181.

Figure 2.1: Portrait of Wang Chuanshan. Wang Chuanshan (1619–1692) was an eminent scholar and Ming loyalist. His thoughts had greatly influenced revolutionary intellectuals. In this portrait, Wang Chuanshan is still dressed in Chinese costume and has not shaved his head as the Manchu Court ordered. It is said that of the three greatest Ming loyalists – Wang Chuanshan, Huang Zongxi (1610–1695), and Gu Yanwu (1613–1682) – Wang was the only one that resisted the Manchu queue, which is not completely plausible since Wang had lived under the Manchu rule for more than 50 years. However, it seems that the National Essence School believed this anti-Manchu story and aimed to propagandize it. This might also be the reason why Huang and Gu's portraits were not published, despite their heavy intellectual influence on the National Essence School. From *Guocui xuebao*, Vol. 1 (1905), 45.

thoughts that made up the national essence were subjected to a heterodox exegesis to comply with the Western modern political system and to fit a republican political construct. Thus, it is not surprising that radical culturalist nationalism of the revolutionaries provoked storms of critical reactions, including reformers whose cultural ideas were no less radical. The above-mentioned Song Shu, for example, spurned the *Guocui xuebao*'s political orientation, which he reproached as perfidious and rebellious.[59]

Sino-Babylonianism and the Western Origin of the Chinese Nation

Revolutionary intellectuals' discourse on culturalist nationalism, which at first glance might sound like the expression of a tradition on the defensive, was largely grounded on Sino-Babylonianism, a theory put forward by the French orientalist Terrien de Lacouperie. Originally from Normandie, Terrien de Lacouperie lived as a child in Hong Kong with his parents who did business there. An enthusiast of Chinese culture, he abandoned the family business to be devoted entirely to the study of Chinese and Chinese civilization. He later moved to London, where he spent his life as an orientalist and a phonologist.[60] The *Western Origin of the Early Chinese Civilization* was published in 1894. It was poorly received in Europe but captured the attention of Japanese historians Shirakawa Jirō (1874–1919) and Kokubu Tanenori (1873–1950), who declared this work to be a major breakthrough in the study of Chinese history, and in 1900 presented it in their book *Shina bunmeishi* (*History of Chinese Civilization*). It was through this book that Chinese intellectuals and Japan students accepted Terrien de Lacouperie's hypothesis without criticism.[61]

Since Chinese civilization appeared to have been surprisingly mature from the very beginning, Terrien de Lacouperie presented Chinese civilization as an importation and extension of a much more archaic civilizational entity.[62] The striking similarities that he discovered between Chinese civilization and Mesopo-

---

59 Song, "Shang Dongfu qingzou chuang Cuihuaxuetang yi," 149.
60 Richard Rutt, *Zhouyi: A New Translation with Commentary of the Book of Change* (Abingdon and New York: Routledge, 2013), 72–3.
61 Ishikawa Yoshihiro, "Ershi shiji chunian Zhongguo liuri xuesheng "huangdi" zhi zaizao – paiman, xiaoxiang, xifang qiyuan lun" (The Rebuilding of "Huang Di" in the 20th Century: Excluding the Manchu, Portraits and Western Originality Theory), *Qingshi yanjiu* 4 (2005): 55–7.
62 Terrien de Lacouperie, *Western Origin of the Early Chinese Civilisation, from 2,300 B.C. to 200 A.D.* (London: Asher & Co., 1894), 3.

tamian civilization in terms of science, art, literature, and political system led him to propose that the Chinese people and their civilization originated in Mesopotamia.[63] He reasoned that the ancestors of Chinese, which he named the Bak tribe, emigrated from the region situated to the west of the Hindu Kush, to the southeast of the Caspian Sea, and in proximity to Elam. Every place whose name contained the syllable *bak* marked the passage of this emigration: Bactria, Bagdad, Bagistan, etc. They traversed the Pamir, Kashgar, and Khotan to finally arrive in Shanxi. The date of this emigration is around 2285–2282 BC.[64] *Bak* itself later became the expression *baixing* (hundred surnames) that is still used today to designate the population.

There is no doubt that Terrien de Lacouperie was a great scholar of Chinese civilization. He drew greatly from *Yijing* (*Book of Changes*), *Shiji* (*Records of the Grand Historian*), *Zhuangzi*, and even *Shanhaijing* (*Classic of Mountains and Seas*) to support his hypothesis. Although rich in sources, his work suffers from many inconsistencies and flaws. Many paragraphs seem quite arbitrary, or even whimsical. For instance, without giving any reliable explanation, he wrote: "In ancient Chinese *nam* is the South, and with the common equivalence *n=l*, it sounds like an old souvenir of *Elam* as a southern country."[65] This type of argument is hardly convincing, but revolutionaries took his theory for granted. The Southern Society member Lin Xie (1874–1926) once argued:

> The word "baixing" has a long history. The Han Chinese used to live in Pamir, not far away from the Kunlun Mountains. They lived together and called themselves "bakuxun." The Yellow Emperor had later guided his numerous descendants to emigrate to China. This nation was still called "bakuxun." One thousand years later, people's accent changed. Consequently, the "ba" became "bai" and the "kuxun," when pronounced quickly, became "xing." Despite the difference in writings, the pronunciations are close. Since our ancestors feared that we, the Han Chinese, mingled with foreigners and forgot our roots, they repeatedly called: "Bakuxun! Bakuxun!" Pronounced quickly, the "bakuxun" became "baixing." And they continued to call themselves "Baixing! Baixing!"

Reasoning of this type, with little scientific rigor, nevertheless appeared repeatedly in revolutionaries' writing. Even Liu Shipei and Zhang Taiyan, arguably two of the most eloquent classical scholars, had a role in disseminating this hypothesis. Liu Shipei pointed out that *bak* was also a homophone of Pangu, the first

---

[63] Ibid., 9–27.
[64] Henri Maspero, "Les origines de la civilisation chinoise," *Annales de géographie* 35, no. 194 (1926): 136.
[65] Terrien de Lacouperie, *Western Origin of the Early Chinese Civilisation*, 26.

living being and the creator in Chinese mythology.⁶⁶ The second edition of *Qiushu* published in Japan in 1904 evokes Terrien de Lacouperie's theory to argue for the superiority of the Han Chinese compared to their "barbarian" neighbors.⁶⁷ Even after the foundation of the Republic, and despite the archeologic and historical discoveries that exploded this theory, Chen Qubing still insisted upon Sino-Babylonianism in a 1929 text on the interpretation of the nomination of the Republic of China.⁶⁸

Since the ancestors of Europeans departed from Mesopotamia in the opposite direction to the Yellow Emperor's tribe, China and the West shared the same racial and civilizational origin. Hence, learning from Western social and political systems and institutions was no longer seen as an importation of foreign culture that risked deteriorating the purity and integrity of Chinese culture. Instead, it was viewed as a rediscovery of authentic Chinese teaching that had been overshadowed by Confucianism, and as such was established as the orthodox state ideology to rationalize the dictatorship of absolute monarchy. Revolutionaries turned this theory into an ideological weapon not only to legitimize audacious cultural and political remodeling, but also to delegitimize the Manchu's rule over China on the grounds of the racial difference between the Han Chinese and the Manchu. As Liu Shipei stated:

> The name of Kunlun is known by all the Han Chinese. Their nostalgia for their natal lands perpetuates, even after the Zhou and the Qin dynasties. Under the Qin and the Han dynasties, their nostalgia for Kunlun was transfigured into a nostalgia for the Yellow Emperor. Hence, the radical thought (*zhongzu sixiang*) was born. This is the proof of the nostalgia of the Han Chinese for the antiquity... The Chinese [authentic] politics and tradition correspond very often to that of the Western nations. There is little doubt that the Han Chinese are originated from the West. Today, Chinese are subjected to a foreign race. If we do not know how to defend our race, how can we restore our nation?⁶⁹

While the concept of national essence was first advocated by reformers, it took on a new meaning and political implication under the pens of revolutionary intellectuals. Since Dong Zhongshu (179–104 BC) convinced Emperor Wu of Han

---

66 Liu Shipei, "Zhongguo minzuzhi" (History of the Chinese nation), in *Liu Shenshu xiansheng yishu* (Works of Liu Shipei), Vol. 17 (Edition of Ningwu Nanshi, 1936), 2.
67 Zhang Taiyan, *Qiu shu chongdingben* (Revised Version of Urgency), in *Zhang Taiyan quanji* (Complete Works of Zhang Taiyan), Vol. 3 (Shanghai: Shanghai renmin chubanshe, 1984), 170.
68 Chen Qubing, "Zhonghua minguo shiyi" (On the Name of the Republic of China) in *Chen Qubing quanji* (Complete Works of Chen Qubing), Vol. 2, ed. Zhang Yi (Shanghai: Shanghai guji chubanshe, 2009), 590.
69 Liu Shipei, "Si zuguo pian" (Nostalgia for the Fatherland), in *Liu Shipei nianpu changbian*, 86.

(157–87 BC) to venerate Confucianism as the state ideology, new ideas had to be grounded in Confucian classics if they were to be accepted.[70] To revolutionaries, this symbolized the moment when "national learnings" (*guoxue*) was replaced and overwhelmed by "imperial learnings" (*junxue*), in the sense that knowledge served the imperial house and not the nation. The glorification of national learnings and the condemnation of imperial learnings led to a differentiation between the Chinese nation and the Manchu government.[71] National essence now needed to be sought in traditional teachings other than Confucianism, so that the Chinese nation could be restored both politically and culturally.

The First Revolution of Social Liberation

In this ideological context, branches of the Hundred Schools of Thought (*zhuzi baijia*) other than Confucianism, such as Mohism and Legalism, were rediscovered. These alternative traditions made up an important part of China's national essence, as Deng Shi remarked:

> Books of the Hundred Schools of Thought covered psychology, ethics, rhetoric, sociology, history, politics, jurisprudence, physics, and chemistry of the West... Since the Han Dynasty, Confucianism has dominated our country for more than one thousand years. When a new foreign knowledge is introduced in China, we find it strange. However, once we examine it closely, we become aware of its grand utility in the West... [and] realize that there are other teachings other than Confucianism and there are, not only the Six Confucian Classics, but also the Hundred Schools of Thought... Even if these ancient teachings are little-known, they also constitute our national essence.[72]

The reason Deng Shi gave for reviving ancient learnings was not for their own sake, but for their concordance with Western learnings. Deng Shi compared the "the burning of books and burying of scholars" with the Seven Sages, whose philosophy fell down with the rise of the Roman Empire that exalted Christianity. Christianity and Confucianism were both established as the state religion or ideology, to the detriment of ancient Greek and other traditional Chinese teachings. He believed that China and the West followed the same historical

---

[70] Feng Youlan, *Zhongguo zhexueshi* (A History of Chinese Philosophy) (Beijing: Zhonghua shuju, 1961), 485.
[71] See, for example, Deng Shi, "Guoxue zhenlun" (On the Truth of the National Learning), *Guocui xuebao* 7, no. 28 (1907), 3025.
[72] Deng Shi, "Guxue fuxing lun" (On the Restoration of Ancient Learnings), *Guocui xuebao* 1, no. 9 (1905): 115–6.

trajectory. Hence, not unlike the European Renaissance, the twentieth century would witness the same rediscovery and re-lighting of the authentic ancient learnings.[73] The rediscovery of the other branches of the Hundred Schools of Thought also led to a rehabilitation of more recent ideas that challenged the dictatorship of absolute monarchy, notably the work of the Ming loyalists Wang Chuanshan, Gu Yanwu, and Huang Zongxi. Furthermore, since Chinese history now came to be understood as following the same path of progress as the West, progressive political remodeling could be justified at both the intellectual and political level.

The culturalist nationalism structured around this logic aimed first to promote what I call the first revolution of social liberation. As mentioned earlier, the strengthening of state power was no longer regarded as a governmental affair, rather it was based on the social restructuring that bolstered the nation. Dong Zhongshu's political view that "the sovereign should win the people" was revived along with the awareness of the importance of the people's role in modeling the country's future.[74] The growing emphasis on popular power challenged the foundation of the imperial government, since an excess of the sovereign's power was believed to be one of the key impediments to the people's ability to exercise their political rights and duties.[75] Reformers placed the popular power and the sovereign's power on the two sides of a seesaw – their mission was to balance each other. For revolutionaries however, this would never happen. Despite the profound economic, educational, and industrial reforms that were undertaken during the New Policies, the Qing Court was reluctant to establish a constitution and a parliament. Thus, revolutionaries took social congregation as the viable force to reverse the central government, with the first step towards this to lift traditional moral codes and practices that impeded individual emancipation. As such, the implementation of the first revolution of social liberation was accompanied by a reflection on the question of citizenship. Individuals should no longer be passive subjects, but citizens, and citizens held rights, not just with respect to each other, but with respect to the very government itself. It follows that the translation of the term society also changed from *qun* to *shehui* – the latter was a neologism of the Japanese term *shakai* and designates what were originally traditional communitarian activities. Although *shehui* has an in-

---

73 Ibid., 112–3.
74 The original text reads: 君子, 群也。 This phase appeared in many revolutionary and reformer magazines and journals of the time; see, for example, Deng Shi, "Junzhu" (Sovereign), in *Jindai Zhongguo shiliao congkan xupian diershiqiji*, 107.
75 See, for example, Jiang Fangzhen, "Minzu zhuyi lun" (On Nationalism), in *Xinhai geming qianshinianjian shilun xuanji*, Vol. 1, 49.

digenous origin, the term was imbued with a type of secret association with a hidden agenda in late Qing China. In contrast, *qun* implied unity and solidarity, as well as a certain level of loyalty towards the emperor. With the rise of the revolutionary force, *qun*, understood in this way, became obsolete.[76] Furthermore, as Kai Vogelsang points out, in comparison with the collective *qun*, *shehui* "could be reformed, reconstructed, even revolutionized."[77] This is exactly what revolutionaries and reformers were aiming for.

Revolutionaries and reformers shared the view that a new society should be characterized by a cohesive and organic social structure. The theory of social organicism was borrowed from Herbert Spencer (1820–1903). To him, the social body was "a commonwealth of monads, each of which has independent powers of life, growth and reproduction; each of which unites with a number of others to perform some function needful for supporting itself and all the rest."[78] Yan Fu interpreted the benefits of an organic society in Xun Zi's words (cf. *supra*). Society was defined as an organism built upon the harmony between the part and the whole. The lack of such a harmony suggests the existence of social pathology. It was common at the time to divide society into three hierarchical levels: the superior society (*shangceng/shangdeng shehui*), the middle society (*zhongceng/zhongdeng shehui*), and the inferior society (*xiaceng/xiadeng shehui*). Liang Qichao defined the superior society as the dominant social group who used to have the capacity and measures to control other people, the middle society as composed of honest citizens, while people belonging to the inferior society were in general thugs and thieves whose existence and activities posed a threat to the interests of the middle class.[79] More commonly, the inferior society was defined as the ensemble of individuals massively affected by illiteracy who had a limited knowledge of Chinese history and geography and remained indifferent towards China's crisis. In 1903, Li Shucheng (1882–1965) described the Chinese inferior society as one that reassembled in general the crowds who were "rude, unreasonable, aggressive, mercilessly stupid, impulsive without re-

---

[76] Jin Guantao and Liu Qinfeng, "Cong 'qun' dao 'shehui', 'shehuizhuyi' – Zhongguo jindai gonggong lingyu bianqian de sixiangshi yanjiu" (From "Grouping" to "Society" and "Socialism" – An Intellectual History on the Evolution of Modern Chinese Public Sphere), *Bulletin of the Institute of Modern History, Academia Sinica* 35 (2011): 7, 15 and 20–21.
[77] Kai Vogelsang, "Chinese 'Society': History of a Troublesome Concept," *Oriens Extremus* 51 (2012): 183.
[78] James Elwick, "Herbert Spencer and the Disunity of the Social Organism", *History of Science* 41 (2003): 48.
[79] Liang Qichao, "Zhongguo lishishang geming zhi yanjiu" (A Research on the Revolution Throughout the Chinese History), in *Xinmin congbao* 46–48 (1904).

flection, hostile to foreigners and contemptuous of law."⁸⁰ But if society can be compared to a living body that can fall ill, its pathologies can also be "cured."⁸¹

An organic system sustaining a society will not be effective unless the underprivileged social stratum becomes an active participant in promoting general interests. Thus, the first step towards the construction of an organic society is to grant all individuals the same conditions to fully develop their talents to become independent. To this end, Li Hsiao-ti shows that there was a historical period in the 1890s and 1900s when progressive intellectuals were actively involved in the social enlightenment movement.⁸² For revolutionaries, the movement of the awakening of the popular spirit included the publication of vernacular journals, the speech given in local languages, the performance of the so-called "improved Peking Opera," the promotion of women's education, and the ban on foot binding.

The Southern Society created several women's schools and "associations for natural foot" (*tianzuhui*). Spencer's influence here was also evident. In his 1851 *Social Statics*, he denounced the unequal treatment that women suffered and stated that the status of women indicated the level of civilization of a society.⁸³ In 1902, the chapter *Rights of Women* was translated by the Southern Society and the Tongmenghui member Ma Junwu (1880–1940), who exhorted China to "join the civilization" and to grant women the same rights as men.⁸⁴ At the cultural level, Liu Yazi blamed Neo-Confucianism squarely for the atrocious familial and social practices that enslaved women.⁸⁵ Huang Jie undermined Neo-Confucianism by referring to more ancient traditional learning, arguing that the word "wife" (*qi*) signified an equal (*qi*) to her husband in the *Shuowen jiezi* (*Explaining Graphs and Analyzing Characters*).⁸⁶

---

80 Li Shucheng, "Xuesheng zhi jingzheng" (The Concurrence Among Students), *Hubei xueshengjiei*, 2 (1903).
81 Claude Blanckert, La nature de la société: organicisme et sciences sociales au XIXᵉsiècle (Paris: L'Harmattan, 2004), 8–17.
82 Li Hsiao-t'i, *Qingmo de xiacengshehui qimeng yundong* (Movement of Social Enlightenment of the Inferior Society at the end of the Qing) (Taipei: Institute of Modern History, Academia Sinica, 1998).
83 Spencer Herbert, *Social Statics, or The Conditions Essential to Human Happiness Specified, and the First of Them Developed* (London: John Chapman, 1851), 161.
84 Ma Junwu, "Sibinsai nüquan pain" (Spencer on Rights of Women), in *Ma Junwu ji* (Works of Ma Junwu), ed. Mo Shixiang (Wuhan: Huazhong shifandaxue chubanshe, 1991), 16–27.
85 Liu Yazi, "Ti Liuxi Qinming nüxiao xiezhen wei Huiyun zuo" (Poem on the Photography of the Women's School of Qinming in Liuxi for Huiyun), *Nanshe congke* 1, no. 1 (1909): 202.
86 Huang Jie, "Huangshi lunli shu" (On Ethics, the Yellow History), *Guocui xuebao* 1, no. 8 (1905): 535.

The social enlightenment movement did not constitute an end itself but aimed to essentially mobilize the whole population to bring about an anti-Manchu republican revolution. Southern Society's *Zhongguo baihuabao* (*Chinese Vernacular Journal*) and *Ershi shiji dawutai* are both cases in point. The latter audaciously published a play named *Anlewo* (*Cozy Nest*), in which the Empress Dowager Cixi was depicted as a dissolute woman who was ruling China as a shameless Manchu dictator.[87] These plays were actually put on the stage in Shanghai.[88]

To believe that the organic system would only be effective in sustaining a society if this underprivileged social stratum became independent, one must make no mistake about the nature of this kind of individualism. In such an organic society, individuals are invited to become more autonomous and at the same time more implicated in the social interaction. By citing Guan Zhong (ca. 720–645 BC), Huang Jie argued that by assigning each individual a specialized task, a sense of social responsibility would naturally accompany individualism:

> Everyone has his own profession (*zhiye*). Guan Zi said: if there is a farmer who does not cultivate his field, all the others will be hungry; if there is a woman who doesn't weave the fabric, all the others will feel cold... Qiu Jun said: "people should have a profession, they will then be able to feed and take care of each other. If one does not take his work seriously, someone else's life will suffer." So, for me, once everyone adheres to his responsibility and to his proper work, the condition and the order of the society (*qunzhi*) will be greatly improved.[89]

In a different context, his optimism on the positive effects of society is quite similar to that of Durkheim (1858–1917) for whom modern individualism does not necessarily lead to the destruction of all social bonds nor contribute to narrow utilitarianism.[90] Society will be in order if individuals can be allowed to freely pursue their own interests. In 1903, an article published in *Zhengyi tongbao* maintained that "altruism (*lita*) is nothing but a form of egoism (*liji*)."[91] By citing

---

[87] Jing'an (Sun Huangjing), "Anlewo" (Cozy Nest), *Ershi shiji dawutai* 2 (1904).
[88] The *Shenbao* (Shanghai News) published publicities of two plays in 1904. See *Shenbao* (September 26, 1904; October 14, 1904).
[89] Huang Jie, "Huangshi lunli shu" (On Ethnics, the Yellow History), *Guocu ixuebao* 1, no. 6 (1905): 510–1.
[90] Sylvie Mesure, "Durkheim et Tönnies: regards croisés sur la société et sur sa connaissance," *Sociologie* 4, no. 2 (2013).
[91] "Shenghuo shehui zhi erfangmian" (Two Aspects of the Social Life), in *Jindai Zhongguo shiliao congkan xupian diershibaji*, 133.

Yang Zhu's (440–360 BC) words, Gao Xie also affirmed: "if no one pulls out even just one hair of his own to benefit the world, the world will be regulated."[92]

In the traditional hierarchical structure, the Chinese population was divided into four categories in order of decreasing social status: the *shi* (gentry), *nong* (farmers), *gong* (artisans), and *shang* (merchants). In the late Qing context, this structure was less about social hierarchy and more about a collective awareness of the importance of the division of labor in a multicellular social organism. The traditional stratified society was now transformed into a society characterized by functional differentiation.[93]

Revolutionaries urged people from all walks of life to organize their own associations to manage the interests and requests of a wide range of individuals and group members. This advice itself was nothing new, since Kang Youwei had already proposed a similar idea during the Hundred Days' reform. The idea of *xinmin* (new people) that Liang Qichao developed in 1902 and 1903 was of particular importance for the social associations advocated by the reformers. At the core of the politics of *xinmin* is the idea of grouping, articulated around *side* (private morality) and *gongde* (public morality). While *side* requires that everyone pays careful attention to the improvement of their morality and capacity (*dushan qishen*), *gongde* is the moral quality that drives each member of the society to improve the well-being of the community to which he or she belongs (*xiangshan qiqun*).[94]

To revolutionaries, Liang was too naïve. An article published in *Zhejiang chao* (*The Zhejiang Trend*) accused Liang Qichao of being excessively idealistic in placing his hopes for reform solely in the Chinese people. The political indifference of the people was a direct consequence of the absolutism imposed by the Manchu Empire that Liang was not prepared to overturn.[95] The sociopolitical remodeling came to be seen not as a result of the sovereign's or the people's willingness to reform, but rather as the direct confrontation and revolution against the Manchu Court. Thus, revolutionaries' appeal implied an assault on the Manchu government from the liberated society, which proved to be effective. Take merchants' associations' manifestations as an example. The Qing government's plan to raise funds from foreign banks to develop railways and consequently give

---

92 Gao Xie, "Yang Zhu xueshuo zhi gaizao" (The Transformation of Yang Zhu's Doctrines), in *Gao Xie Ji* (Works of Gao Xie), ed. Gao Tian et al. (Beijing: Zhongguo renmindaxue chubanshe, 1999), 6.
93 Kai Vogelsang, "Chinese 'Society'," 60.
94 Liang Qichao, "Xinmin shuo", 123.
95 Feisheng, "Jinshi erda xueshuo zhi pinglun" (Comments on Two Recent Popular Doctrines), in *Xinhai geming qianshinianjian shilun xuanji*, Vol. 1, 519–21.

foreigners control of local railways fueled the nationalist discontent with the local gentry, who condemned foreign intervention in China's internal economy and the invasion of Western capital.[96] Using their influence and networks, businessmen marshalled large strikes and demonstrations to pressure the Qing government to abort the plan to sell the local ventures to foreigners.[97] Their assemblies in Shanghai in 1907 mounted an articulate challenge to the Manchu government's reluctance to establish a constitution. In a 1904 article written for the drapers of the Jiangnan area, the author motivated his target readers to use their economic clout for political gains, asking seditiously why they did not dare to revolt when "all these princes and eunuchs were indulging in extravagance and smoking opium" paid for with their taxes.[98] In another article, the author, full of indignation, assailed the Manchu government for its anti-popular policies that did not allow any social associations, public speech or press to be freely arranged to hinder people's rights, equality, independence, and freedom, without which they were doomed to be slaves and not nationals.[99]

This type of revolt against the state from society in the name of the nation was welcomed by the revolutionaries. For them, the central message of Dong Zhongshu indicated that without the people's contentment, any form of government would be illegal. Liu Shipei called this political theory *minyue* – the Chinese translation of "social contract" borrowed from the Japanese term *minyaku*. The Rousseauist social contract was misread in Japan, which directly influenced its reception in China. Whereas Rousseau's original idea rested upon the homogeneity between people and the sovereign, both the Japanese and Chinese translators failed to seize this sensibility and resorted to Rousseau's theory to legitimize popular revolution against the sovereign.[100] However, this misinterpretation allowed them to see a way to fight against the central government and overthrow the Manchu Dynasty. To Liu Shipei, social contract was closely intertwined with the tradi-

---

[96] Zheng Xiaowen, *The Politics of Rights and the 1911 Revolution in China* (Stanford: Stanford University Press, 2018), 2.
[97] Zhu Ying, "Qingmo xinxing shangren ji minjian shehui" (New Businessmen and the Popular Society at the End of Qing), *Twentieth-First Century* 3 (1990): 42.
[98] Pinglushi, "Baohu chouduan de fazi" (The Way to Protect Silk), in *Xinhai geming qianshinianjian shilun xuanji*, Vol. 1, 878.
[99] "Shuo Guomin" (On the Nation), in *Xinhai geming qianshinianjian shilun xuanji*, Vol. 1, 72–7.
[100] See, Wang Xiaoling, "Lusuo "pubian yizhi" gainian zai Zhongguo de yinjie jiqi lishi zuoyong" (On Rousseau's "general will" in China and its historical function), *Sixiangshi* 3 (2014): 1–66; idem, "Liu Shipei et son concept de *contrat social chinois*," *Études chinoises* XVII, no. 1–2 (1998): 155–90.

tional Chinese political notion of *minben* (people as root).¹⁰¹ The Manchu government prohibited interethnic marriage between Manchu and Han who assumed different social status and roles. Wang Jingwei indicated that this practice of racial segregation opened the way for Han people to set up associations to take over local affairs in which Manchu people were not usually involved, such as agriculture and commerce, in order to affirm and consolidate their political rights vis-à-vis the state.¹⁰²

The Chinese concept of society was also deeply embedded in social evolutionism. This theory was popularized by Yan Fu as well. Yan was inspired by Edward Jenks's (1861–1939) *A History of Politics*, which Yan translated in 1903 under the title of *Shehui tongquan* (*A general interpretation of society*). Jenks resumed his theory in these words:

> Nowadays, the principle which binds together these communities of the modern type is the tie of *military allegiance*. In the States which practice conscription, or universal military service, this is very obvious. The most heinous political offence which a Frenchman or a German can commit is attempting to evade military service; or, possibly worse, taking part in military service against his own country... In the older conditions of society, however, to which allusion has been made, the tie was not that of military allegiance, but *kinship*, which... we may call *patriarchal* or *tribal*. Until quite recently it was believed that this *patriarchal* type was the oldest type of human community... But the brilliant discoveries of the last half century have revealed to us a still more primitive type of society... Its scientific name of *Totemistic* is too elaborate and technical for popular use. Perhaps it will be best to call it the *savage* type, though it must be clearly understood that the term implies neither contempt nor blame. It merely signifies that the type in question is very *primitive* or *rudimentary*. Here, then, we have our three historical types of human society – the *savage*, the *patriarchal* and the *military* (or "political" in the modern sense).¹⁰³

Yan Fu translated savage, patriarchal, and military society respectively as *manyi/tuteng shehui*, *zongfa shehui*, and *junguo/guojia shehui*.¹⁰⁴ The social evolution was believed to give rise to a military nation (*jun guomin*). In 1906, "military veneration" was officially codified in the objectives of the national education poli-

---

101 Liu Shipei, "Zhongguo minyue jingyi" (The Quintessence of the Chinese Social Contract), in *Liu Shenshu xiansheng yishu 16*, Vol. 3.
102 Wang Jingwei, "Manzhou lixian yu guomin geming" (The Constitution of the Manchu Government and the Nationalist Revolution), *Minbao* 8 (1906): 48.
103 Edward Jenks, *A History of Politics* (London: J. M. Dent, 1900), 2–4.
104 Yan Fu, *Shehui tongquan* (A General Interpretation of Society) (Beijing: Commercial Press, 1981), 4.

cy,¹⁰⁵ but private schools and associations took the lead in military education. The Anti-Russia Volunteers (*Ju'e yiyong dui*) created in the context of the Anti-Russia Movement is an example.¹⁰⁶ Although military officials enjoyed a lower status than literary officials, and military personnel were not regarded highly in ancient China, Yao Guang, Liu Shipei, and Song Jiaoren all argued that the Han Chinese had a glorious military tradition.¹⁰⁷ Following this logic, even the image of the Yellow Emperor had been radically modified. Represented in military uniform, the Yellow Emperor was imbued with a Caucasian physiognomy, which alludes to Sino-Babylonianism.

The contradiction remains that, on the one hand, revolutionaries appealed for an individual emancipation from traditional moral codes and practices, while on the other hand, liberated individuals seemed to be forced into a social collectivism, with a military-like structure. Indeed, the revolutionaries' fight for the first revolution of social liberation in the 1900s was never about personal gain. Individuals were expected to break the traditional chains that once enslaved them, so that they would be ready to be devoted to nation-building. Liu Shipei, for one, glorified the ultimate sacrifice in the battlefield.¹⁰⁸ The Southern Society members Zhou Xiangjun (1870–1914) and Zhou Shi (1885–1911) had deplored the foot binding and lauded women's education, not because women with no political rights or little education hampered the national progress, but because the oppression imposed on the women was simply inhuman and diabolic.¹⁰⁹ However, this kind of voice was too weak not to crumble under the nationalist conception of societal construction.

---

105 Hwang Jin-lin, "Jindai Zhongguo de junshi shenti jian'gou, 1895–1949" (The Construction of Military Body in Modern China 1895–1949), *Bulletin of the Institute of Modern History, Academia Sinica* 43 (2004): 185.
106 Chang Yu-fa, *Qingji de geming tuanti* (Revolutionary Associations at the End of the Qing) (Taipei: Institute of Modern History, Academia Sinica, 1982), 659.
107 Jacques Gernet, *La vie quotidienne en Chine: à la veille de l'invasion mongole (1250–1276)* (Paris: Hachette, 1978), 75; Yao Guang, "Shangwu jiuguo lun" (Military Veneration and Salvation of the Nation), in *Yao Guang quanji* (Complete Works of Yao Guang), ed. Yao kunqun et al. (Beijing: Shehui kexue wenxian chubanshe, 2007), 37; Gongming (Song Jiaoren), "Hanzu qinlue shi" (A History of Intrusion of the Han Chinese), *Ershi shiji zhi zhina* 1 (1905): 25; Liu Guanghan, "Lun gudai renmin yi shangwu liguo" (Ancient People Constructed the Country with Military Power), *Guocui xuebao* 1, no. 2 (1905): 175.
108 Liu Shipei, "Jun guomin de jiaoyu" (Education of a Military Nation), in *Liu Shipei nianpu changbian*, 75.
109 Zhou Xiangjun, "Chanzu yin" (On Foot Binding), *Nanshe congke* 1, no. 2 (1910): 241; Zhou Shi, "Tangyin nüshi xiaozhuan" (A Short Biography of Ms. Tangyin), *Nanshe congke* 1, no. 3 (1910): 304–7.

Figure 2.2: Portrait of the Ancestor of China, the Yellow Emperor. From *Ershi shiji zhi zhina* 1 (1905).

Figure 2.3: The Great Nationalist Men in the World, the Yellow Emperor. The ancestor of China. From *Minbao* 1 (1905).

Cultural and Ethnical Homogenization of the Chinese Society

Society offered a viable space in which people's political appeals vis-à-vis the state could be heard. Membership of a nation was acquired by one's implications in various "social forms" through which nationalist expectations were expressed. Political nation and society were homogenized in this way. However, the question remained as to whether this social space was reserved for the Han Chinese or all those who were living within the borders of the Qing dynasty. This question was fundamental to the revolutionaries. The culturalist nationalism that they had inherited from Japan and the concrete political situation made it imperative that individuals forming a political nation were culturally and ethnically the same. As argued at the beginning of this chapter, the concept of society that served as the justification for revolution and was the linchpin of the revolutionaries' understanding of political nation did not provide a useful conceptual framework to differentiate people of different ethnical origins.

The unlimitedness of the political reference of society is reflected in socialism that had been constructed in its entirety in the concept of society. Socialism originally negated any modular factors that particularized society and opposed therefore nationalism. When Eric Hobsbawm introduced the influential idea of "invented tradition" (i.e., traditions which appear to be old but are not as old as their proponents claim), he suggested that what was real and substantial was not nation, but social classes and struggle.¹¹⁰ As Engels (1820–1895) and Bakunin (1814–1876) stated, socialists and, in a wider sense, Marxists, could be led to support the fight for national liberation only if the combat of an oppressed nation for independence formed part of international social revolution.¹¹¹ However, despite Chinese intellectuals' widely shared dream of world pacificism and grand unity, they understood that international socialism could only be realized after the period of nationalism. Due to the threat posed to China by foreign powers, it was essential to differentiate "we" from "they" within the society. While socialism and the accompanying concept of society were nationalized by Stalin (1878–1953) and the CCP,¹¹² revolutionaries had culturally nationalized the political reference of society two decades earlier.

On the one hand, general strikes, mobilizations for workers' rights, and factory occupations in the name of the socialist movement were also considered to be struggles for citizens' rights during the revolutionary decade. The Southern Society member Jin Tianhe (1873–1947) proposed a view whereby those who oppressed the destitute part of the population, politically or economically, would be excluded from national society and the nation.¹¹³ This view, which became a reality in Communist China 60 years later, was informed by the awareness that imperialism not only took the form of military action against China; it was also expressed as foreign nations' economic dominance, as Lenin (1870–1924) would characterize it in 1916.¹¹⁴ One example is the Railway Protection Movement that erupted in 1911 mentioned above. For this reason, revolutionaries embraced the Three Principles of the People – nationalism, people's power (*minquan zhuyi*), and socialism (*minsheng zhuyi*) – and called for a parallel socialist and nationalist revolution. Since the majority of the Chinese population belonged to the proletariat and were Han Chinese, the socialist movement and

---

110 Eric Hobsbawm and Terence Ranger (ed.), *The Invention of Tradition* (Cambridge: Cambridge University Press, 1983); Roger Scruton, *Fools, Frauds and Firebrands: Thinkers of the New Left* (London and New York: Bloomsbury Continuum, 2015), 17–38.
111 Karl Marx, *Revolution and Counter-Revolution* (Chicago: Charles H. Kerr & Company, 1907); Michel Bakounine, "Fédéralisme, socialisme et antithéologisme, Proposition motive au Comité central de la Ligue de la paix et de la liberté", in *Œuvres*, Vol. 1 (Paris: P. V. Stock, 1895), 19–20.
112 Douglas Howland, "The Dialectics of Chauvinism: Minority Nationalities and Territorial Sovereignty in Mao Zedong's New Democracy," *Modern China* 37, no. 2 (2011): 178–82.
113 Jin Tianhe, "Guomin xin linghun" (New Soul of the Nation), *Jiangsu* 5 (1903).
114 Jiang Yihua, *Shehui zhuyi xueshuo zai Zhongguo de chuqi chuanbo* (Initial Diffusion of Socialism in China) (Shanghai: Fudan daxue chubanshe, 1984), 224–5; Viladimir Lenin. *L'impérialisme, stade suprême du capitalisme*, 1916, http://marxiste.fr/lenine/imp.pdf. (accessed October 21, 2017).

the nationalist movement could join forces to create the political identity of the Han Chinese nation.[115]

On the other hand, revolutionaries fabricated narratives that excluded the Manchu from the Chinese society. For them, if it could be proved that the Manchu was a foreign nation, then their rule over China would be illegitimate and they could not claim political rights within the nationalized society. While the Sino-Barbarian dichotomy (*huayi zhi bian*) was a major theme in China's ancient historical and political debates, China had been repeatedly conquered by those whom the Chinese disdained as barbarians. To reformers, the incorporation of the Manchu into the Chinese nation was a self-evident truth, since *Chunqiu*, according to the Tang scholar Han Yu (768–824), stipulated that "barbarians who are assimilated to China became Chinese."[116] To Kang Youwei, the Manchu government's policy of "wearing the Manchu queue and changing the Chinese clothes" (*tifa yifu*) conformed to Confucius' ritual teaching.[117]

Revolutionaries found reformers' defense of the Manchu's place in the Chinese nation nonsense, arguing that even if Confucius had blurred the racial boundaries between barbarians and Chinese, the Manchu was never naturalized (*guihua*) or assimilated into (*tonghua*) Chinese culture. The racial segregation ordered by the Manchu government, such as the ban on the interracial marriage, clearly marked their reticence to be integrated into Chinese society.[118] Their veneration of Confucianism was for the sole purpose of consolidating their illegitimate rule over China, while imposing their own customs and traditions.[119]

Moreover, the racial difference between the Manchu and the Han Chinese was too great to ignore. Revolutionaries believed that the ethnic origin of the Chinese nation was located in the West. The Manchu, on the other hand, were barbarians from northeastern China. In the revolutionary pamphlet *Geming jun (Revolutionary Army)*, Zou Rong subdivided the "yellow race" into the Chinese race and the Siberian race. The Chinese race was composed of the Han Chinese, Tibe-

---

115 Zhu Zhixin, "Lun shehui geming dang yu zhengzhi geming bingxing" (Social Revolution Should be Launched Simultaneously with Political Revolution), *Minbao* 5 (1906): 13–35.
116 The original text reads: "夷而進於中國則中國之。"
117 Wong Young-tsu, *Cong chuantong zhong qiubian – Wanqing sixiangshi yanjiu* (Seeing Change in Tradition: Study on Late Qing Intellectual History) (Nanchang: Baihuazhou wenyi chubanshe, 2001), 210; De Bary and Lufrano, *Sources of Chinese Tradition*, Vol. 2, 267–8.
118 Wang Jingwei, "Chi wei Manzhou bianhuzhe zhi wuchi" (On the Shamelessness of the Defenders of the Manchu), *Minbao* 12 (1907): 159–60.
119 See, for example, Baihua daoren (Lin Xie), "Gao guomin shu" (A letter to the Nation), in *Xinhai geming qinshinianjian shilun xuanji*, Vol. 1, 901; Zheshen (Zhu Zhixin), "Lun Manzhou sui yu lixian er buneng" (The Manchu Cannot Establish a Constitution as They Wish), *Minbao* 1 (1905): 37–8.

tans, and Vietnamese. The Siberian race was Mongolian, Turkish, and Tungstic. The Manchu were part of the Tungstic people.[120] Each race had its own place of origin. Although the Han Chinese and the Manchu were both Asian groups, the racial line between them was simply nonexistent. To Zou, the racial segregation imposed by the Manchu government allowed the Chinese to maintain the "purity of blood."[121]

Ethnic nationalism demonstrated a profound "racism" during the first years of the 1900s. In 1777, the Qianlong Emperor (1711–1799) ordered research into the origin of the Manchu. Like almost every nation, the *Manzhou yuanliu kao* (*On the Origin of the Manchu*) recounted a tale in which the Manchu descended from the child of a girl who was impregnated by the red food that a common magpie left on her robe while she was showering in Bulhūri Lake, under Mont Bukuuri.[122] Wang Jingwei cited this legend in *Minbao*, stating that the Manchu were in fact descendants of animals in the Mont Changbai. This is why they were barbarians, or even worse, animals.[123] Huang Jie added to Wang's argument, saying that there were six races in Asia; the Han Chinese was the most noble yellow race, while amongst the other five races, only those living in the regions to the east of China were part of the human race. The others, including the Manchu, were descendants of animals.[124]

Zhang Taiyan was more "rational." Although he also maintained that the Manchu was part of the Tungstic people, he acknowledged that all races were descendants of one common ancestor. This did not mean that he accepted that the Manchu were the same as the Han Chinese. Rather, because of millions of years of racial cohabitation and integration, the "natural nation" (*tianran minzu*) was not the criteria to differentiate the Manchu from the Han. Instead, he argued for what he called the "historical nation" (*lishi minzu*) as the reliable racial boundary. Indeed, history constituted the fundamental dimension of Zhang's view on national essence.

---

120 Zou Rong, "Geming jun" (Revolutionary Army), in *Meng huitou: Chen Tianhua Zou Rong ji*, 40–1.
121 Ibid., 44.
122 For an important research on *Manzhou yuanliu kao*, see Pamela Kyle Crossley, "Manzhou Yuanliu Kao and the Formalization of the Manchu Heritage," *The Journal of Asian Studies* 46, no. 4 (1987): 761–90.
123 Wang Jingwei, "Minzu de guomin" (The Political Nation Grew out of the Culture Nation), *Minbao* 1 (1905): 9–10.
124 Huang Jie, "Huang shi zhongzu shu" (On Race, the Yellow History), *Guocui xuebao* 1, no. 24 (1905): 425.

The history of the Han race, as Zhang termed it, was first and foremost a history that stimulated hatred towards the Manchu. During the revolutionary decades, two forbidden books were discovered in Japan that registered the massacre of the Han by the Manchu army in the mid-seventeenth century – *Yangzhou tucheng qinli* (*Lived Experience of the Yangzhou Massacre*) and *Jiading tucheng jilue* (*On the Massacre of Jiading*). These books were widely circulated in the revolutionary network and greatly agitated anti-Manchu sentiment. Feng Yuxiang (1882–1948), who at the time was a soldier under Yuan Shikai's army, recalled that his attitude towards the Manchu Court drastically changed after reading the *Jiading tucheng jilue*. He had been secretly given the book in 1905 by his superior, Sun Jiansheng (1883–1912).[125] Many plays in the *Ershi shiji dawutai* were inspired by these tragedies. One theme of the magazine was to put on stage "the massacre of Jiading and the misery of the families in Jiading" to "re-forge the Sino-Barbarian dichotomy and provoke a spiteful spirit."[126]

In his study on revolutionaries' "historical trauma," Peter Zarrow remarks that "anti-Manchuism was expressed in extravagant, passionate language, not the reasoned logic of ideological argument (which, if not always reasonable, at least uses logical forms)."[127] However, the reasoned logic of ideological argument and irrational language are not mutually exclusive in revolutionaries' anti-Manchu propaganda. From an historical perspective, they launched a debate with reformers concerning whether the Manchu's place of origin was part of China's territory. In other words, was the Manchu already a sovereign nation before it ascended to the throne in China? If the answer was yes, then the struggle between the Han Chinese and the Manchu would be an international affair, which would delegitimize the Manchu's colonial rule over China. To Liang Qichao, the Manchu originated in Jianzhou, which was a *jimizhoui* of the Ming dynasty. *Jimizhou* should not be confused with the tributary states, the territory of which was not incorporated into China.[128] Rather, *jimizhou* referred to the autonomous administrative unity of ethnic minorities situated at the borders of China. Despite the autonomy, *jimizhou* was considered to be part of China. Therefore, Liang Qichao argued that the rise of the Manchu could be compared with the overturn of

---

125 Feng Yuxiang, *Wode shenghuo* (My Life) (Shanghai: Jiaoyu shudian, 1947), 120.
126 Yalu, "Ershi shiji dawutai fankanci" (Manifesto of the Magazine *20th-Century Stage*), *Ershi shiji dawutai* 1 (1904): 3.
127 Peter Zarrow, "Historical Trauma: Anti-Manchuism and Memories of Atrocity in Late Qing China," *History and Memory* 16, no. 2 (2004): 72.
128 Liu Tong, *Tangdai jimizhou yanjiu* (On *Jimizhou* of the Tang Dynasty) (Xi'an: Xibei daxue chubanshe, 1998), 51–5.

the Qin dynasty by the Han dynasty. This constituted an internal regime change.¹²⁹

Zhu Zhixin (1885–19820) reminded reformers of the fundamental difference between nationalism as a sentiment of allegiance to a community and statism. The awkward situation in which China found herself was comparable to that of Alsace and Lorraine, two French regions ceded to the German Empire following the defeat of France in the 1870 Franco-German War. According to the reformers' logic, people in Alsace and in Lorraine should love Germany and not France, since these two regions were ceded to Germany and the people there were granted German citizenship. He reasoned that the reformers' nationalism dangerously entangled the Chinese in slavery, while being completely blind to their willingness, suffering, and sense of allegiance.¹³⁰

From an historical and philological perspective, Liu Shipei stated that *jimi* should be apprehended in the sense of constraint. By citing several ancient texts, he argued that the aim of the implementation of *jimizhou* was to avoid being involved in barbarians' internal affairs and to avoid all military conflicts with the latter by according to them some economic benefits; people living within each *jimizhou* did not assume any responsibility with regard to China and they were not counted in the census. Therefore, *jimizhou* was neither a territory nor a protectorate of China.¹³¹ He also pointed out that when the Manchu established their own state in Jianzhou, it was registered that the Ming dynasty considered it to be an enemy state.¹³² Tao Chengzhang (1878–1912) further developed Liu's argument, affirming that Jianzhou had been recognized as an independent state long before the collapse of the Ming dynasty, and the Manchu used the Jin as the name of the state while the Ming dynasty still existed. For this reason, "the rule of the Manchu over China marked the victory of a foreign nation-state against Chinese nation-state, and not the substitution of the Ming dynasty by the Qing dynasty... since the Qing Court had been established long before the end of the Ming dynasty."¹³³

While the revolutionaries' exclusion of the Manchu from the Chinese nation was condemned by reformers, the latter differentiated foreigners from the Chi-

---

129 Liang Qichao, "Da mou zabao" (Response to a Certain Journal), *Xinmin congbao* 84 (1906).
130 Xuanjie (Zhu Zhixin), "Xinli de guojiazhuyi" (Psychological Nationalism), *Minbao* 21 (1908): 13–35.
131 Weiyi (Liu Shipei), "Bian Manren fei Zhongguo zhi chenmin" (Manchu Was not Chinese Subjects), *Minbao* 14 (1907): 97–111.
132 Ibid., 40.
133 Sigu (Tao Chengzhang), "Lun Manzhou dang Mingmo shidai yu Zhongguo wei diguo" (The Manchuria was an Enemy State of China at the End of the Ming dynasty), *Minbao* 21 (1908): 4.

nese in a way that pulled reformers and revolutionaries into the same ideological pattern. Liang Qichao advocated the "great nationalism" (*da minzu zhuyi*) to challenge revolutionaries' "small nationalism" (*xiao minzu zhuyi*). While the "small nationalism" was the Han-nationalism, "great nationalism" describes the feeling towards foreign nations of all ethnic groups living within the borders of the Qing dynasty. Acknowledging the defining role played by culture in the formation of a nation, Liang maintained that hundreds of years of cohabitation had amalgamated all of these ethnic groups into what he called a unified Chinese nation (*Zhonghua minzu*).[134] In this way, the political nation and cultural nation remained one under Liang's pen, albeit with an objective diametrically opposite to that of the revolutionaries.

With the advent of the Revolution of 1911, revolutionaries had to come up with a new framework to include all the ethnic groups living within the country's borders in the nation. The Republic of China claimed sovereignty over the entire territorial boundaries of the Qing Empire. While recognizing the ethnic diversity of China, revolutionaries denied their ability to form an independent political entity. To Wang Jingwei, the revolutionaries' nationalism aimed to assimilate different cultural nations into one political nation. Assimilation did not entail a fusion of nation A and nation B to create nation C, but rather the integration of nation B into nation A. In China, this "apex nation" was the Han, so that "the nationalism promoted by the state makes the political nation and the cultural nation a unified entity."[135] Zhang Taiyan proposed to implement autonomous regions, in which ethnic minorities would be "purified," before granting them the totality of political rights.[136] Revolutionaries' radicalism terrified Manchu communities in Japan, who even planned to instigate a Japanese intervention in the Revolution of 1911.[137] This forced Zhang Taiyan to address them in a letter, assuring them that the revolutionaries' enemy was the Manchu government, and they could live with full political rights and equality in the future Republic.[138]

In 1912, the Republic of China was founded upon the slogan "Five People Under One Union" (*wuzu gonghe*). This included the Han, the Hui (Muslim), the Tibetans, the Manchus, and the Mongolians. However, the Republic ultimately revealed a great-Hanism (*da hanzu zhuyi*) disguised as a unified *Zhonghua*

---

134 Liang Qichao, "Zhengzhixue dajia Bolunzhili zhi xueshuo" (Political Theory of Bluntschli), *Liang Qichao quanji*, 1069.
135 Wang Jingwei, "Minzu de guomin," 31.
136 Zhang Taiyan, "Zhonghua minguo jie" (On the Republic of China), *Minbao* 15 (1907): 13–4.
137 Tang Zhijun, *Zhang Taiyan zhenglun xuanji*, Vol. 1, 520.
138 Ibid.

minzu. Liang's term invented to defend the Qing was now appropriated by the revolutionaries as fundamental to the Republic of China.[139] As a late Qing revolutionary, Sun Yat-sen openly interpreted "Five People Under One Union" as the Sinicization of non-Han.[140] This insistence upon a unified *Zhonghua minzu* intensified after the Manchurian Incident which took place on September 18, 1931 and led to the Japanese seizure of Manchuria. Manchukuo was founded in Manchuria as a puppet state in February 1932 and Japan appealed to non-Han to foster their own nationalism and liberate themselves from the Han chauvinism. In response, nationalist intellectuals affiliated with the KMT turned to history, archaeology, and ethnology for scientific proof that all ethnic groups in China shared the same bloodline. For them, differences in terms of habits and customs were simply consequences of different environmental conditions and it was only a matter of time before all ethnic minorities were integrated into the *Zhonghua minzu*.[141]

The voice of the ethnic minorities was rarely expressed in the revolutionaries' print media. Wang Xiaonong (1858–1918), a Manchu Peking Opera actor and editor of the *Ershi shiji dawutai*, was an exception. In *Minbao*, there is only one anti-Manchu text written by a non-Han writer. He was a Mongolian. In this article, he fully supported the Han revolutionaries' political radicalism, and it seems that he was willing to be "purified," supporting that, according to the law of social evolution, Mongolian nomad culture was inferior and backward.[142] But the reality was quite different. After the foundation of the Republic, both Tibet and Mongolia asserted their ancient independence. In January 1913, the two sides signed a treaty of friendship and alliance and declared mutual recognition, although the treaty failed to obtain international recognition and Tibet remained nominally part of China.

Revolution was restoration of the Chinese sovereignty, usurped by a foreign nation that colonized China and refused any form of assimilation. But there was a contradiction in the revolutionaries' discourses. Since the ancestors of the Chi-

---

139 Liu Xiaoyuan, "From Five "Imperial Domains" to a "Chinese Nation": A Perceptual and Political Transformation in Recent History", in *Ethnic China: Identity, Assimilation, and Resistance*, ed. Li Xiaobing and Patrick Fuliang Shan (Lanham: Lexington Books, 2005), 26.
140 Sun Yat-sen. "Zai Zhongguo Guomindang benbu teshe zhu Yue banshichu de yanshuo" (Speech Given at the Special Office of the Headquarters of the Kuomintang in Guangdong), in *Sun Zhongshan quanji* (Complete Works of Sun Yat-sen), ed. Sun Yat-sen Research Institute, Department of History, Sun Yat-sen University, Vol. 5 (Beijing: Zhonghua shuju, 1985), 474.
141 James Leibold, "Competing Narratives in Republican China: From the Yellow Emperor to Peking Man," *Modern China* 32, no. 2 (2006): 181, 201–2.
142 Mengyi zhi duofenzi, "Menggu yu Hanzu jiehe gongshen taoman fuchou dayi zhi xuanyanshu" (Declaration on the Cooperation Between Mongolians and the Hans for the Just Revenge Against the Manchu), *Minbao* 20 (1908): 113–9.

nese nation also came to China from the West as immigrants and conquerors, should the Chinese not give back the territory to its original owner? Zhang Taiyan tackled this contradiction by eliminating the paragraphs concerning Sino-Babylonianism in the 1915 version of *Qiushu*, stating that there was no substantial evidence to support this hypothesis.[143] Others were content to resort to social evolutionism, arguing that the expulsion of the indigenous people demonstrated the "survival of the fittest."[144] Could the same thing be said about the Manchu's conquest of China? Jing Yaoyue (1881–1944) was one of the few who considered this question and simply said that it was an exception to the universal law of social evolutionism.[145] Such rhetoric seems unconvincing, but it was how the political and cultural preponderance was endowed to the Han Chinese.

Origins of Political Reforms

The cohesion between politics and culture translated not only into the homogenization of the diverse ethnic identities of the Chinese population within a highly politicized society, but also into attempts to combine Chinese tradition with Western modern politics. National essence included Chinese language, history, as well as ancient political theories and institutions. Whereas the acquisition of the language demonstrated one's belonging to the national community, the preservation of ancient political theories and institutions attested revolutionaries' willingness to transform tradition into a living reality, rather than a cluster of relics. As I have demonstrated, national essence as cultural politics did not automatically entail radicalism or conservatism; it all depended on which cultural elements were preserved to inform which type of political order. Revolutionaries' idea of national essence was highly interchangeable with that of national learnings.[146] National learnings, or *guoxue*, originally meant state school. It acquired its modern signification in Japan (*kokugaku*) and joined force with national essence to resist the imperial learnings that served the imperial house.[147]

---

143 Zhang Taiyan, *Jianlun* (*Essay of revisions*), in *Zhang Taiyan quanji*, 360–77.
144 Yang Zhiqiang, "Cong 'Miao' dao 'Miaozu' – lun jindai minzu jituan xingcheng de 'tazhexing' wenti" (From "Miao" to the "Miao Nation" – On the Question of the "Otherness" in the Formation of Modern Nation), *Xinan minzu daxue xuebao* 6 (2010): 5.
145 Taiyuangongzi (Jing Yaoyue), "Shanxi xuangao tao Manzhou xi" (Anti-Manchu Declaration of Shanxi), *Minbao* 21 (1908): 102.
146 Zheng Shiqu, *Wanqing guocui pai*, 114.
147 Ibid., 118.

Xu Zhiheng (1877–1935), a contributor to *Guocui xuebao*, once remarked: "When the government evokes national essence, the purpose is to control the intellectual field and to stop people from taking the initiative to renew [the politics and the culture]." In this sense, the traditional culture defended by reformers and officials of the Qing Court, however culturally radical it might have been, was geared towards the objective of preserving the Qing dynasty.[148] Although imperial learnings mainly comprised of Confucianism and reformers cherished Confucianism as the nation's national essence, it would be misleading to suggest that revolutionaries rejected Confucianism all together. While Liu Shipei, for instance, condemned Confucian scholars as opportunists who "retreated from political life at sensitive times and took the chance whenever it arose,"[149] he spoke highly of those who refused to serve the succeeding dynasties that they criticized on the ground of Confucian classics.[150] To him, only by refraining from the political realm could Confucian scholars fully focus on their profession as a creator of national learnings, whose values and foundation opposed the indoctrination of the imperial learnings. It was also for this reason that Zhang Taiyan rejected national education sanctioned by the state's educational system.[151] Their view implicitly attacked the educational reform of the New Policies, which aimed to abolish Imperial examination and select officials into state bureaucracy in state-owned Western-style schools.

The Japanese *kokugaku* academic movement that began in the seventeenth and eighteenth centuries drew heavily from Shinto and Japan's early classics by discarding the influence of the then-dominant Chinese, Confucian, as well as Buddhist teachings.[152] As in the case of national essence, revolutionaries found Japanese national learnings too narrow. Huang Jie reasoned that by excluding foreign cultural elements, neither national learnings nor national essence would be brought to light.[153] Destruction was the keyword here, without which it was impossible to preserve anything, since preservation was based on an insightful selection and reinterpretation of the valuable elements in the cul-

---

148 Xu Zhiheng, "Du Guocui xuebao ganyan" (Remarks on the *Guocui xuebao*), *Guocui xuebao* 1, no. 6 (1905): 89.
149 The original text from Xun Zi reads: "時絀而絀, 時伸而伸。"
150 Liu Shipei, "Jinru xueshu tongxi lun" (Confucianism in Recent Times), *Guocui xuebao* 7, no. 27 (1907): 3019–23.
151 Zhang Taiyan, "Yu Wang Heming shu" (Letter to Wang Heming) in *Zhang Taiyan quanji*, Vol. 4, 151–3.
152 David Magarey Earl, *Emperor and Nation in Japan, Political Thinkers and the Tokugawa Period* (Seattle: University of Washington Press, 1964), 66–7.
153 See, Luo Zhitian, "Qingji baocun guocui de chaoye nuli ji guannian yitong," 55.

tural tradition.¹⁵⁴ Therefore, revolutionary intellectuals welcomed the new knowledge from the West to "complete" the national learnings and believed that national essence should be constantly changing to respond to the demand of political circumstances and match their evolution.¹⁵⁵ That is why Xu Shouwei (?–?), another writer of the *Guocui xuebao*, stressed the theme of rebirth and claimed that "national essence was not a barrier to Europeanization."¹⁵⁶

Hon Tze-ki remarks that the political culture of national essence was to a certain extent anti-Western, but it pointed to the same evolutionary end as the West. Borrowing from Wang Fan-sen's thesis, Hon maintains that in this "anti-Western Westernization," the National Essence School accepted "the global pattern defined by Western power," while insisting upon China's "separate route to achieve the same goal." He concludes that their political blueprint focused on "an idealized political system of the Western Zhou period," whose system of local autonomy greatly inspired the members' political outlook for a modern China.¹⁵⁷ Nonetheless, the West in the eyes of the National Essence School had two facets. While its members abhorred the colonial and imperialist West, they greatly appreciated the Western civilization that the West brought into China. To revolutionary intellectuals, the possibility of transposing foreign political theories and practices, such as social organicism and social contract, was based on the belief that traditional learnings echoed the latter. Nevertheless, contrary to the preceding generation of literati, they admitted that the West was clearly ahead of China. However, the parallel between the Renaissance and the Chinese revival of ancient learnings meant that China could join the modern world and catch up with the West by reviving national learnings and shaping national essence, as Hon argues.¹⁵⁸

It should also be noted that the traditional cultural elements that revolutionaries mobilized to theorize China's national essence were not limited to those of the Western Zhou period, and national essence was further forged and harmonized with Western modern political themes of democracy, constitutionalism, federalism, division of powers, parliamentarism, the motto of the French Republic,

---

154 Ibid., 56.
155 Shijiang (Gao Xu), "Xueshu yan'ge zhi ganlun" (A General Discussion on the Evolution of Knowledge), *Xingshi* 1 (1905): 33.
156 Xu Shouwei, "Lun guocui wuzu yu ouhua" (On National Essential not Being a Barrier to Europeanization), *Guocui xuebao* 1, no. 7 (1905): 100.
157 Hon, *Revolution as Restoration*, 118–119.
158 Hon Tze-ki, "Revolution as Restoration: Meanings of 'National Essence' and 'National Learning' in Guocui Xuebao", in *The Challenge of Linear Time: Nationhood and the Politics of History in East Asia*, ed. Viren Murthy and Axel Schneider (Leiden: Brill, 2014), 265.

and even Western science.¹⁵⁹ Scientific spirit was reflected in a more rational organization of ancient learnings. Traditionally, ancient books were categorized in four thematic groups: *jing* (Confucian classics), *shi* (history), *zi* (Hundred Schools of Thought and religion), and *ji* (literature). Deng Shi and Liu Shipei were among the first to break this tradition by proposing a categorization of ancient learning according to Western disciplines, such as psychology, ethic, rhetoric, sociology, religion, politics, law, economy, military science, pedagogy, physics, mathematics, and philosophy.¹⁶⁰

While the French Revolution remained a source of inspiration for revolutionaries, reformers evoked the disastrous Jacobin period to warn against revolution. For reformers, the French Revolution symbolized the absolute failure of radical actions, the bloody consequence of which ironically led to the restoration of monarchy. Not unlike English conservatives, Liang Qichao reasoned that the French Revolution was launched by the invention of liberty, which was absent in the French political tradition. Therefore, the illusion of liberty incited the mob to create a savage despotism. Consequently, the liberty enjoyed by the French during the post-revolutionary period was considerably restricted and incomplete in comparison to nations where liberty had been obtained through a long historical course and in a non-violent manner.¹⁶¹

For revolutionaries, the Chinese quest for liberty did not suggest an ideological invention; the spirit was believed to be rooted in traditional political culture, which was later crushed by imperial dictatorship in favor of the oppressive state Confucianism. Chen Tianhua (1875–1905), a well-known revolutionary, uncovered the vestige of liberty of expression, publication, and assembly during the Three Dynasties (*Sandai*), while the burgeoning Hundred Schools of Thought and the reunion between Confucius and the 3,000 students who accompanied him to spread Confucianism in various states also attested to the existence of the liberty of expression and movement in ancient China.¹⁶²

---

159 For a discussion on Liu Shipei, see Joachim Kurtz, "Disciplining the National Essence: Liu Shipei and the Reinvention of Ancient China's Intellectual History," in *Science and Technology in Modern China, 1880s–1940s*, ed. Jing Tsu and Benjamin A. Elman (Leiden: Brill, 2014), 67–91.
160 Liu Shipei, "Zhoumo xueshushi xu" (Preface to the History of Knowledge of the End of the Zhou Dynasty), in *Liu Shipei shixue lunzhu xuanji* (Selective Historical Articles of Liu Shipei), ed. Wu Guoyi and Wu Xiuyi (Shanghai: Shanghai guji chubanshe, 2006), 58–95.
161 Liang Qichao, "Mieguo zhengzhi pinglue" (Comments on American Politics), in *Xinhai geming qianshinianjian shilun xuanji*, Vol. 2, 793.
162 Sihuang (Chen Tianhua), "Zhongguo geming shilun" (On the History of Chinese Revolutions), *Minbao* 1 (1905): 55.

As to equality and fraternity, Mohism was often evoked as the philosophy in which the Chinese origin of these spirits were rooted. Intellectually opposing Confucianism, Mohism questions the Confucian ethical doctrine in favor of a materialist vision of life and politics.[163] Unlike Confucianism, in which the concept of "love" is fully incorporated and hierarchized into the social responsibility and role of each person, Mohism promotes the so-called *jian'ai* (universal love). Gao Xu actively promoted Mohism as the intellectual pioneer of egalitarian society and even suggested replacing Confucianism with Mohism as the state ideology.[164] The economic vision and "logic" mode of thinking of Mohism were further compared with Western materialism and scientific spirit.[165]

To Wang Jingwei, "given the fact that the nation has the ideals of liberty, equality and fraternity and that the people's power and constitution are all founded on these ideals, these political themes certainly suit the nation."[166] Kang Youwei did not deny that some of these values existed in China. However, it is exactly because the Chinese enjoyed these rights under the Manchu government that China did not need a revolution.[167] Ironically, this was also the reason evoked by Zhang Zhidong during the Hundred Days' Reform to refute Kang's reformist constitutional monarchy.

It is clear that a republic cannot be founded solely upon values. Revolutionaries had repeatedly ensured that their activism would lead to a "civilized revolution" rather than a "barbarian one." Zou Rong explained: "Barbarian revolution destroys without construction... which leads to terror, as the Boxer Rebellion and [the movement guided] by Garibaldi... Civilized revolution destroys and constructs at the same time... and aims to bring about liberty, equality, independence and individual autonomy."[168] Thus, once these indigenous values were set in law, constitution, parliament, and republican government, a democratic regime could be fully concretized.

None of these institutions was considered to be foreign to China. Instead, they were fully developed in an archaic past. Liu Shipei spoke highly of the age before the Three Dynasties. To him, the sovereign was elected by the people

---

163 Feng Youlan, *Zhongguo zhexueshi*, 307–48.
164 Yang and Wang, *Nansheshi changbian*, 301.
165 Sun Zhimei, *Nanshe yanjiu* (A Study on the Southern Society) (Beijing: Renmin wenxue, 2003), 266.
166 Wang Jingwei, "Bo xinmin congbao zuijin zhi fei geminglun" (Rebuttal of the Recent Anti-Revolutionary Argument of *New People Journal*), *Minbao* 1, no. 4 (1905): 29.
167 See, for example, Mingyi (Kang Youwei), "Faguo gemingshi lun" (On the History of the French Revolution), in *Xinhai geming qianshinianjian shilun xuanji*, Vol. 2, 330.
168 Zou Rong, "Geming jun," 35.

at the time, and the hereditary system was unheard of.[169] Thus, a sovereign's mandate was fixed, just like in the Western republican nations; when the mandate expired, another leader should be elected.[170] Huang Jie further stated that not only was the leader elected, but he also only had an accessory role in the political decision-making process. Everything was transparent and open to debate.[171] Even if the state Confucianism destroyed this political system, Liu Shipei perceived in the concept of Mandate of the Heaven a relic of the ancient politics, stipulating that the power of the sovereign came from and was supervised by Heaven. Thus, it could not be abused.[172]

The establishment of rational laws was also frequently debated. This recalls the dichotomy between the Confucian rule of man (*renzhi*) and the Legalist rule of law (*fazhi*). Confucianism emphasizes the virtuous nature of men and believes that the value of a community depends on education instead of punishment, while for Legalism, the vicious human nature makes discipline, intimidation, and rigid laws possible and indispensable. Legalism associated the concept of law with justice, individual dignity, and protection from oppression. However, since Legalism is, after all, an authoritarian form of thought, it may seem problematic that the inclusion of the emperor in the juridical order was sufficient for many revolutionaries to support Legalism as the ideological foundation of the juridical system of the regime that they were engaged in constructing.[173] This preference for Legalism needs to be contextualized in the late Qing reformist period, in which the preponderance of the rule of man was deemed to be the consequence of the expansive power of the emperor, which resulted in the lack of rational law and dictatorship.[174] At the intellectual level, the return to the rule of law, as with all the other political forms, recalled the duality between national learning/national essence and imperial studies.

Local self-government was another institution that revolutionaries aimed to restore, as Hon has shown. This political form of social organization was implemented by the Qing Court as part of the New Policies. However, far from decen-

---

169 Liu Guanghan, "Guzheng yuanshi lun" (On the Real image of Ancient Politics), *Guocui xuebao*, 1, no. 2 (1905): 221.
170 Ibid., 222.
171 Huang Jie, "Huang Shi juaner" (The Yellow History, Volume 2), *Guocui xuebao* 1, no. 3 (1905): 450.
172 Liu Guanghan, "Lianghan zhengzhixue faweilun diyi" (First Discussion on the Politics of the Two Han Dynasties), *Guocui xuebao* 10 (1905): 837.
173 Liu Shipei, "Zhoumo xueshushi xu," 112–3.
174 Yongan, "Xingshi susongfa zhi yu Zhongguo" (The Penal Code in China), *Xingshi* 1 (1905): 29.

tralizing the central government, the Manchu Court wished to integrate local elites into the bureaucratic structure and take charge more firmly of local affairs by penetrating even the smallest administrative division.[175] Thus, local self-government was restricted to the activities promoting local public warfare (*difang gongyi shiye*). Even this was closely supervised by the government, making people reluctant to participate.[176] Revolutionaries believed that local self-government could be effectively used as a platform to nourish the people's nationalism, arguing that this was exactly the form of the state during the Three Dynasties.[177]

In a wider context, local self-government was intimately related to federalism, which revolutionaries related to the ancient form of government *fengjian* that could be compared with European feudalism.[178] *Fengjian* refers to the political organization of the Zhou dynasty, in which centralization was relatively weak. The organization allowed region states to have their own government to manage civil, financial, juridical, and military affairs. The head of each regional state was subject to the sovereign of the Zhou dynasty.[179] After the collapse of the Zhou dynasty, the succeeding Qin dynasty adopted the political system of *junxian* – prefectures and counties, which has permeated the Chinese political system to the present day. Although *fengjian* only enjoyed a momentary existence, Confucian scholars always felt nostalgic for it, since the Qin dynasty, which adopted Legalism as the state ideology, was judged arbitrary and tyrannical.

Gu Yanwu, for example, estimated that *junxian* reflected the absolutism of the central government.[180] While acknowledging the impossibility to return to *fengjian*, he was persuaded that the spirit of *fengjian* (that the emperor did not treat the country as his own property) could be combined with *junxian* by allowing officials to govern his place of origin. To him, the *si* (egoism, private) could stimulate the officials' willingness to enhance local warfare. The emperor, on the other hand, should demonstrate the *gong* (public) by being altruistic and disinterested. Huang Zongxi had a similar thought. He called for a reimplementation of *fangzhen* (or *fanzhen*). This was a military city situated at the borders, created during the second half of the Tang dynasty to resist foreign intrusion. Local af-

---

175 Hon, *Revolution as Restoration*, 91–2.
176 Niu Ming-shi, "Wanqing difang zizhi yundong de fansi" (Reflection on the Local Self-Government at the End of the Qing Dynasty), *Twenty-First Century* 98 (2006): 44.
177 Hon, *Revolution as Restoration*, 83–112.
178 Li Feng, "'Feudalism' and Western Zhou China: A Criticism," *Harvard Journal of Asiatic Studies*, 63, no. 1 (2003): 115–44.
179 Ibid., 142.
180 Gu Yanwu, "Junxian lun" (On *Junxian*), in *Gu Yanwu wenji* (Works of Gu Yanwu), ed. Qian Zhonglian (Suzhou: Suzhou daxue chubanshe, 2001), 1–11.

fairs were managed by officials called *jiedushi*. To Huang, the same politics should be applied to all regions.

Zhang Taiyan compared the decentralization with *fangzhen* that Huang Zongxi had evoked. He believed that this system incarnated the spirit of *gong*, as long as *jiedushi* did not privatize his territory. In 1899, he asserted that the time had come for the *fangzhen* to be revived.[181] In the ensuing decade, federalism was considered by revolutionaries to be the only feasible solution in order to maintain national unity and counter the omnipresence of the central government. This explains why many editors denominated revolutionary journals by using the name of their native provinces: *Xin Guangdong* (New Guangdong), *Xin Hunan* (New Hunan), *Jiangsu*, and *Jiangxi*, to name a few.

This type of localism was not considered to be in conflict with nationalism. As mentioned earlier, love for the nation began with love for one's native place. This native place was generally regarded as the province where one came from. To Ou Jujia (1870–1911), given the diversity of language, culture, and customs of each province, it was simply nonsense to require the Chinese to love the nation. He called the love for province "small nationalism." Only after fostering this nationalism could the "great nationalism" come about.[182] Hou Hongjian (1872–1961) even argued that China only had two options: she could be subjugated as Poland and India, or she could choose to become as powerful as Germany and the United States.[183] Revolutionaries placed China in a new spatial-temporal context, supporting whatever form of state derived from the new political reality of the world. Evidently, this was a simplistic view that reduced the prosperity or the sufferance of a nation to its political structures. But as I shall show in the following chapters, the question of *fengjian*, federalism and local self-government continued to haunt the political debates during the Republican era.

Ideas and Counter Ideas

In a text published in the 1990s, Lucian Pye states that nationalism in China jeopardized the nation's modernization and the insistence on the preservation of Chinese cultural tradition only reinforced the opposition between Chinese na-

---

181 Zhang Taiyan, "Fenzhen," in *Zhang Taiyan zhenglun xuanji*, 106.
182 Ou Jujia, "Xin Guangdong" (New Guangdong), in *Xinhaigeming qianshinianjian shilun xuanji*, Vol. 1, 270.
183 Hou Sheng (Hou Hongjian), "Ai Jiangnan" (Mourning Jiangnan), in *Xinhaigeming qianshinianjian shilun xuanji*, Vol. 1, 538.

tionalism and modernization.[184] While this observation might be partially true in the case of the Court's officials, Pye fails to recognize the constitution of political projects that integrated the reformulation of traditional teachings and modern Western politics. A culture both traditional and unconventional but also open to interpretation and flexible enough to respond and adapt to the new political realities did emerge in the 1900s. Borrowing from tradition does not make one automatically traditionalist or conservative, since the culturalist nationalism during the revolutionary decade was a highly politicized culture that gave rise to different or even opposite political aims. This chapter explored this give-and-take relationship between the conservative and the radical culturalist nationalists against the backdrop of evolving political circumstances of the 1890s and 1900s. In the same culturalist language, two forms of political future were conceptualized and confronted with each other. This debate concluded with the Revolution of 1911 that saw the end of China's dynastic system and the establishment of the first republican regime in Asia. The next question is, how did the late Qing culturalist nationalism, which shaped both conservatism and radicalism of this period, come to be dismissed as conservative altogether after the foundation of the Republic?

---

[184] Lucian Pye, "Zhongguo minzuzhuyi yu xiandaihua" (Chinese Nationalism and Modernization), *Twenty-First Century* 9 (1991): 23.

3 From Culturalist Nationalism to Nationalist Conservatism

Culturalist nationalism informed both late Qing conservatism and radicalism. At what point in time and in what sociopolitical configuration was culturalist nationalism indiscriminately judged as conservative? Conservatism does not exist without its opposite. In the eyes of the younger generation, even the political and cultural claims of the late Qing radical culturalist nationalism appeared to be demoded, outdated, and backward-looking. This process did not happen overnight but occurred over an extended period from the foundation of the Republic in 1912 to the May Fourth New Culture Movement. Despite the intellectual challenges that defenders of the late Qing culturalist nationalism confronted immediately after the establishment of the Republic, a real reaction from them was only triggered with the rise of the New Culturalists.

On the eve of the victory of the Revolution of 1911, *Guocui xuebao* published its last issue. Perhaps Deng Shi and Huang Jie were out of funds. After all, its publication was sponsored by donations and sales of ancient fine art reproductions. At the intellectual level, *Guocui xuebao* was replaced by *Guxue huikan* (*Journal of Ancient Learning*), the manifesto of which reads:

> The publication of the *Guocui xuebao* has lasted for seven years. Thanks to the Yellow Emperor, the politics of China is now restored. The foundation of the Republic is inseparable from public opinion and citizens' spirit... Today we have succeeded in retrieving our usurped political power... What matters now, is thus no longer the [shaping of] public opinions, but rather the authentic knowledge. Hence, we have decided to cease the publication of the *Guocui xuebao* and created the *Guxue huikan*, so that we can be concentrated on research, instead of continuing to incite or provoke public reactions by utilizing classic learning.[1]

As this citation shows, the voluntary retreat from the political realm characterized the intellectual trajectory of many former revolutionaries, including numerous members of the Southern Society who, for the most part, were "old-style" poets. The founders of the Society for the Preservation of National Learnings were no longer politically active. Deng Shi put an end to his political life and devoted himself entirely to the collection and preservation of ancient books, paintings, and antiquities. Huang Jie found a teaching position at the Peking University. The founder of the Critical Review Group, Wu Mi (1894–1978), mentioned in his 1927 diary that Huang was so frustrated by the decadence of ancient learning

---

[1] "Guxue huikan fakanci" (Manifesto of the *Guxue huikan*), *Guxue huikan* 1 (1912).

that he cried in front of him.² The culture that Huang admired no longer encapsulated the political utility that it had during the revolutionary decade. Ma Xulun once recalled that when the Republic was founded, he believed that revolutionary intellectuals had completed their mission and political construction should be on the shoulders of politicians.³

Meanwhile, the market-oriented Southern Society writers, notably Wang Dungen (1888–1951), Bao Tianxiao (1876–1973), Fan Yanqiao (1894–1967), Zhou Shoujuan (1895–1968), Yao Hechu (1893– ?), and Xu Zhenya (1889–1937), provided popular entertainment with love stories, adventures or political scandals for urban audiences in the traditional novel form of storytelling. This kind of writing is known as the Mandarin Ducks and Butterflies literature (*Yuanyang hudie pai*). Although recent research has shown that this type of literature melded well with the progressive dimensions of the intellectual sphere at the time, it was harshly dismissed as nothing more than escapist entertainment by the New Culturalists.⁴

This political laxity, expressed sometimes in the traditional form of escapist literature, served as a pretext for the New Culturalists to attack the intellectuals of the previous generation for selling "fake antiques," an apt name for the commodified traditional culture. However, it must be stressed that the New Culturalists did not obtain the discursive authority within the intellectual sphere overnight. In fact, their initial attacks on the defenders of traditional culture were not so readily noticeable that Qian Xuantong had to invent a fictional conservative Wang Jingxuan in a 1918 issue of *New Youth* to rouse the attention of the public to the "struggle between traditional culture and new culture."⁵ It was those who later unwittingly became conservatives that dominated the intellectual sphere before it was taken over by the New Culturalists from the late 1910s onwards. Despite the retirement of some conservatives from the political scene at the beginning of the 1910s, many, including Zhang Taiyan, Liang Qichao, and Liu Yazi, still assumed the political role that China's politico-cultural tradition bestowed on educated elites.

Their political allegiance was quite different at the very beginning. Zhang Taiyan and Liang Qichao served the Yuan Shikai government, whereas Liu Yazi firmly opposed the 1911 North-South Conference (*Nanbei yihe*), in which Sun Yat-sen agreed to cede the presidency to Yuan Shikai on the condition

---

2 Wu Mi, *Wu Mi riji 1925–1927* (Wu Mi's Diary, 1925–1927) (Beijing: Sanlian shudian, 1998), 353.
3 Ma, Wo zai liushisui yiqian, 44.
4 See, for example, Chen Jianhua, "Minguo chuqi xiaoxian zazhi yu nuxing huayu de zhuanxing" (The Magazines of Leisure and the Transition of the Female Discourses at the Early Republican Era), *Zhongzheng hanxue yanjiu* 22 (2013): 355–386.
5 Hill, *Lin Shu, Inc.*, 192–230.

that Yuan would accept the principle of the republican government and persuade the Xuantong Emperor to abdicate.[6] Although culture still had a role to play in intellectuals' political choices, it seemed that political construction took precedence over cultural concerns. But Yuan's restoration and abrupt death in 1916 ended this situation. Intellectuals were disappointed by the chaotic political reality, and Lu Xun's motto that medicine could not save Chinese, but literature could, resurfaced. Intellectuals' attention was refocused towards an organic society in which free individuals support one another and share a real conception of collective political life. This idea emerged once again as the spearhead of nationalist movements and expectations.

To shape this society, New Culturalists appealed for a scientific and democratic enlightenment. This was believed to be totally incompatible with the Confucianism of which Yuan took advantage to rationalize his restoration. While late Qing political activists also criticized many aspects of tradition without losing sight of the way in which traditional culture could be revamped through a new vision to inform modern politics, the New Culturalists' "holistic mode of thinking" rejected traditional learning as a whole. To save China as a political entity, the residue of traditional culture had to be abandoned in favor of a "New Culture." As such, those who attempted to preserve culturalist nationalism as the potent intellectual tool to institutionalize the nation-state came to be seen as conservative.

The New Culturalists did not silence the conservatives, but the conservatives failed to emulate their opponents' success. In comparison with the revolutionary decade, their discourse became increasingly limited to a defensive cultural expression. Consider the *Critical Review* magazine in which the re-politicization of Confucianism, capitalism, socialism, industrialization, democracy, and the religious questions were rigorously debated in the first few issues. However, it quickly lost its political vigor and became an almost purely cultural magazine.[7] In her study on *Shenbao* (*Shanghai News*), Barbara Mittler asked "*how* the Chinese press functioned and thus *how* and *why* it could be powerful."[8] In this chapter, I emphasize the multiple forces that shaped the Chinese intellectual sphere from the mid-1910s to the 1920s and ask how and why the conservative voice became weak and trailed off. Since language reform lay at the heart of the cultural debate, I will focus mainly on this issue to depict the stratagems deployed in the

---

6 Liu, *Nanshe jilue*, 39–40.
7 Luo Gang, "Lishi zhong de Xueheng" (The Critical Review Group in History), *Twenty-First Century* 28 (1995): 41.
8 Barbara Mittler, *A Newspaper for China? Power, Identity, and Change in Shanghai's News Media, 1872–1912* (Cambridge, Mass.: Harvard University Press, 2004), 413–4.

struggle between conservatives and New Culturalists for the domination of the intellectual field to tackle this question.

The Polemic of the Vernacular Language

The New Culturalists blamed the decadence of the nation squarely on the residue of traditional learnings. Apart from Confucianism, classical Chinese (*wenyan*) was another main target. The national language movement (*Guoyu yundong*) was not an invention of the New Culturalists. Before 1900, the movement launched by the Qing government aimed to unify the spoken language and the written language, while between 1901 and 1911, the government attempted to standardize the spoken language.[9] The language of Beijing – or Mandarin – was chosen as the standard language, and classical Chinese remained the official written language. Although the New Politics quickly suffered defeat with the collapse of the Qing Court, the republican government pursued this language movement. A proposition presented to the Third Assembly of the National Federation of Education (*Quanguo jiaoyu lianhehui*) reads: "The national language designated Mandarin (*guanhua*) that we now call *Putonghua*."[10] In 1913, the government organized the Committee for the Unification of Pronunciation (*Duyin tongyihui*) in order to standardize the national language. In 1916, the Association for the Research on National Language (*Guoyu yanjiuhui*) was created by the Ministry of Education. The National Language Unification Preparation Committee (*Guoyu tongyi choubeihui*), founded in 1919 under the Ministry of Education, successfully codified the vernacular language of Mandarin (*baihua*) as the national language. In the following year, vernacular language was promoted by the government as the teaching language of all primary and secondary schools.[11]

If the New Culturalists had achieved their goal, why did they persist in attacking intellectuals who continued to write in classical Chinese? Self-proclaimed as the leaders and the enlighteners of society, perhaps they were infuriated by the popular Mandarin Ducks and Butterflies literature. But the use of

---

9 Li Jinxi, *Guoyu yundong shigang* (History of the Movement of National Language) (Shanghai: Commercial Press, 1935), 10.
10 Quanguo jiaoyu lianhehui disanjie huiyi qingding guoyu biaozhun tuixing zhuyinzimu tian (Proposition on the Promulgation the *Bopomofo* and on the Standardization of the National Language During the 3rd Assembly of the National Federation of Education), in *Zhonghua minguo shi dang'an shiliao huibian•disanji•jiaoyu* (Archives of the Republic of China, Volume 3, Education), ed. Second Historical Archives of China (Nanjing: Jiangsu guji chubanshe, 1991), 770.
11 Li Jinxi, *Guoyu yundong shigang*, 109–10.

classical Chinese did not end there, and conservatives had come up with reasonable arguments against the adoption of Mandarin as the national language.

Unlike the culturalist nationalists of the 1900s, the New Culturalists not only wished to popularize the vernacular language to facilitate the education of the inferior society, but also aimed to invent a new nation by drastically changing its language. By remodeling the new vernacular language according to Western grammar, New Culturalists aspired to a new nation that broke away from its cultural heritage and fell in line with Western nations. Such a political and intellectual orientation was unacceptable to Zhang Taiyan. To him, vernacular language should take root in the nation's linguistic tradition. From this point of view, Mandarin did not conform to the archaic Chinese language nor represent the authentic Chinese vernacular language since the nation was not defined by the North.[12] The advantage of classical Chinese rested in the fact that it enabled Chinese who spoke different local languages to understand each other by writing without being forced to learn a "foreign" language. This is one of the arguments advanced by *Guogu yuekan* (*National Heritage Monthly*), a journal published by former members of the National Essence School:

> What is this so-called national language that they [the New Culturalists] talk about? Is it not the Mandarin of the Qing? Is it not a dialect of Beijing? But how many Chinese speak this dialect? How many educated Chinese speak it?... If the Beijing dialect is indeed superior to others and allows us to express things that we cannot express in others, Mandarin should be the national language, but this is not the case... Mandarin was spoken by the emperor and Qing officials and is now used by officials of the government. This is its only advantage. But forcing Chinese to learn Mandarin, just because it was used by officials is possible only in a regime of dictatorship. We are now in an egalitarian country and Chinese do not have this obligation to learn it.[13]

Thus, language is democracy. For the author, a government that imposed a language to the detriment of others was no different to a dictatorial regime. From an evolutionist perspective, the author reasoned that all languages were vulgar at the very beginning and became more and more refined. Thus, he did not understand the replacement of an elegant language – classical Chinese – by a vulgar vernacular language. For him, a return to the vernacular language was as retro-

---

12 Luo Zhitian, *Guojia yu xueshu: Qingji minchu guanyu "guoxue" de sixiang lunzheng* (The State and Academia: Debates Over the Thought of National Learning at the End of the Qing and the Beginning of the Republic) (Beijing: Sanlian shudian, 2003), 185.
13 Zhang Xuan, "Yanwen heyi pingyi" (On the Unification Between the Spoken Language and the Written Language), *Guogu yuekan* 1 (1919): 3.

grade as the restoration of monarchy.¹⁴ He further argued that it would be fatuous to adopt the vernacular language while the classical language was already the language of teaching:

> Education is necessary to popularize the vernacular language. All teachers today know how to write classical Chinese in a simple and clear manner, while few among them speak the dialect of Beijing. If we use the vernacular language at a national level, local languages will have to disappear. To this end, all the teachers should quit their post of teaching and learn the national language first. Three years are minimum for them to be capable of writing and speaking it fluently. During these three years, will these teachers be all replaced by those from Beijing? [In this case], would it not be easier to keep the way as it is?¹⁵

Mostly importantly, the establishment of the vernacular language as the national language jeopardized the natural evolution of the language itself. Like Zhang Taiyan and Liu Shipei who wished for a vernacular language based upon the fusion of local languages, the author argued that a national language that would serve the whole population should be the culmination and outcome of the natural amalgamation of local languages, instead of an imposed Northern dialect.¹⁶

On the political level, conservatives' cultural orientation echoed their refusal to accept preconceived political reforms or to adopt a Western political system that countered local identities and customs. The polemic of the vernacular language reappeared regularly in the ensuing years, the controversial point of which did not always relate to the question of literary style. In his study of Lin Shu, Michael Gibbs Hill discusses Lin's "becoming" a conservative during the New Culture Movement by drawing from Bourdieu's theory on the "field of cultural production," according to which the prize in the struggle between writers is "the monopoly of the power to consecrate producers of products."¹⁷ I further argue here that this struggle within the field of cultural production should be analyzed first and foremost within the historical context of the birth of the profession of the modern intellectual. What legitimized intellectuals' discursive authority within the society is the professional value of "disinterestedness" (i.e., independence vis-à-vis the state, political apparatus, as well as the market). Inherent to the cultural debates between conservatives and New Culturalists was the forceful effort to appropriate this symbolic capital that enabled them to univer-

---

14 Ibid.
15 Ibid., 4.
16 Yao Dianzhong and Dong Guoyan, *Zhang Taiyan xueshu nianpu* (Academic Biographical Chronology of Zhang Taiyan) (Taiyuan: Jiangxi guji chubanshe, 1996), 161.
17 Hill, *Lin Shu, Inc.*, 193–4.

salize their own particularistic politico-cultural standpoints as acting in general interest.

Classical Chinese and Depoliticized Literature

Traditional literati became intellectual in China with the abolition of the Imperial examination in 1905. Cultural elites deprived of the traditional access to the state bureaucracy were forced to compete for a job outside the political realm and find a career corresponding to their specific expertise. Haunted by the politico-cultural tradition that knowledge should be put in use for the public good, the educated elite in general earned a living as translators, newspapermen, journalists, lecturers, publishers, and writers – professions that enabled them to continue to influence society. Moreover, through collective actions and associations, they got involved in cultural debates and carried out collective actions. As Michel Hockx points out, culture, as the specific expertise of intellectuals, was used to redefine their identity from traditional literati to modern intellectuals. Cultural activities were thus expected to be "associated with the well-being of their country and their people."[18] In this context, educated elites' cultural capital became a source of charisma and a tool of political engagement.

This was exactly what late Qing reformers and revolutionaries did in the 1900s. Propaganda through printed media was one of the strategies of the Tongmenghui to foster and coagulate the readers' anti-Manchu nationalism. In the early Republic, the late Qing culturalist nationalists attempted to emulate their success in the 1900s. Liu Yazi recalled that major newspapers of the 1910s and 1920s, such as the *Taipingyang bao* (*Pacific News*), *Tianduo bao* (*Heavenly Bell News*), *Shibao* (*Times Newspaper*), *Minquanbao* (*People's Rights*), *Minlibao* (*People's Stand*), *Shenzhou ribao* (*China Daily*), *Dagonghe bao* (*The Great Republic*), *Minguo xinwen* (*Republic News*), and *Minsheng ribao* (*People's Life*) were almost monopolized by writers of the Southern Society.[19] During Yuan's reign, the Southern Society assumed their professional responsibility of engagement by collectively denouncing Yuan's authoritarianism and defending justice, liberty, and freedom of speech through these publications

Despite continuous political commitment, the reality was harsh. Not only were intellectuals deprived of access to the state bureaucracy to directly influence political decision-making, but they were also confronted with the newly

---

18 Hockx, "Introduction," in *The Literary Field of Twentieth-Century China*, 12.
19 Liu, *Nanshe jilue*, 42–3.

emerging social groups that challenged their traditional prestige. Intellectuals had to accept the fact that they were no longer the leaders of a society whose discursive authority was taken for granted.[20] To legitimize their claim to represent the people, intellectuals had to gain the symbolic capital of prestige, distinction, and fame by practicing their profession in the spirit of disinterestedness, a strategy that consisted of "refusing at the same time subordination to the state and submission to the market, intended to be linked only with the public."[21] *Zhonghua xinbao* (*Chinese New Journal*), edited by members of the Southern Society, for example, defined its objective in these terms in 1915: "We wish to expose social and political situations justly. Our sole goal is to search for universal values by rising up against dependence on money or political power."[22] Miao Fenglin (1899–1959), a member of the Critical Review Group, spoke of disinterestedness by citing Matthew Arnold (1822–1888):[23]

> It is of the last importance that English criticism should clearly discern what rule for its course, in order to avail itself of the field now opening to it, and to produce fruit for the future, it ought to take. The rule may be summed up in one word, – *disinterestedness*. And how is criticism to show disinterestedness? By keeping aloof from practice; by resolutely following the law of its own nature, which is to be a free play of the mind on all subjects which it touches; by steadily refusing to lend itself to any of those ulterior, political, practical considerations about ideas [...] which criticism has really nothing to do with.[24]

Under this professional value of disinterestedness, intellectuals claimed their independence with regard to the political field and the market and rationalized the universalization of their politico-cultural standpoints as acting in the general interest. The autonomy of the intellectual sphere certainly did not prevent intellectuals from renouncing their critical independence by adhering to political parties or individuals holding political power. While it was almost impossible for intellectuals to maintain political neutrality, some of them went so far as to be dominated voluntarily by the ideological apparatus of the state or political parties, to the detriment of what is commonly regarded as the general interest. One case in point is that of Liu Shipei. Supporting Yuan's restoration, he exercised his profession as an intellectual by repudiating critical freedom, creating poems, essays,

---

20 Hon, *Revolution as Restoration*, 114–6.
21 Lucien Karpik, *Les avocats. Entre l'État, le public et le marché, XIIIe–XXe siècle* (Paris: Gallimard, 1995), 90–1.
22 Yang and Wang, *Nansheshi changbian*, 402.
23 Miao Fenglin, "Wende pian" (On the Moral of Literature), *Critical Review* 3 (1922).
24 Matthew Arnold, "The Function of Criticism at the Present Time," in *Essays in Criticism* (London and Cambridge: Macmillan and Co., 1865), 18–9.

and political comments deemed by mainstream intellectuals to be contrary to the public interest. Literature of the left wing (*zuoyi wenxue*) of the CCP and the art of the Three Principles of the People (*sanminzhuyi wenyi*) of the KMT were other examples. Although this type of subordination to the state's political demands exempted them from repression, censorship or even prosecution as a result of dissidence, boycotting of these intellectuals often took place within the intellectual field. One example is Gao Xu's editing and 1914 publication of poems composed over the previous 30 years, from which the writers who had supported Yuan and the Manchu Court were deliberately excluded.[25]

Another famous affair is that of the scandalous election of Cao Kun (1862–1938) as the president in 1923. Cao Kun won the majority in the Parliament by bribery. On October 20, 1923, *Shenbao* published a list of 550 members of parliament who voted that day.[26] Although this list did not reveal who voted for Cao Kun, the public believed that all of them were bribed by the latter. On the list were the names of 19 members of the Southern Society. Tian Tong (1879–1930) called for an immediate exclusion of these members from the Society in the *Minguo ribao*.[27] The exclusion was later confirmed in the same newspaper.

Politicians could also take advantage of intellectuals' discourses against their will to gain legitimacy of their politics. Zhang Taiyan's case offers a telling example. His status as an erudite scholar and an emblematic figure of the Revolution of 1911 granted him such charisma that Yuan Shikai did not dare to assassinate him, despite his open contestation with the latter.[28] It was also because of his charisma that politicians took advantage of his political standpoints for their own benefit. Zhang supported the argument that Beijing should be the capital of the Republic and that federalism was an effective means to unify the nation. Nevertheless, these proposals were picked up by Yuan Shikai and warlords to consolidate their own political interests and powers.

The differentiation between the political sphere and the intellectual sphere forced intellectuals to seize culture as the foundation of their intellectual autonomy. The strategy of refusal of subordination to the state took several forms. The first consisted of reclaiming the autonomy of criticism and role of spokesmen for public interests. However, given the government's censorship, it was hard to ignore the potential consequences of political criticism. The independence of the intellectual field was, in this regard, only relative. Although

---

25 Yang and Wang, *Nanshe shi changbian*, 360.
26 "Beijing toupiao yiyuan mingdan daohu" (List of the Members of the Parliament Who Voted), *Shenbao* (October 20, 1923).
27 Yang et Wang, *Nansheshi changbian*, 582.
28 Yao and Dong, *Zhang Taiyan xueshu nianpu*, 220.

many still assumed the professional responsibility of engagement, some writers chose to refrain from commenting on political issues. This silence towards politics also resulted from frustration with the ongoing political chaos. The Republic was plunged into instability from the first year of its foundation. After Yuan's death, China declined into a period of warlordism. Yuan's death allowed Sun Yat-sen to return to China from exile in Japan. However, the military government that he established in Guangdong in 1917 to challenge the Beiyang government failed to gain international recognition and unify China. China would remain a disintegrated nation for more than a decade to come. For some intellectuals, this was proving to be the final straw for political activism.

Disappointed by the political turmoil that unfolded across the country after 1912, Liu Yazi recalled that during this period, his literary works were nothing but the "sound of decadence" (*mimi zhi yin*).[29] Between 1915 and 1917, he founded at least three associations for classical poetry whose activities expressed the manifest determination to take refuge in an intellectual environment disconnected from reality. One of these literary associations was the "Alcohol Society" (*Jiushe*). Indeed, alcohol and poetry permeated the history of the Southern Society. Its members were known to indulge in the lifestyle of the so-called "literary celebrity" (*mingshi*) of ancient times. The "elegant assemblies" (*yaji*) they regularly attended not only offered an occasion to discuss politics, but also constituted a "romantic" event, held in restaurants and gardens in Shanghai, Suzhou, and Hangzhou, where the participants improvised classical poems in an atmosphere of conviviality produced by alcohol.

Writers of the Mandarin Ducks and Butterflies school took this lifestyle to an extreme. This literary movement started in the 1920s and included all types of popular old-style novels deemed by New Culturalists to be morally reprehensible and artistically irrelevant, such as scandalous novels (*heimu xiaoshuo*), love stories (*aiqing xiaoshuo*), detective histories (*zhentan xiaoshuo*), and martial heroes (*wuxia xiaoshuo*). To augment the profits, literary works from the Mandarin Ducks and Butterflies in classical Chinese were usually published as a serial novel in magazines and tabloids (*xiaobao*). Their cultural activities were carried out in the spirit of the Shanghainese notion of *bahsian*,[30] a polysemic word denoting "play," "leisure", and "funny". Old-style literature and tabloids were forms of *bahsian*. Titles of the major tabloids of the Mandarin Ducks and Butterflies school, such as *Kuaihuo* (*Pleasure*), *Youxi zazhi* (*Game Magazine*), and *Libai-*

---

29 Liu, *Nanshe jilue*, 61.
30 Li Nan, *Wanqing, Minguo shiqi Shanghai xiaobao yanjiu* (Study on Tabloids in Shanghai from the Late Qing Period to the Republican Era) (Beijing: Renmin wenxue chubanshe, 2005), 198–220.

liu (Saturday) all carried a sense of the hedonism implied in *bahsian*. To attract readers in search of novelty and pleasure, photographs of famous courtesans, anecdotes, political scandals, and news of leisure resorts in Shanghai were regularly published. However, it was their lifestyle that embodied the best of *bahsian*. Incorporating the habits of ancient literary celebrities, they indulged themselves in alcohol, classical literature, theatre, leisure resorts, and prostitution. Brothels, euphemistically called *huachang* (flower sites), where they socialized in the presence of courtesans, constituted an essential component of their sociocultural life.[31] In this way, the habits of literary celebrities was combined with economic success within the market of cultural consumption. These practices ran against intellectuals' pretentions of literary engagement. These depoliticized activities dispossessed intellectuals of their charisma and authority associated with the responsibility of engagement within the social space.

The independence that these intellectuals gained vis-à-vis the political field came at a price. On the one hand, the literary creations of writers such as Liu Yazi retreated from the public to the private sphere and took the form of escapism. On the other hand, writers of the Mandarin Ducks and Butterflies school submitted themselves to the market. All of them abdicated from the professional value of disinterestedness and the responsibility of engagement. Traditional culture became an object of leisure, a tool for profit, rather than an ideological instrument for institutionalizing the nation-state. All these factors made them an easy target for the emerging New Culturalists.

The Stigmatization of Conservatism

The political indifference resulting from disillusionment, the habits of literary celebrities, and the loss of the professional value of disinterestedness of certain members of the Southern Society allowed the leaders of the New Culture Movement to pose as bulwarks against conservative intellectuals. In a speech given at Yenching University in 1929, Lu Xun mentioned the Southern Society as a counterexample of literary revolution:

> The so-called revolutionary writers of yesterday should disappear; new literary revolutions cannot appear only when (the previous) revolution begins to pay off and allows us to reflect on it. Why? Because seemingly revolutionary works emerge from time to time when the society of yesterday is about to collapse, but in the end, these works do not constitute a rev-

---

31 Zhang Yongjiu, *Yuanyang hudie pai wenren* (Writers of the Mandarin Ducks and Butterflies school) (Taipei: Showwe, 2011), 6.

olutionary literature. [...] (Some writers of yesterday) hated the old society, but they did not have a project for the future; they called for a transformation of society, but their ideal was nothing but a utopia. [...] In China, there were writers desirous of a revolution who remained silent when the revolution came. The Southern Society of the late Qing is an example. It is a literary society that advocated revolution. Its members mourned the oppression that the Han Chinese were subjected to, abhorred the Manchu's dictatorship and called for a restoration (of a Han regime). However, they became silent after the foundation of the Republic. In my opinion, that is because the only thing they wanted was to see the renaissance of the stately manner of the Han official (*hanguan weiyi*) and people walking on the street in traditional costumes, after the Revolution of 1911. But when things did not turn out as they wished them to be, they became bored (with politics) and no longer created anything.[32]

While true to some extent, Lu Xun's view is biased. The "stately manner of the Han official" probably refers to one of Chen Qubing's 1904 poems,[33] but this locution should be contextualized. Written during the revolutionary decade, it was used as a metaphor for the orthodox culture and the political legitimacy of the Han Chinese. Furthermore, Lu Xun ignored the role that the Southern Society played in the struggle against Yuan Shikai. Publications of the Society had a significant voice in the anti-Yuan Shikai campaign, and a significant number of members sacrificed their lives during the Second Revolution.[34] Accentuating some writers' choice to disengage from politics, Lu Xun's criticism annulled the intellectual and political contributions of the Southern Society as a whole.

The political indifference of certain members of the Southern Society was called into question, as was the cultural system sustaining this indifference and the habits of the literary celebrity. As an association of traditional literati, the Southern Society was attached to old-style literature in classical Chinese and many members were Confucian scholars who loathed the loss of traditional values. They believed in the plasticity of tradition, which was not only capable of, but also necessary for, educating the masses and guiding the nation's pursuit of political modernity. During the 1900s, it was this conviction that incited the Southern Society to be involved in the movement of social enlightenment and to find a path for political reforms in the ancient sources that they defended as the national essence.

---

[32] Lu Xun, *Lu Xun quanji* (Complete Works of Lu Xun), Vol. 4 (Beijing: Renmin wenxue chubanshe, 2005), 134–5.
[33] Zhang, *Chen Qubing quanji*, 23.
[34] Jiang Guo, *Nanshe xiaoshuo yanjiu chutan* (Preliminary Study on the Southern Society's Novels) (Changchun: Jilin daxue chubanshe, 2012), 25.

For the New Culturalists, who saw in the persistence of traditional culture the root of the evil of the country's past and future, such a politico-cultural inclination was simply inconsequential. Their aversion to traditional culture was certainly related to Yuan's regime under which attempts to restore the cult of Confucius took place to rationalize his restoration. Contrary to the conservatives' attempts to insert elements of traditional culture into political reforms, the New Culture activists refuted such a logic of selective cultural preservation. They argued that modern political systems could only be concretized by the "museumification" of traditional culture, transformed from a living reality and creative force to a relic, wherein tradition was not considered for its political uses but for its value as a museum artifact.[35] The May Fourth intellectuals' "holistic mode of thinking" gradually turned the rejection of Confucianism into a revolt against traditional culture as a whole.[36] Thus, their criticism of certain traditional-style writers, such as those of the Southern Society, tended to generalize and degrade all intellectuals with conservative leaning.

This struggle between the conservatives and the progressives over cultural issues reveals a fight for legitimacy. As Gisèle Sapiro points out, "The principle that the newcomers challenge the literary orthodoxy and assert themselves against their elders… was one way by which the literary field fought the potential routinization of the charismatic figure of the writer."[37] The New Culturalists' rejection of "traditional culture" in favor of the "New Culture" therefore posed a challenge to the existing power structure within the intellectual field.

Like the Southern Society, the New Culture activists were also culturalist in their insistence on the priority of cultural over sociopolitical changes.[38] While there was nothing innovative in their culturalist approach to sociopolitical reforms with respect to the late Qing intellectuals, they proposed a new framework – the "New Culture" – within which this culturalist approach operated. If sociopolitical transformations were underpinned by cultural revolution, they asserted that only their own was able to complete this task successfully. One example in this regard are the words of Zheng Zhenduo (1898–1958), criticizing the Mandarin Ducks and Butterflies school: "the thoughts of the writers of *Libailiu* are purely demoded but they pretend to be modern… by adding new concepts,

---

[35] Luo Zhitian, "Songjin bowuyuan: Qingji minchu quxin renshi cong 'xiandai' li quchu 'gudai' de yingxiang" (Sending the Tradition to the Museum: Tendency of the Proponents of Novelty to Push the Tradition out of Modernity) in *Liebian zhong de chuancheng*, 91–130.
[36] Lin, *The Crisis of Chinese Consciousness*, 29.
[37] Gisèle Sapiro, "Forms of politicization in the French literary field," *Theory and Society* 32 (2003): 640.
[38] Lin, *The Crisis of Chinese Consciousness*, 29.

such as 'liberation' or 'reforms of family', into their stereotyped novels."³⁹ By appropriating the role of social enlighteners with their own cultural values, the New Culturalists reduced traditional-style writers and their ideas to a simple and irrelevant conservatism. Conservatism was then viewed as a pejorative judgement, associated with the nation's cultural tradition and rejected as ignorable.

Obviously, the New Culturalists' cultural ideas and political sensibilities cannot be bracketed together. Nevertheless, they all shared the same desire to replace the cultural tradition of the nation with the "New Culture" and deliberately created the conservative/progressive polarity to position themselves as the true defenders of the general interest and to disqualify the conservatives from having any say in the shaping of the nation's political future.⁴⁰ In 1924, Liu Yazi, the linchpin of the Southern Society, converted to the New Culture camp. He began to denounce Shanghainese novelists, many of whom were his comrades of the Southern Society, as literary beggars (cf. *infra*), interested only in money and in insulting the New Culturalists.⁴¹ Reacting to his remarks, Wang Dungen, Liu's former colleague in the Southern Society, questioned what really differentiated the progressives from the conservatives:

> New Culture writers accuse the "literary beggars" of writing for money, and they find romance fiction immoral. However, they are also paid for their works […] and most of the vernacular fiction and poems are stories of a man and a woman. Their words (in these novels) are sometimes more audacious than those in traditional-style fiction. When it comes to the vernacular fiction of writers of the previous generation, they also slander them as salacious and obscene, only because of the lack of modern punctuation in these works. However, once punctuation is used and a charismatic writer's comments are added (their works of the same genre) are admired and used as materials in textbooks.⁴²

For Wang, whether a writer is progressive or conservative is not so much determined by the literary style and language, but rather the camp that one chooses. His observation underlines an important aspect of the intellectual field of the early twentieth century: it was usually through associations that intellectuals became involved in cultural debates and carried out collective actions. This was the most effective way to establish discursive power and occupy the dominant posi-

---

39 Xidi, "Sixiang de fanliu" (Reactionary Orientation of Thoughts), *Wenxue xunkan* 4 (1921).
40 Kuo, "Polarities and the May Fourth Polemical Culture."
41 Zhang Chuntian, Geming yu shuqing: Nanshe de wenhua zhengzhi yu Zhongguo xiandaixing (1903–1923) (Revolution and Lyricism: Cultural Politics of the Southern Society and Chinese Modernity, 1903–1923) (Shanghai: Shanghai renmin chubanshe, 2015), 347.
42 Zhang, *Yuanyang hudie pai wenren*, 56–7.

tion within the intellectual field,[43] as shown in the simplistic polarity created by the leaders of the New Culture Movement between progressives and conservatives mentioned above.

Exactly what "New Culture" was believed to entail varied from one person to another, but the denial of the utility of tradition in bringing Chinese society and politics into line with the modern world persisted. However, like "traditional culture," the "New Culture," promoted by the progressives as *the* ideological instrument of sociopolitical transformation, reveals only their own vision of what the nation's future should be like and by no means automatically constitutes a normative frame of reference for reform. To universalize the "New Culture" as acting in the general interest, the New Culturalists had to form their identity as true disinterested intellectuals. Too often however, such a task appeared to be homologous with dispossessing the conservatives of this notion.

In the context in which intellectuals were excluded from the state bureaucracy and forced to compete with other social groups for symbolic power, the professional value of disinterestedness allowed intellectuals to gain popularity, hold on to the traditional social status of cultural elites, and self-qualify as "public intellectuals" with the mission and legitimacy to advise the government and educate society. Desirous of replacing the traditional intellectuals of the previous generation and becoming the new leaders in the intellectual field, the New Culture activists faulted the traditional intellectuals for their failure to defend the value of disinterestedness and questioned directly their symbolic power and discursive legitimacy.

Their first strategy consisted of relating traditional culture to monarchial restoration. Such a correlation is not entirely wrong. Reactionaries, including Kang Youwei and Liu Shipei, exploited Confucianism to rationalize their wishes and efforts to overthrow the Republic. However, reactionaryism must not be confused with conservatism, as I have shown in the first chapter. Thus, it is simply misleading to reduce the conservative logic of using traditional culture as a strategic resource for political change of the Republic to restoration. Chen Duxiu's challenge to *Dongfang zazhi* (*The Eastern Miscellany*) illustrates that the New Culturalists sometimes formulated totally groundless accusations to provoke public reactions against the conservatives. *Dongfang zazhi*, the flagship magazine of the Commercial Press edited by Du Yaquan, published a text on June 15, 1918, entitled *Zhongxi wenming zhi pingpan* (*Criticism on Chinese and Western civilizations*). The article is the Chinese translation of a text by a certain German writer, which

---

[43] Michel Hockx, "Playing the Field: Aspects of Chinese Literary Life in the 1920s," in *The Literary Field of Twentieth-Century China*, 72–6.

appeared in the Japanese magazine *Tōa no hikari* (*Light of East Asia*). In this text, the author discusses two of Gu Hongming's articles on the particularities of the Chinese civilization. He agrees with Gu's point of view and underlines that the Chinese civilization is a spiritual one and that Germany should learn the lessons of the First World War and abandon the spirits of cupidity and materialism.

In the Chinese intellectual landscape, the duality between the Chinese "spiritual" civilization and the Western "materialistic" civilization was no more than a platitude. The outbreak of the First World War, viewed by some conservatives as the inevitable consequence of the development of a deficient civilization, provided yet another occasion for them to reaffirm Chinese civilization as an indispensable and integral part of the nation's road to modernity. However, this article, which presented nothing particularly innovative on this topic, provoked controversy. Chen issued a critique in the *New Youth*, still a relatively insignificant magazine at the time in comparison to *Dongfang zazhi*, which was later regarded as conservative. He asked rhetorically, if *Dongfang zazhi*, by publishing this review, aligned itself with Gu's support for the restoration of the Manchu Court.[44] This accusation was baseless, but Gu, a royalist, was an easy target for Chen to use to generalize and relate traditional culture to restoration. Du Yaquan responded and urged Chen to ensure that his accusation was grounded in fact and not based on his imagination and an intent to defame.[45] The damage, however, had already been done. The magazine's sales decreased dramatically due to Chen's accusation, forcing the Commercial Press to lower the price in order to make sales.[46]

Regarding the Mandarin Ducks and Butterflies school, Zhou Zuoren (1885–1967) and Qian Xuantong repeatedly linked their literary creations with Yuan's regime. To them, the rise of these traditional-style popular novels in classical Chinese was due to the conservative political environment of the 1910s, to which these "regressive" literary works also contributed. Zhou explained: "During Yuan's restoration, the nation was plunged into a sort of revivalism (*fugu*) and foreign thoughts came to be considered worthless. Consequently, they

---

[44] Chen Duxiu, "Zhiwen Dongfang zazhi jizhe – Dongfang zazhi yu fupi wenti" (Questioning the Journalists of *Dongfang zazhi* – *Dongfang zazhi* and the Question of Restoration), *New Youth* 5, no. 3 (1918): 206.
[45] Cangfu (Du Yaquan), "Da Xinqingnian zazhi jizhe zhi zhiwen" (Response to the Interrogation of the Journalist of *New Youth*), *Dongfang zazhi* 15, no. 12 (1918): 15.
[46] Wang Qisheng, *Geming yu fangeming – shehui wenhua shiye xia de Minguo zhengzhi* (Revolution and Counter-revolution, Republican Politics from a Sociocultural Perspective) (Beijing: Shehui kexue wenxian chubanshe, 2010), 20–1.

began to create novels of ancient style and... became popular."⁴⁷ Conservatism in the cultural field was in this way slammed as the accomplice of political conservatism, or even reactionaryism.

The Mandarin Ducks and Butterflies school was further subjected to the New Culturalists' disdain and mockery for their market-oriented novels. Lu Xun described these writers as "contracted literary giants" (*shangding wenhao*) who collaborated with publishers for better sales but pretended to be indifferent to fame.⁴⁸ They were belittled as "literary beggars" (*wengai*) and "literary prostitutes" (*wenchang*) who wrote for money and provided readers with nothing more than escapist entertainment.⁴⁹ Some progressives even designated themselves educators responsible for "purifying" the traditional-style writers. Cheng Fangwu (1897–1984) said in 1922: "We feel very ashamed... [that] we do not know what to do to help these traditional writers who refuse to progress."⁵⁰ In 1921, Luo Jialun (1897–1969) professed that though these traditional writers "were not worth even his slurs," he felt "obliged to spare two minutes" to inveigh against them to fulfil his duty towards "public critique" and to prevent their fictions from "polluting" the minds of innocent readers.⁵¹

Not only were the traditional-style writers depicted as culturally regressive and politically reactionary, but it also became the progressives' duty as public intellectuals to expose the traditional-style writers' loss of the professional value of disinterestedness. However, the traditional-style writers would not concede without a struggle, claiming that the so-called progressives were simply a little band of self-righteous hypocrites.

Conservatives' Responses

The New Culturalists took the moral high ground and reproached the conservatives for repudiating the professional value of disinterestedness in order to dis-

---

47 Chen Jianhua, "Minguo chuqi Zhou Shoujuan de xinli xiaoshuo – jianlun "Libailiu pai" yu 'Yuanyang hudie pai' zhi bie" (Zhou Shoujuan's Psychological Fiction in the Early Republican Era and the Difference Between the *Saturday* School and the Mandarin Ducks and Butterflies School), *Xiandai zhongwen xuekan* 11 (2011): 37–49.
48 Lu Xun, "Shangding wenhao" (Contracted Literary Giants), *Shenbao Ziyoutan* (November 11, 1933).
49 C.S., "Wenchang" (Literary Prostitutes), *Wenxue xunkan* 38 (1922).
50 Housheng (Cheng Fangwu), "Tiaohe xinjiu wenxue jinyi jie" (On the Harmonization of the Old and the New Literatures) *Wenxue xunkan* 6 (1921).
51 Zhixi (Luo Jialun), "Jinri Zhongguo zhi xiaoshuojie" (Today's Field of Fictions), *Xinchao* 1 (1919).

parage traditional culture in favor of the New Culture. This said, one would have expected them to engage in activities that were the exact opposite of what they condemned, which, according to the conservatives, was not always the case. While Zhou Zuoren and others linked the traditional intellectuals' cultural activities with Yuan's restoration, the progressives could also be accused of maintaining an ambiguous relationship with Yuan. In fact, Zhuo's brother, Lu Xun, had even worked within Yuan's government. In 1915, Lu Xun was appointed chief of the fiction section of the Research Committee for Popular Education (*Tongsu jiaoyu yanjiuhui*), affiliated with the Ministry of Education. Lu Xun assumed this position when the Committee was given the mission of shaping a favorable public opinion towards Yuan's restoration. This political aim translated into an ideological control of literary production, required to reflect the Confucian values of loyalty (*zhong*), filial piety (*xiao*), contingency (*jie*), and righteousness (*yi*).[52] Under these guidelines, Lu Xun censored 32 works of fictions from the Mandarin Ducks and Butterflies school, which, ironically, his brother accused of contributing to Yuan's restoration.[53]

For conservatives, the New Culturalists' success in disseminating their ideas among the younger generation was largely due to their manipulation of institutional power, intentional defamation of the conservatives, and exclusion of the latter from the discursive space. Mei Guangdi (1890–1945), a member of the Southern Society and the Critical Review Group, condemned the leaders of the New Culture Movement as self-promoters and savvy politicians (*zhengke*).[54] For Wu Mi, the very reason why the vernacular language replaced classical Chinese and became the national language and the language of instruction was that its promoters had infiltrated the government; Hu Shi, Qian Xuantong, Zhou Zuoren, and Liu Bannong (1891–1934) were all involved in the Preparatory Commission for the National Language Unification Preparation Committee of the Ministry of Education.[55] Hence, the victory of the vernacular language did not reflect the natural evolution of the Chinese language as Hu Shi suggested, but the decadence of the intellectual field whose autonomy was jeopardized by intellectuals who resorted to political power to impose their personal cultural viewpoints.

---

52 Zhou Huimei, *Minzhong jiaoyuguan yu Zhongguo shehui bianqian* (Popular Schools and the Evolution of the Chinese Society) (Taipei: Showwe, 2013), 88–9.
53 Qian Liqun et al., *Zhongguo xiandai wenxue sanshinian* (30 Years of Chinese Literature) (Taipei: Wunan, 2002), 100–1.
54 Mei Guangdi, "Ping tichang xinwenhuazhe" (Comments on the Promoters of the New Culture Movement), *Critical Review* 1 (1922).
55 Chen Chun-chi, "Wu Mi yu Xinwenhua yundong" (Wu Mi and the New Culture Movement), *Bulletin of the Institute of Modern History, Academia Sinica* 56 (2007): 74.

For conservatives, the New Culturalists' hegemonic position within the intellectual field also impeded freedom of expression to the point where the cultural tradition had no place in such an intellectual environment. Following the controversy created by Chen Duxiu over the role *Dongfang zazhi* played in the Manchu restoration, Du Yaquan urged the editors of *New Youth* "not to restrict freedom of speech like the authoritarian officials of the Qing."[56] Mei Guangdi had expressed the same concern, arguing that the only objective of these "academic politicians" of the New Culture Movement was to silence the traditional-style writers. They did this by creating a dualism that he judged groundless between the so-called "aristocrat's literature" (*guizu wenxue*) and "commoner's literature" (*pingmin wenxue*), or even between "dead literature" (*siwenxue*) and "living literature" (*huowenxue*). He said: "Those who do not agree with them are called 'old,' 'dead,' 'aristocratic' or 'against the global trend,' so their opponents are bound to fail before even having a chance to debate."[57]

While the progressives ridiculed the Mandarin Ducks and Butterflies school for their pursuit of economic profits, this accusation sounded ironic to the very writers they condemned. Several writers from the Mandarin Ducks and Butterflies school sneered at what they perceived as pure hypocrisy in the progressives' comments, remarking that Hu Shi and Lu Xun's books were priced so high that they could not afford them.[58] In 1928, Wu Mi observed that New Culture activists enjoyed not only a tremendous reputation, but also huge sums of money: "Zhou Shuren [Lu Xun] earned more than 2,000 yuan for *Nahan* [*Cries*] [while] [...] Zhang Ziping [1893–1959] and Yu Dafu [1896–1945]... are paid more than 20 yuan per 1,000 words."[59] If what Wu Mi said was correct, the income of the famous New Culture writers was quite considerable – in comparison, a Shanghai worker earned no more than 0.408 yuan per day on average in the 1920s and 1930s.[60]

---

56 Cangfu, "Da 'Xin Qingnian' zazhi jizhe zhi zhiwen," 15.
57 Mei Guangdi, "Ping jinren tichang xueshu zhi fangfa" (On the Way in Which Intellectuals Promote Knowledge Today), *Critical Review* 2 (1922).
58 Hu Shi's books in the 1920s could cost as much as 3 yuan and Lu Xun's 1927 *Eryiji* (*And That's That*) cost 0.65 yuan. In comparison, magazines of the Mandarin Ducks and Butterflies school were priced in general between 0.1–0.4 yuan. See Rui Heshi et al., *Yuanyang hudie pai wenxue ziliao* (Literary Sources of the Mandarin Ducks and Butterflies School) (Fuzhou: Fujian renmin chubanshe, 1984), 108 and 189; Perry Link, *Mandarin Ducks and Butterflies. Popular Fiction in Early Twentieth-Century Chinese Cities* (Berkeley: University of California Press, 1991), 249–60.
59 Wu Mi, *Wu Mi riji 1928–1929*, 17.
60 Xu, *Chinese Professionals and the Republican State*, 56–60.

The progressive writers might claim that their works contributed to the shaping of modern Chinese society and were by no means market oriented. Conservatives would not agree with such a statement. Mei Guangdi argued that their iconoclastic cultural politics, set against the classical tradition representing thousands of years of accumulated design knowledge, was formulated to pander to the vulgar taste of the masses and the radicalism of young students.[61] He concluded that the leaders of the New Culture Movement could not even be called intellectuals or academics because their cultural activities were biased, fame-motivated, and far from disinterested. One magazine of the Mandarin Ducks and Butterflies school satirically remarked:

> The ninth issue of the 13th volume of *Xiaoshuo yuebao* [*Fiction Monthly*][62] of the Commercial Press attacked magazines of other publishing houses, such as *Libailiu* [...], etc. *Xiaoshuo yuebao* considers these magazines extremely obscene and the writers highly frivolous. Some believe that this literary debate is a competition between new thought and old thought. In my opinion, such a point of view is too bombastic; this debate is only about the question of livelihood – that is, the question of jobs (*fanwan*). I admit that those who write traditional-style fiction are extremely frivolous, but will these new writers admit that they are all wealthy capitalists? When two unworthy persons meet one another, they fight each other for bread and butter.[63]

The author acknowledges that the traditional-style writers like himself fail to act in the true spirit of disinterestedness, but at the same time he highlights the insincerity of the New Culture writers, who treat the traditional market-oriented writers with contempt while imposing their own cultural views for the same purpose. The New Culturalists' "ambition" to save the nation with the New Culture was judged to be nothing more than a pretext for fame and wealth. Another text, entitled *Literary Beggar* and signed with the penname of Literary Beggar, remarks that, once paid for the words written, a writer becomes a literary beggar; if a certain writer belongs to the camp of the New Culture in Beijing, this does not make his income more noble than that of a traditional-style writer of popular fiction in Shanghai.[64]

The tactics used by both the conservatives and the progressives to rebut, or even silence each other's voices, is quite similar. The focal point of the dispute

---

61 Mei, "Ping tichang xinwenhuazhe."
62 Although created by writers of the Mandarin Ducks and Butterflies school, progressive writers began to edit *Xiaoshuo yuebao* in 1921.
63 Xingxing, "Shangwu yinshuguan de xianyi" (Suspicion of the Commercial Press), *Jingbao* (September 21, 1922).
64 Wengai, "Wengai" (Literary Beggars), *Jingbao* (October 21, 1922).

was identifying the guardians of the professional value of disinterestedness with the legitimacy to universalize their particularistic ideas as acting in the general interest. While the progressives reproached the conservatives for giving publicity to restoration and obtaining profit through cultural activities, the conservatives attacked the progressives' intentional defamation and, above all, their hypocrisy. In the name of saving the nation, they imposed the New Culture by political power and earned no less than those they despised.

However, despite all their efforts to redeem their role as social leaders and educators, the conservatives were fighting a losing battle. The struggle between the old faction and the new faction is certainly more complicated than can be conveyed in this chapter. Beyond the literary field, conservatives did manage to oust their opponents from time to time. Chen Duxiu, for example, was forced to resign from Peking University in 1919 after being accused of leading a questionable lifestyle involving brothels.[65] Nevertheless, through personal networks and advertisements, the "New Culture," as Forster shows, became the hegemonic buzzword with which political and cultural programs had to be identified if they were to be considered legitimate.[66] Even for some conservatives, "traditional culture" needed to be justified within the framework of the "New Culture."[67]

This obligatory identification with the new faction's particularistic cultural standpoints did not silence the conservatives, but they never managed to emulate the New Culturalists' success.[68] From the late 1910s onward, a substantial number of magazines and literary supplements edited by members of the Southern Society, including *Xiaoshuo yuebao* and *Shenbao*'s *Ziyoutan* (*Free Talk*), began to recruit New Culture writers as chief editors. Other major periodicals of the Southern Society, such as *Chenbao* (*Morning Post*), *Guomin* (*Nation*), *Taipingyang bao*, and *Minguo ribao*, removed, one after another, sections devoted to classical literature in favor of literary works of New Culture.

The Southern Society, the biggest literary association of the late Qing and the early Republican era, finally dissolved in 1923. According to Liu Yazi, one of the reasons behind the dissolution of the Southern Society was the gap between what the young generation wished to read and what the Southern Society produced.[69] In 1923, the New Southern Society was founded in Shanghai. With no political agenda, it was only a commemorative literary association and did not

---

65 Forster, *1919 – The Year That Changed China*, 50.
66 Forster, *1919 – The Year That Changed China*, 144–9; 195.
67 See for example Wu Mi, "Lun Xinwenhua yundong" (On the New Culture Movement), *Critical Review* 4 (1922).
68 Forster, *1919 – The Year That Changed China*, 139.
69 Liu, *Nanshe jilue*, 153.

last long. With regard to the New Southern Society's appeal to "preserve national learnings," Liu responded that "organizing national learnings resembles looking for fragrance in dung – it will result in nothing at all."[70] While he agreed that those who wished should be granted the freedom to conduct research on traditional culture, Liu affirmed that this kind of research should only be carried out for its own sake.[71] As condoned by his former opponents in the New Culture camp, traditional culture was "museumified" and denied any political significance.

The First World War and the Diversification of Conservative Thought

Taking the Southern Society as a case study, this chapter has shown that while disappointment with the political crisis did lead several conservatives to voluntarily reject political issues after the foundation of the Republic, the depoliticization of modern Chinese conservatism during the New Culture era cannot be attributed merely to personal choice. Essentially, this process is intertwined with the rise of the profession of the intellectual and the struggle for legitimacy within the intellectual field. In the period where cultural reforms were regarded as the first step towards the creation of a new political order, both the conservatives and the progressives attempted to bring political issues to their cultural foundation. The conservative logic was based on the strategic inventory of traditional culture as a resource for political change, while for the progressives, political construction required the establishment of a solid moral transformation through the "New Culture." The "traditional culture" and the "New Culture" that the conservatives and the progressives respectively defended nevertheless merely represented their own particularistic views. The universalization of one's particularistic view as acting in the general interest required the symbolic capital associated with the intellectual's professional value of disinterestedness, which legitimized the claim to discursive authority. In this struggle between the conservatives and the progressives it was the latter, with an abundance of media coverage, personal networks, and successful marketing, who managed to prescribe their particularistic cultural version of New Culture as the legitimate representation of public interest and relegate the conservatives to the marginalized periphery of the intellectual field. Thus, conservatism became both an intellectual disposition and an

---

70 Ibid., 102.
71 Ibid., 103.

epithet reserved for those who disagreed with the New Culturalists, and conservatives were ousted by the New Culturalists.⁷²

The "depoliticization" of conservatism as an intellectual movement is not to be understood strictly as rejection of politics; more importantly it is the loss of political influence and control. In such an intellectual environment, not only was the discursive space where the conservative intellectuals' voices could be heard considerably reduced, their political agendas were also easily dismissed as inconsequential. Consequently, instead of finding an effective way to inject traditional culture into the political institutionalization, conservatism became increasingly an intellectual disposition centered on the defense of cultural heritage. It was now possible to speak of cultural conservatism as defined by Schwartz – an insistence upon the enduring value of traditional culture and the link between the nation's survival and her cultural heritage. This insistence, it should be noted, failed to be politically concretized.

At approximately the same time in the West, conservatism enjoyed such a great popular support that conservative parties even supported universal suffrage.⁷³ Why was the same phenomenon not produced in China? Language seems to be one of the reasons that obstructed them in the effective dissemination of their ideas. Conservatism in Republican China was mainly an urban movement that targeted urban readership. Since conservatives were in general writing in classical Chinese, they were unable to make themselves understood in a country with a largely illiterate population. The editors of the *Critical Review* magazine, for example, still published articles in classical Chinese several years after the vernacular language was codified as the national language. Liang Shih-chiu (1903–1987), the only intellectual involved in the translation and dissemination of the philosophy of New Humanism that the Critical Review Group advocated (cf. Chapter 5), recalled that the *Critical Review* magazine elicited little public interest and failed to spread New Humanism effectively because of the classical language that the magazine adopted; even Liang, who was a university student in the early 1920s, was first repelled by the language style of the magazine.⁷⁴

Nevertheless, although Liu Yazi and some members of the Southern Society surrendered in this struggle between the old and the new, other defenders of culturalist nationalism would not concede to the New Culturalists' domination with-

---

72 Kuo, "Polarities and the May Fourth Polemical Culture."
73 Arthur Aughey at al., *The Conservative Political Tradition in Britain and the United States* (London: Pinter Publishers, 1992), 129–30.
74 Liang Shih-chiu, "Guanyu Baibide xiansheng ji qi sixiang" (On Mr. Babbitt and His Thought), in *Liang Shiqiu piping wenji* (Liang Shiqiu's Critiques), ed. Xu Jingbo (Zhuhai: Zhuhai chubanshe, 1998), 212.

out a fight. They led the fight by directly attacking the two bastions of the New Culture project – science and democracy. The First World War radically changed conservatives' and New Culturalists' outlook on nationalism. While the latter castigated nationalism that risked falling under the sway of folly, conservatives began to question the merit of Western civilization that they had previously admired. Certainly, during the revolutionary decades, late Qing conservatives and radicals had drastically different political aims, but they all resorted to Western political systems to conceptualize either the constitutional monarchy or the republic. Nevertheless, the atrocities of the Great War led many to come up with a new model for China. As a consequence, traditional learnings or national essence that shaped culturalist nationalism during the Republican era further diversified. As such, Chinese conservatism should be understood as a polysemic intellectual and political current.

4 Liberal Conservatism and Anti-modern Conservatism (mid-1910s–1930s)

In his classic study of French right-wing politics, René Rémond points out that the French right comprised conservatives, royalists, counterrevolutionaries, reactionaries, and extremists. More specifically, the French right could be divided into *légitimiste*, *orléaniste*, and *bonapartiste*.[1] Pluralism has been the principle of the French right. The same thing can be said about conservatism. Karen Stenner divides contemporary conservatism into authoritarianism, status quo conservatism, and laissez-faire conservatism.[2] In a more recent research, Jan-Werner Müller evokes four dimensions of conservatism: sociological conservatism that represents the ideology of a social group seeking the preservation of their prerogatives, methodological conservatism that insists upon the continuation of tradition in reforms, aesthetic conservatism that encapsulates a nostalgia for the past, and philosophical conservatism that defends a series of immutable social values, notably social hierarchy and natural inequality.[3] The aleatory combination of these dimensions generates a diverse landscape of conservative politics.

In Chinese historiography, pluralism in conservative thought has yet to be given due attention. Chinese conservatism in Republican China took root in the late Qing culturalist nationalism that built upon the conviction that the political nation and the cultural nation should be unified, while implying that concrete political reforms resonated with China's cultural tradition. The New Culturalists' successful domination of the intellectual field, which led to an invalidation of political reforms based on traditional culture, did not expel conservatives from the realm of political discourse. The continued political crisis and the First World War led a considerable number of conservatives to question the merits of the West that they had admired during the revolutionary decade. In response to this, the meanings attributed to traditional culture that shaped culturalist nationalism diversified. In this regard, conservatism in Republican China should be understood in plural terms.

At this point history repeated itself in the sense that, much like the late Qing period, culturalist nationalism gave rise to different and contradictory political agendas. These depended on the cultural elements selected by different intellectuals and politicians to inform political reforms. The coherence of Chinese con-

---

1 René Rémond, *Les trois droites en France* (Paris: Aubiers, 2014).
2 Karen Stenner, "Three kinds of 'Conservatism'," *Psychological Inquiry* 20 (2009): 142–59.
3 Jan-Werner Müller, "Comprehending Conservatism: A New Framework for Analysis," *Journal of Political Ideologies* 11, no. 3 (2006): 359–365.

servatism of the Republican era is manifest in the commitment to defend pre-1912 culturalist nationalism which insisted upon the integration of the political and the cultural nation. However, conservative ideas differed with respect to the political, social, and economic systems in which the Republic was to function. Despite this, their political aim was ultimately the same: the preservation of the present, that is, the Republic of China. This chapter surveys two typologies of Chinese conservatism that appeared in the mid-1910s and 1920s – liberal conservatism and antimodern conservatism, which attributed different values and judgements to liberty, capitalism, and industrialization, as well as Westernization.

In the West, conservatism is not necessarily in opposition to liberalism or liberty. In fact, the question of liberty occupies the central place in the thoughts of a number of Western conservatives, for example, Edmund Burke, Tocqueville (1805–1859), and Benjamin Constant (1767–1839), to name a few. Western classical conservatives' efforts to conserve liberty was provoked by what Hayek (1899–1992) called "false individualism" that prevailed during the French Revolution.[4] To classical conservatives, liberty is associated with order that "put a limit on the proliferation of subjective rights that lead to anomie," as well as the predominance of community and social institutions to which an individual belongs.[5] In contrast, the liberty of the French Revolution enveloped a disguised tyranny. The Declaration of the Rights of Man and of the Citizen of 1789, which stipulates that "men are born and remain free and equal in rights," resulted in a total dissolution of internal and external limits that impeded liberty. It was in this spirit that Rabaut Saint-Étienne (1743–1793) proclaimed to the National Assembly: "To make people happy, they should be renewed, their ideas should be changed, their law should be changed, their customs should be changed, change the men, change the thing. Destroy everything, yes, everything should be destroyed because everything should be recreated."[6] When Disraeli stated that English society was sustained by the equality of liberty, he referred to what Isaiah Berlin would call "negative liberty," in contrast to the French "positive liberty:"[7]

---

[4] F.A. Hayek, "Individualism: True and False," in *Individualism and Economic Order* (Chicago: The University of Chicago Press, 1948), 1–32.

[5] Vincent, *Qu'est-ce que le conservatisme?*, 20.

[6] Gilles Lebrenton, "Liberté, libertés," in *Le dictionnaire du conservatisme*, ed. Frédéric Rouvillois et al. (Paris: Les Édition du Cerf, 2017), 563.

[7] Isaiah Berlin, "Two Concepts of Liberty," in *Four Essays on Liberty* (Oxford: Oxford University Press, 1969), 118–72.

> The basis of English society is Equality… there are two kinds of equality; there is the equality that levels and destroys, and the equality that elevates and creates. It is this last, this sublime, this celestial equality, that animates the laws of England. The principle of the first equality, base, terrestrial, Gallic, and groveling, is that no one should be privileged; the principle of English equality is that everyone should be privileged. Thus the meanest subject of our King is born to great and important privileges; an Englishman, however humble may be his birth, whether he be doomed to the plough or destined to the loom, is born to the noblest of all inheritances, the equality of civil rights; he is born to freedom, he is born to justice, and he is born to property. There is no station to which he may not aspire; there is no master whom he is obliged to serve; there is no magistrate who dares imprison him against the law; and the soil on which he labors must supply him with an honest and decorous maintenance.[8]

In order that liberty acquired a meaningful sense, Disraeli appealed for a maintenance of institutions such as family, local communities, and church that made liberty possible in the first place. That is to say, liberty is not an abstract idea which can only be concretized through the sharing of a collective life. The destruction of intermediate institutions (*corps intermédiaire*) which led to a vacuum between people and the state during the Jacobin time ran counter to the realization of the liberty defended by classical conservatives. To recreate the social links between individuals, Tocqueville felt the need to restore the intermediate institutions of the *Ancien Régime* in order to reinforce the dependence and hierarchy among individuals, who would act collectively and spontaneously in regard to the state's power.[9] The collectivities to which the individual belongs and that place duties on him have an "agent-relative and unconditional, irreducible, non-instrumental" value.[10] Order is not exclusive of liberty and must not be confused with arbitrariness or coercion. If order is to be reconciled with liberty, social cohesion must start from below, ensuring the organic integrity of free and self-reliant men. In other words, liberty is not solely a matter of individual action; it must involve restraint and be established at an institutional level. As I shall show, Chinese liberal conservatives shared Western classical conservatives' concern about individual liberty and its relation to institutional liberty.

To understand anti-modernism, it is necessary to first retrace the constitutive elements of modernity. The genesis of modernity dates back to seventeenth cen-

---

[8] Benjamin Disraeli, *Vindication of the English Constitution* (London: Saunders and Otley, 1835), 204–5.
[9] Pierre Manent, *Tocqueville et la nature de la démocratie* (Paris: Gallimard, 2012), 44–6.
[10] John Skorupski, "The Conservative Critique of Liberalism," in *The Cambridge Companion to Liberalism*, ed. Steven Wall (Cambridge: Cambridge University Press, 2015), 409.

tury Europe.¹¹ It is the product of an interaction between a set of political, economic, social, and cultural processes. Marx (1818–1883) interpreted it as a general movement of commercialization and banalization of all forms of human creativity, which were fatally condemned to become standardized and economic commodities.¹² To Weber (1864–1920), modern society clearly provides for a rationality that does not limit itself to capitalism as Marx argued, but also concerns bureaucracy and religion.¹³ Durkheim, as I have mentioned in Chapter 2, saw in modern society the advent of industrialization and an intensification of the division of labor, bringing individuals to automation and differentiation.¹⁴ These classical works left an influential legacy for today's understanding of modernity. Stuart Hall summarized four characteristic traits of modernity: laicity, capitalism, division of labor, and decline of religion.¹⁵ At the philosophical level, modernity is characterized by a temporal rupture, in the sense that tradition is seen as suspended in the past and that collective consciousness is not capable of being integrated as a continuity.¹⁶ This does not mean that tradition has faded away in the modern age. As Baudelaire (1821–1867) stated, modernity is an attitude "allowing to seize what is heroic in the present moment."¹⁷

Antoine Compagnon reasons that antimodern thinkers do not establish their ideas as the antithesis of modernity. On the contrary, these are the moderns "full of delicateness with modern time, modernism or modernity" or the moderns who "became moderns reluctantly, torn up moderns or even untimely moderns" – a modernity without modernism.¹⁸ Compagnon decomposed anti-modernism into six dimensions: anti-revolution, anti-Enlightenment, pessimism, original sin, sublime, and vituperations. Under these labels, Compagnon categorizes a number of French thinkers, intellectuals, and writers from the fourteenth to the twentieth century, whose resistance to modernity is characterized by the suspicion of or hostility towards the myth of progress, the optimism of the Enlightenment, the

---

11 Anthony Giddens, *The Consequences of Modernity* (Stanford: Stanford University Press, 1990), 1.
12 Judy Cox, "An Introduction to Marx's Theory of Alienation," *International Socialism, Quarterly Journal of the Socialist Workers Party (Britain)* 79 (1998), http://pubs.socialistreviewindex.org.uk/isj79/cox.htm, accessed February 8, 2018.
13 Stephan Kalberg, "Max Weber's Types of Rationality: Cornerstones for the Analysis of Rationalization Processes in History," *The American Journal of Sociology* 85, no. 5 (1980): 1145–1179.
14 Mesure, "Durkheim et Tönnies."
15 Hall, *Formations of Modernity*, 6.
16 Matei Calinescu, *Five Faces of Modernity: Modernism, Avant-garde, Decadence, Kitsch, Postmodernism* (Durham: Duke University Press, 1987), 3–34.
17 Michel Foucault, *Qu'est-ce que les Lumières?* (Rosny: Bréal, 2004), 72.
18 Compagnon, *Les antimodernes de Joseph de Maistre à Roland Barthes*, 7.

turmoil of industrialization, the destruction of the past, and the standardization of mentality. All these ideas, hostile but consubstantial to modernity, make antimoderns moderns who are not duped by a modernism "naïve and zealous of progress." Thus, antimoderns for Compagnon are not conservatives, traditionalists or reactionaries.[19] Although it remains unconvincing to put Charles Maurras (1868–1951), Joseph de Maistre, Chateaubriand (1768–1848), and Roland Barthes (1915–1980) under the same category, while claiming none of them was conservative, traditionalist or reactionary, Compagnon's thesis provides us with an open sesame to examine at the theoretical level anti-modernism in the Chinese Republican context.

Liberal Conservatism and the Second Revolution of Social Liberation

Although the struggle for individual liberation and the social enlightenment movement were led by culturalist nationalists during the revolutionary decade, it seems that these movements really came into their own during the May Fourth New Culture period. As with many neologisms, the notion of individual that was translated in Chinese as *geren* originated from the Japanese term *kojin*. The word first entered the Chinese vocabulary in 1898, and the rise of the discourses on individualism was engendered with an awareness of individual independence and political rights. Nevertheless, once individuals obtained freedom by discarding the traditional moral coercion that was imposed on them, they were expected to reinforce their sense of responsibility vis-à-vis a larger community – that of society and nation. It was within this rhetoric that individualism, organic society, and military nation, though seemingly paradoxical, managed to become integrated with each other. For intellectuals of the 1900s, order and a sense of collectivity posed a necessary limit to the proliferation of individual rights.

The same can be said about liberty (*ziyou*). Also borrowed from the Japanese language (*jiyū*), the concept was charged with economic preoccupations which permeated European liberalism. In contrast, liberalism in China at the end of the nineteenth century put the sole emphasis on the adherence to the national and societal-building projects and public morality.[20] Individual freedom was deeply embedded in institutional freedom. But this idea of individualism and

---

19 Ibid., 8.
20 Edmund S.K. Fung, "The Idea of Freedom in Modern China Revisited: Plural Conceptions and Dual Responsibility," *Modern China* 32, no. 4 (2006): 473.

freedom became increasingly unattractive among the younger generation. The American sociologist Edward A. Ross (1866–1951) carried out six months of field work in China. On his return to the United States, he published a book describing the change in mentality operating in China. In 1911, *New York Times* published a review of his book. Its observations depict the situation that alarmed Chinese conservatives at the time. To Ross, China was incapable of endorsing moral responsibility towards social well-being:

> Right here we come upon the gravest problem arising from China's change of base; whence will come the morality of tomorrow? In the reaction against the old classical tradition with its emphasis on ethics there has been a tendency to neglect instruction in morals. Though they must do homage once a month to Confucius's tablet, the young men are inwardly scoffing. Confucius! He never rode on a train or used the telephone or sent a wireless. What did he know of science? He is only an old fogy! And so the Sage, whose teachings have kept myriads within the safe way, has little authority over the educated part of the rising generation. What they covet is riches and power, and perceiving that the wealth and martial prowess of the West rests immediately upon exact knowledge, the students are all for science. The hidden moral foundations of Western success they are apt to overlook. Neglecting their own idealism and missing ours, they may develop a selfish materialistic character which will make the awakening of China a curse instead of a blessing.[21]

Conservatives were worried that the abandonment of morality in the quest for individualism would lead to a legitimized egoism, built upon materialist satisfaction. Hence, late Qing social enlighteners harshly condemned the excessive individual liberty submerged in what they saw as irrational temptation and disorder. For example, *Minhu ribao* (*People's Call Daily*) published an illustration portraying two females dressed in modern outfits. The lady on the left wore a dress with a logo saying, "money supremacy" (*jinqian zhishang*); the lady on the right wore a dress with a logo saying, "civilized marriage" (*wenming jiehun*). The author asked rhetorically which one was a prostitute and which one was a student. Here, he satirized female students in big cities, who, in the name of freedom, ignored social responsibility in favor of a narcissistic quest for an ego-ideal identity.

Compared with the 1900s, the first years of the Republican era witnessed a much more courageous and pronounced spiritual and social liberation, which the conservatives perceived as erecting an insupportable fanatical and egoist individualism. The second revolution of social liberation was both a further development and a denial of the first revolution of social liberation of the 1900s. Its

---

[21] "The Chinaman's Destiny is that of the White Man," *New York Times* (October 29, 1911).

proponents castigated not only Confucianism, but also encouraged an individualism that prevailed over nationalism and social collectivism.

The first revolution of social liberation advocated a sexual division of labor. Ideas of Yao Guang, a fervent defender of women's rights at the end of the Qing dynasty, perfectly illustrated this tendency. Although he slayed male chauvinism and supported women's education and sexual equality, he reasoned that women's education should emphasize homemaking (*jiazheng*): "Even if women engage themselves in politics and become citizens, they cannot liberate themselves from the three steps of their life: daughter, wife and mother."[22] This type of argument was largely inspired by Nakamura Masanao's (1832–1891) theory of "wise wife, good mother" (*xianqi liangmu*).[23]

Figure 4.1: Who is prostitute? Who is student? From Chen Pingyuan, *Wanqing tuxiang:* Dianshizhai huabao *zhiwai* [Illustrating the Late Qing: Besides Dianshizhai Illustrated Newspaper] (Hong Kong: Open Page, 2015), 317.

The slogan was very popular during the 1900s and only became obsolete during the New Culture Movement. Hu Shi, for one, called for his female readers to establish a role outside the house and follow the example of American women who

---

[22] Yao Guang, "Nüzi yi zhuzhong jiazheng shuo" (Women Should Pay Attention to Homemaking), *Yao Guang quanji*, 19.
[23] Joan Judge, "Talent, Virtue, and the Nation: Chinese Nationalisms and Female Subjectivities in the Early Twenty Century," *The American Historical Review* 106, no. 3 (2001): 771.

were devoted entirely to a professional career and remained single.[24] To Chen Duxiu, the kind of family revolution that he endeavored to achieve aimed at doing away with the moral foundation of family in order to recast the human relation solely on juridical and economical levels.[25] To the New Culturalists, traditional morality embodied Confucian credo, which was associated with extremely rigid and inhuman ethic practices that impeded individual emancipation. Filial piety (*xiao*), which is the foundation of Confucian morality, was condemned harshly by New Culturalists.[26] Chen even ridiculed filial piety in a very provocative manner, stating that he saw no reason for children to be grateful towards parents, since they were only products of their parents' carnal desire.[27]

This determination to break with all existing moral values and hostility towards the institution of family captured the attention of those who claimed to be its victim, notably Wu Yu (1872–1949), who had lost his job as a teacher following a legal dispute with his father.[28] Wu Yu later joined the editorial group of *New Youth*. To him, the Chinese institution of family that kept women and children in submission vis-à-vis husbands and parents resembled a miniature of political dictatorship that should be brought down.[29]

From a Marxist perspective, Li Dazhao (1888–1927) remarked that China had always been an agricultural society. For this reason, its system of "big family" that engendered the traditional customs and virtues rested upon its social structure that required filial piety and patriarchy. However, the bitter failure of China towards the West demonstrated the necessity to transit from an agricultural society sustained by Confucian morality to a democratic, industrialized, and individualist society that made Europe the continent of progress and prosperity.[30]

---

[24] Hu Shi, "Meiguo de funü" (American Women), *New Youth* 5, no. 3 (1918).
[25] Chen Duxiu, "Dongxi minzu genben sixiang zhi chayi" (The Fundamental Difference Between the Thoughts of Eastern and Western People), *New Youth* 1, no. 4 (1915): 284
[26] See, for example, Wu Yu, "Chiren yu lijiao" (Cannibalism and Neo-Confucianism), *New Youth* 6 (1919).
[27] Yü Yingshi, "Tan 'Tiandi jun qin shi' de qiyuan" (On the Origin of Heaven-Earth-Sovereign-Ancestors-Masters), in *Xiandai ruxue de huigu yu zhanwang* (A Retrospective and Prospective View on Contemporary Confucianism) (Beijing: Sanlian shudian, 2004), 126–31.
[28] Wang Fan-sen, "Sichao yu shehui tiaojian – xinwenhua yundong zhong de liangge lizi" (Ideas and Social Condition – Two Examples from the New Culture Movement) in *Zhongguo jindai sixiang yu xueshu de xipu*, 255–68.
[29] Wu Yu, "Jiazu zhidu wei zhuanzhi zhuyi genjulun" (Family as the Foundation of Dictatorship), *New Youth* 2, no. 6 (1917).
[30] Li Dazhao, "You jingjishang jieshi Zhongguo jindai sixiang biandong de yuanyin" (The Economic Interpretation of the Changes in Modern Chinese Thought), *New Youth* 7, no. 2 (1920).

The individualism and liberalism promoted during the New Culture period further implied a questioning of the primacy of society and nation. In the progressive and radical vocabulary of the early 1920s, self-fulfillment outweighed responsibility with regard to nation and society. The writer of *New Youth,* Gao Yihan (1885–1968), claimed that the idea that individuals should sacrifice themselves for national causes was outdated and obsolete. Each individual possessed an intrinsic value independent of the nation, whereas the latter had no right to limit or pose an obstacle to the liberty that people were entitled to fully enjoy.[31] This argument was formulated on the eve of the end of the First World War, when Chinese intellectuals became increasingly interested in a political form other than nationalism that had lapsed into madness in Europe. Gao Yihan's statement resonated with that of Chen Duxiu cited in the introductory chapter. Loyalty towards the nation should not be confused with an obligation to obey in a situation where the nation failed to act as the guardian of people's rights.[32] As Yi Baisha (1886–1921) proclaimed: "It is better to love the world, rather than the nation... the world is the destiny that I share with others."[33]

It bears stressing that I am not suggesting that New Culturalists were not nationalist in the sense of standing for the independence and self-determination of China and forming a mental project for the nation. Yi Bashai, for example, was a nationalist. He drowned himself in 1921 after a failed attempt to assassinate a local warlord. However, as I have explained in the introductory chapter, their nationalism was conditional and was "situated in a cross-cultural context as a fusion of personal, national, social, civic, and moral freedoms."[34] Also, I do not intend to challenge Lydia Liu's thesis that the discourse of individualism "contributed to the process of inventing *geren* for the goals of liberation and national revolution."[35] Nevertheless, this temporary historical moment, in which the question of the individual took precedence over that of the society and the nation, and the social consequences this viewpoint generated need to be emphasized.

The ideal political reference that Yi Baisha shared with his fellow New Culturalists at the beginning of the New Culture period was society, a society that operated in a centripetal manner by the aggregation of individuals sharing the same rights. In the chaotic situation of the new Republic, in which people's

---

31 Li Jing, Xin Qingnian Zazhi huayu yanjiu (On the Discourse of the *New Youth*) (Tianjin: Tianjin daxue chubanshe, 2010), 92.
32 Ibid.
33 Ibid., 95.
34 Fung, "The Idea of Freedom in Modern China Revisited," 455.
35 Liu, *Translingual Practice*, 91.

rights and liberty were undermined by constant political struggles among warlords, it was little wonder that the nation, which had failed in its task of bringing about stability, gave way to the primacy of the individual and the personal satisfaction which constituted the first step towards a real harmonious and integrated society. It was by this logic that Li Dazhao reasoned: "What we expect now is a liberated and free ego and a world in which everyone loves each other. Everything that makes a barrier between me and the world, be it family, social class or nation, impedes the evolution and should be progressively aborted."[36]

It was regarding this situation that conservatives spoke against the second revolution of social liberation and attempted to restore the balance between authority and freedom. To ensure that liberal conservatives were for individual freedom and political liberty, Xu Zhenya stated that he preferred death to a life without freedom.[37] But he insisted upon the integration of personal liberty and national liberty; the first served the latter.[38] An editorial in *Minquan bao* affirmed that the right to liberty signified autonomy and could only be exercised in the framework of ethic.[39]

The conservative notion of liberty is always attached to that of authority. Indeed, for conservatives, the operation of the social structure began by the natural affection of each member of the society to their most immediate community. This community imposed on them an authority that they should respect. The slogan of "love for the nation begins with love for one's native place" was embedded in this ideology. The overhaul of the society during the second revolution of liberation was based on the engenderment of a new moral code. This code was not structured around the selected and reinterpreted tradition, but around the ideal of humanism (*rendao zhuyi*), which coincided with New Culturalists' aspiration for the advent of a denationalized society.[40] But the foundation of a society upon the new moral of "humanism" was considered to be illusionary by conservatives. For them, the engagement of each individual in the community did not constitute an instrumental choice, but a consciousness of duty and virtuous service. This is why a clear barrier did not exist between the public sphere and the private sphere, as Gao Xie argued: "Mencius once said: 'each one should love his relatives and respect his superiors, in this way, the world will be well organ-

---

36 Li Jing, Xin Qingnian *Zazhi huayu yanjiu*, 96.
37 Xu Zhenya, "Si yu ziyou" (Death and Freedom), *Minquan bao* (July 2, 1912).
38 Xu Zhenya, "Yinshi yu zi you" (Hermite and Freedom), *Minquan bao* (March 2, 1916).
39 Lianjiang, "Mengzi zhi ziyou" (Mencius and Liberty), *Minquan bao* (April 11, 1912).
40 See, for example, Zhongmi (Zhou Zuoren), "Zailun 'heimu'" (Revisiting the Conspiracies), *Dongfang zazhi* 15, no. 9 (1919).

ized.' Therefore, private sentiment and public sentiment are in fact the same thing."[41]

The liberty conceived by conservatives was unique in the sense that it involved an institutional dimension. Personal liberty was inserted into the liberty of the community to which one belonged. But personal liberty also constituted the first step towards communitarian liberty. Therefore, the authority of a community could not impose itself as a domination that finished by betraying the principle of liberty. This is the reason why conservatives expressed outrage not only at traditional moral codes that impeded personal liberty, but also at an individualism with little regard for responsibility towards family, society, and, ultimately, the nation. Hence, while New Culturalists denounced Confucianism as a cannibalistic machine and drew a clear line between republicanism and traditional culture, liberal conservatives responded by enumerating elements of the traditional moral values which should be preserved. Du Yaquan wrote:

> The *New Youth* questioned the place of the traditional principle of sovereign (*jundao*), integrity of servants of the state (*chenjie*), moral code (*mingjiao*) and moral order (*gangchang*) in the republican regime. [For them], these themes constitute a rebellion and a betrayal of the Republic and an anti-constitutional crime. I am convinced that republicanism is in no case incompatible with indigenous civilization. [...] If the political regime (of China) has changed, the political principle remains intact. Thus, it is only a matter of integration of civilizations to combine the indigenous civilization, based on the traditional principle of sovereign, integrity of servants of the state, moral code and moral order, and the current republican regime.[42]

The combination of republicanism with these traditional values stressing personal responsibility, social hierarchy, and disciplined human relations as the foundation of a sane political culture underscores the relation between individual liberty and traditional order inherent in conservative liberal thinking. If the Republic was to achieve stability, traditional values should be selectively actualized to allow for individual emancipation, which promoted liberty on an institutional level. The renewal of Confucian moral doctrine in the modern era constituted Du Yaquan's *tiaohe lun* (accommodation) – a search for compromise and moderation that were built upon the harmonization of Chinese and Western civilization.

---

[41] Gao Xie, "Lun xueshu liu" (On Learning, 6), in *Gao Xie ji*, 22.
[42] Cangfu, "Da Xinqingnian zazhi jizhe zhi zhiwen," 14–5.

Lee Ou-fan observes that the May Fourth era witnessed, for the first time in Chinese history, the opposition between the individual and collectivity.[43] This did not imply that the New Culturalists encouraged a total dissolution of morality. To Chen Duxiu, the abandonment of traditional moral principles should be accompanied by an adoption of European moral ethos. However, it was impossible to assimilate the latter overnight. Thus, a moral vacuum emerged in big Chinese cities. It follows that in cases where there was no concrete new moral ethos to replace the discarded old moral doctrine, the New Culturalists' dream of concretizing a healthy society by granting individuals maximum emancipation and freedom appeared to be an illusion. As Xu Jilin points out, "once hedonism and the search for pleasure are legitimized, the social effect will be out of the progressive intellectuals' control."[44] This social environment triggered conservatives to react.

But after all, Republican China remained a traditional country. Even in Shanghai, old moral codes still maintained order in the houses of the gentry and traditional businessmen. It was students, in particular, who identified with progressive and radical ideas. Shanghai was a paradoxical city. On the one hand, female schools prospered, Shanghainese women from rich families or the middle class adored Parisian fashion and regularly frequented theaters, casinos, and bars (access to which was previously prohibited for women), divorce rates rose to 2.382‰ in 1929, and a club for female celibacy was created in 1919. But on the other hand, many women still refused to remarry after the death of their husbands[45] and men and women were forbidden to talk to each other in public spaces such as theatres, even in the big cities.[46] The writer Lu Yin (1895–1934) once recalled that when she was a student at the Beijing Female Normal Superior School (*Beijing nüzi gaodeng shifan xuexiao*) in 1919, a student who gave a speech on "freedom of love" shocked the students who booed her off stage.[47]

Hence, conservatives' aversion to the second revolution of liberation might have been an overreaction. But it is worth mentioning that even some New Cul-

---

[43] Lee Ou-fan, *Zhongguo xiandai wenxue yu xiandaixing shijiang* (Ten Lessons on Chinese Modern Literature and Modernity) (Shanghai: Fudan daxue chuanshe, 2008), 20.
[44] Xu Jilin, "Geren zhuyi de qiyuan – 'Wusi' shiqi de ziwoguan yanjiu" (The Origin of Individualism: A Study of the View about the Self in the May Fourth Period), *Tianjin shehui kexue* 6 (2008): 119.
[45] Ibid., 107.
[46] Li Changli, *Zhongguo jindai shehui shenghuo shi* (History of Chinese Modern Social Life) (Beijing: Zhongguo shehui kexue chubanshe, 2015), chapter 4.
[47] Duan, *Shisu shidai de yiyi tanxun*, 107.

turalists were worried that the social effect produced by the freedom that they appealed for had gone too far. In 1923, the former Chinese anarchist in Paris Wu Zhihui (1865–1953) suggested that Mr. Science (*sai xiansheng*) and Mr. Democracy (*de xiansheng*) – two objectives of the New Culture Movement – should be reinforced with a Miss Moral (*muleer guniang*): "China has welcomed two misters – Mr. Science and Mr. Democracy – of course we should welcome them, but it now becomes clear that China lacks private morality (*side*)."[48]

Also, the majority of conservative intellectuals lived in the big cities and were detached from the rest of the nation where traditional moral principles were still rigorously practiced. Chinese conservatism as an intellectual movement was detached from rural areas. The only exception was the antimodern conservatives, whose engagement in rural reconstruction, as I will show later, produced nevertheless little substantial effects. As a result, conservatives did not disseminate their ideas among the population that might in fact have supported them. But most importantly, conservatives did not tolerate the traditional Confucian moral codes practiced in rural China. The first revolution of social liberation had targeted this very moral principle. Therefore, conservatives faced a dilemma. On the one hand, they were unable to change the traditional mindset in the countryside. On the other hand, with the New Culturalists' ideas being accepted progressively by more and more students and young people who became increasingly disinterested in conservatives' moral lectures, no attractive alternatives in terms of moral ethos that appealed to the younger generation were proposed by conservatives.

Liberal Conservatism and Politics

Liberalism and Chinese traditional politics may seem at odds. The nation's lack of a liberal tradition makes it impossible to conserve something that does not exist. Even the translation of the word "liberty" was a foreign neologism. The seed of liberalism in traditional learning reflected intellectuals' efforts to enable a dialogue between Chinese traditional learning and Western political culture rather than the true spirit of the tradition.

However, as Chapter 2 shows, aspirations for a more liberal regime had fueled progressive intellectuals' political engagements at the end of the Qing dy-

---

[48] Duan Lian, *Shisu shidai de yiyi tanxun – Wusi qimeng sixiang zhong de xindaodeguan yanjiu* (Search for Meaning in a Secular Age – Study of the New Moral in the Thought of Enlightenment of the May Fourth Period) (Taipei: Showwe, 2012), 141.

nasty. Liberal conservatives' optimism regarding the function and the effect of the modern Western liberal political system was accompanied by the willingness to update traditional culture that constituted the national essence in political construction and organization. Indeed, the unique spirit of a cultural nation elaborated by tradition, as Habermas points out, should emerge in the process of forming the political nation, such that the liberal regime remains stable and capable of federating people around a common political end.[49] During the May Fourth period, the place of the cultural tradition in the liberal regime marked the main point of contrast between liberals and liberal conservatives. Despite a shared understanding of liberty with New Culturalists, liberal conservatives refused to blame the Chinese pursuit of a less authoritarian regime in cultural tradition for impeding individual and national emancipation. In this regard, liberal conservatives might be the most loyal heirs of the revolutionary intellectuals' culturalist nationalism of the 1900s.

One of the liberal conservatives was Zhang Shizhao (1881–1973). In the same year that Du Yaquan appealed for the formation of a conservative party, Zhang was studying for his master's degree in political science at the University of Aberdeen. During his stay in Scotland, he regularly sent political commentaries to the newspaper *Diguo ribao* (*Imperial Daily*). Given his educational background, it is not surprising that he borrowed from British thinkers and politicians to theorize his ideal of Chinese parliament. What is interesting is that almost all those who he quoted were conservatives or Tories, notably Walter Bagehot (1826–1877), Albert Venn Dicey (1835–1922), and Edmund Burke.

In the 1900s, Zhang Shizhao was anxious to find a way to build a less authoritarian regime without sacrificing the state's power. He found the answer in the British political system. Inspired by Bagehot, he argued that the cabinet system, in which legislative power and executive power were not strictly separated, made British government strong and powerful, thanks to the superiority of cabinet members' capacities in regard to other members of parliament. The cabinet and the parliament were dominated by the majority party which had to step down if it lost popular support.[50] By citing Dicey, Zhang reasoned that a strict separation of powers resulted in constant quarrels between legislative power and executive power, which weakened the government.[51] Furthermore, the Brit-

---

49 Jürgen Habermas, "The European nation-state. Its achievements and its limitations. On the past and future of sovereignty and citizenship," *Ratio Juris* 19, no. 2 (1996): 130.
50 Zhang Shizhao, "Hewei zhengdang neige" (What is Cabinet), *Diguo ribao* (June 12 and 13, 1911).
51 Zhang Shizhao, "Zhengdang zhengzhi guo shiyu jinri zhi Zhongguo hu" (Is Party Politics Really Suitable for Today's China), *Diguo ribao* (May 29, 1911).

ish political system protected personal liberty and representation was directly linked to taxation. By quoting Burke, Zhang argued that tax revenue always belonged to the people, with their liberty guaranteed by private property and respected by the government.[52] In this way, the British government was able to stay both strong and liberal, and China should emulate this political structure.

Was Zhang aware that his intellectual mentors were conservative, or did he intentionally choose to adopt British conservative views?[53] Zhang did find an echo in British conservatives' skepticism towards universal suffrage. Neither Burke, Dicey nor Bagehot supported democracy in the form of common suffrage. While Zhang wished to ensure a political future in line with the English model, it should be emphasized that liberalism – understood as the protection of civil liberties along with the rule of law – does not necessarily entail democracy.[54] In *The English Constitution*, which Zhang frequently cited, Bagehot maintained that "cabinet government is possible in England because England was a deferential country", and "the masses of the 'ten-pound' householders did not exact of their representatives an obedience to those opinions; that they were in fact guided in their judgement by the better educated classes; that they preferred representatives from those classes, and gave those representatives much license".[55]

To Bagehot, the conservative themes of social hierarchy, natural aristocracy, and the confidence of the people in their betters made the English representative government possible. Zhang argued that even the West was worried that people's morality and capacity were not sufficient for democracy to thrive. So, for China, whose population was too accustomed to authoritarianism, it would be even more problematic for universal suffrage to function immediately after the foundation of the Republic[56]: "We are a country for the people, but we cannot afford an extreme populist government. We are a country for majority rule, but we cannot expense the political spirit of the elites."[57] Thus, Zhang found, in these Brit-

---

52 Zhang Shizhao, "Hewei buchu daiyishi buna zushui" (What is No Taxation Without Representation), *Diguo ribao* (October 22 and 23, 1910).
53 It needs to be underlined that, although self-identified as a Whig, Burke's political philosophy was later interpreted as "the intellectual and moral cornerstone" of English conservatism, when Tories and Conservatives enacted cautious reforms in response to the rise of "English Jacobinism" from the 1880s onwards. Zhang's time in Scotland coincided with the "canonization" of Burke as the founder of modern conservatism. See Emily Jones, *Edmund Burke & The Invention of Modern Conservatism, 1830–1914* (Oxford: Oxford University Press, 2017), 175–209.
54 Skorupski, "The Conservative Critique of Liberalism," 403–4.
55 Walter Bagehot, *The English Constitution* (Boston: Little, Brown, and Company, 1873), 6.
56 Zhang Shizhao, "Guojia yu zeren" (State and Responsibilities), *Jiayin* 2 (1914).
57 Zhang Shizhao, "Lun pingmin zhengzhi" (On Populism), *Minlibao* (March 1, 1912).

ish conservative thoughts, a path not only to establish a liberal political system, but also to maintain and rejuvenate traditional Confucian "paternalism".

After the foundation of the Republic, Zhang Shizhao served in the Yuan Shikai government with Liang Qichao, who, at the end of 1912, returned to China and joined the conservative Chinese Democratic Party (*Zhongguo minzhu dang*) to resist the KMT.[58] In 1913, the Chinese Democratic Party merged with Li Yuanhong's Republican Party to form the Progressive Party (*Jingbu dang*). The members of the Party were largely financed by Yuan Shikai, who counted on the conservative force to confront the KMT.[59] Unlike the former revolutionaries who argued for popular sovereignty, Liang Qichao supported a strong central government to guide the population.[60] As Chang Peng-yuan points out, his support for a strong central government and opposition to decentralization were not only inspired by Western and Japanese political systems, but were also deeply rooted in Confucian teaching of "paternalism" whereby the government should be like a parent and shape politics when individuals acted in self-destructive ways. Given China's long history of absolute monarchy, Liang Qichao believed, much like Zhang Shizhao, that this political system, in which the population was guided by their betters, was the most appropriate to avoid the "rule of the mob."[61]

Liang's alliance with the Yuan Shikai government to fight against the KMT turned out to be a mistake. With unparalleled power, Yuan began to stage his own restoration act after he had suppressed the Second Revolution and purged the KMT. He quickly abandoned the Progressive Party that he had used to resist the KMT.[62] Liang formally broke with Yuan in 1915 and urged each province to declare independence from the central government.[63] But this incident disappointed the public and intellectuals, who now lost confidence in party politics.[64] For the first time since the foundation of the Republic, Liang collaborated with his political opponents to condemn Yuan in the media.

Many members of the Southern Society were killed during the Second Revolution and Yuan heavily censored newspapers that reacted against his politics.

---

58 Ding, *Liang Qichao nianpu changbian*, 647–9.
59 Chang Peng-yuan, *Minguo chunian de zhengdang* (Political Parties of the Early Republic Era) (Changsha: Qiulu shushe, 2004), 444–6.
60 Liang Qichao, "Xianfa zhi sanda jingshen" (Three Major Principles of the Constitution), in *Liang Qichao quanji*, 2560–8.
61 Liang Qichao, "Zhongguo liguo dafangzhen" (Major Guidelines of the Establishment of the Chinese nation), in *Liang Qichao quanji*, 2488–2504.
62 Chang, *Minguo chunian de zhengdang*, 445.
63 Ding, *Liang Qichao nianpu changbian*, 728–9.
64 Chang, *Minguo chunian de zhengdang*, 451.

It suffices to simply scroll through publications after 1912 to appreciate the fury and deception in regard to the Beiyang government under Yuan. Popular power, constitutionalism, democracy, and especially liberty were the keywords of the Southern Society's major newspapers.[65] One of these newspapers was *Minguo ribao*, founded in 1916 in Shanghai. Its principal writers were members of the Southern Society, including the founder of the newspaper Chen Qimei, the chief editor Ye Chucang, and the chief editor of the supplement Shao Lizi (1882–1967). The newspaper, which later became the official body of the KMT, claimed to be the defender of the Republic that was usurped by Yuan Shikai.[66] Destined to promote popular power in disregard of Yuan and the Beiyang government, the newspaper assented that there was a culture characterized by the virtue of Chinese people and capable of "knocking down the current regime, destroying warlords and constructing a real republic."[67]

Questions remained as to what this culture really meant and how could it inform political changes. The archaic political form of *fengjian* resurfaced. To offset the central government, local self-government, which was banned by Yuan Shikai in February 1914, had to be reimplemented.[68] After the death of Yuan, former revolutionaries estimated that the time had come to put forward these political views. Ye Chucang once wrote:

> People's attitude towards the republican regime is very much influenced by the institution of local self-government. In fact, citizens have not still fully understood government's power. So, it is difficult for them to judge if the central administration works well or not. Local self-government, on the other hand, exercises an effect much more immediate. Through local self-government, citizens can directly have a sense of the advantages and benefits brought about by the republican regime. In this way, they will support the Republic.[69]

The political transformation was inseparable not only from traditional culture, but also from industrialization and capitalism. Zhang Dongsun (1886–1973), for one, did not agree with antimodern conservatives' hostility towards industri-

---

65 See, for example, Li Dingyi, "Ziyou zhen" (On Liberty), *Minquanbao* (March 12, 1912).
66 Ai, "Benbao fakanci" (Manifesto), *Minguo ribao* (January 22, 1916).
67 Ai, "Luanshi zhi minde" (Popular Morality in an Era of Chaos), *Minguo ribao* (January 26, 1916).
68 Keith R. Schoppa, "Local Self-Government in Zhejiang, 1909–1927," *Modern China* 2, no. 4 (1976): 514.
69 Ye Chucang, "Shenlun huifu difangzizhi" (On the Reimplementation of Local Self-Government), *Minguo ribao* (October 4, 1917).

alization and capitalism. In a tone that echoed Hu Shi, he observed in 1920 that China was too poorly developed to scoff at Western science and technologies:

> Part of the Chinese population lives in the commercial ports. These people are used to Western material civilization. Hence, they thought that since Westerners criticize their material civilization, that they can do the same. However, they forget the difference between the interior of China and the West... Even if I have yet to visit the interior part of China, I think that the only way out for China is to develop the economy. For this purpose, China should be industrialized because the only illness that Chinese suffer is poverty, which has reached its limit.[70]

Furthermore, he underlined the importance of industrialization and capitalism as the foundation of popular power, arguing that *min* in the West signified *shimin:*

> What is citizenship? Citizens are those who involve themselves in the economic system and possess economic power. I welcome the progressive growth of the power of commercial associations, those which are established in commercial ports are particularly capable. The local self-government today is illustrative proof of the emergence of the middle class. It is pertinent to note that the emergence of the middle class did not come about from nothing but rather relied on power. The strongest powers today are military power and economic power.[71]

In the Western ideological context, the relationship between capitalism, private property, and popular rights constitutes a major theme of conservatism. For the pro-capitalism conservatives, capitalism and industrialization cannot be carried out successfully without the preservation of traditional values. Private property, for example, acts as "a *jus* that is fully integrated... in the virtue of *justitia*. Nevertheless, justice wants the exercise of rights to be backed by benevolence *benignitas*. This is a very conservative theme: rights do not go without duty."[72] As Disraeli professed, to avoid the struggle between classes, aristocracy should assume the responsibility of helping the poor.[73] According to Metzger, certain Chinese literati and intellectuals do not acknowledge the frontier between the Chinese "spiritual" civilization and the Western "material" civilization because the improvement of people's living condition has been regarded as the primary condition of

---

[70] Zhang Dongsun, "You neidi lüxing er dezhi youyi jiaoxun (Reflection After the Visit to the Interior Part of China), in *Zhongguo xiandai sixiangshi ziliao*, 616.
[71] Zhang Dongsun, "Xianzai yu weilai" (Today and Future), in *Zhongguo xiandai sixiangshi zilia*, 618.
[72] Vinent, *Qu'est-ce que le conservatisme?*, 188–9.
[73] Suvanto, *Conservatism from the French Revolution to the 1990s*, 35–6.

individuals' moral perfection since the seventeenth century.[74] Hence, industrialization and capitalism could be well rationalized within the framework of traditional learnings.

Traditional cultural elements that were familiar to the people and capable of arousing their consciousness of solidarity could be appropriated to politically inform concrete reforms. This was why liberal conservatives refused to study traditional learning without value judgement and museumify the national essence as the New Culturalists suggested. The real tradition was not a museum artefact but a source of living inspiration that assumed the pragmatic utility in the political organization.

In this regard, the Constitution of the Republic of China, which divides the power into five Yuan (ministry), is often evoked as the example *par excellence* to illustrate this cohesion between culture and politics.[75] Sun Yat-sen's 1924 speech in Kobe, in which he urged Japan to join China to defend the "royal way" (*wangdao*) against the "hegemonic way" (*badao*), is another example:

> Confucius and Mencius had spoken of popular power more than two thousand years ago. Confucius said: "When the great Way is concretized; the entire world becomes a public propriety." He evoked here an egalitarian world that protected people's political power... [and considered] people as the *raison d'être* of power... Mencius said: "people are more important than the state, and the state is more important than the emperor."... It seemed that at that time, Confucius had already realized that the emperor was not indispensable, and his reign would not last for long... It becomes clear that we Chinese have had the idea of popular power for more than two thousand years.[76]

Another conservative activist, Gao Xu, had several reservations regarding the themes of liberty and Confucianism enacted as the foundation of national education in the 1913 Draft Constitution of the Republic of China. He first urged the government to modify the Constitution to restrain the power of the presidency and remove undue limits on citizens' freedom. He then affirmed that Confucianism was complicit in dictatorship and to make Confucianism the state ideology jeopardized religious freedom. If Confucianism was useful to children, as he maintained, "Confucians would have already created a strong nation."[77] One

---

74 Thomas Metzger, *Escape from Predicament: Neo-Confucianism and China's Evolving Political Culture* (New York: Columbia University Press, 1977), 222
75 Hu Huaichen, "'Guoxue gailun' zonglun" (On the *General Theory of National Learning*), in *Shenwen yu mingbian*, 935.
76 Sun Yat-sen, "Minquan zhuyi Diyijiang" (Principle of People's Power, 1), in *Guofu quanji diyici* (Complete Works of Sun Yat-sen, Volume 1), ed. Chin Hsiao-yi (Taipei: Guofu quanji bianji weiyuanhui, 1989), 60–61.
77 Guo and Jin, *Gao Xu ji*, 523.

of the measures adopted by Yuan Shikai to consolidate his power was the proscription of local self-government. Gao believed in the absolute necessity of protecting the administrative independence of local authorities from the arbitrariness of the central government. He pointed out, as attested in *Zhouli* (*Rites of Zhou*), that China had a long history of decentralization, which reflected "our ancestry's political spirit."[78] While he believed that China could learn from Western political systems, he professed that Western experiences could not be automatically transposed in order to orient the Chinese pursuit of a liberal regime, the concretization of which relied heavily on the nation's archaic political culture.

However, liberal conservatives' political proposals were often met with opposition from the government. Gao Xie, for example, stated that the Chinese culture was structured around Confucianism. He was well aware of the fact that Confucianism had been manipulated to consolidate authoritarian reign and he condemned the inhuman Confucian moral practices.[79] But he argued that the foundation of the Republic marked the ideal moment to rectify the abusive usage of Confucian learning, of which Confucianism was itself a victim, in order to bring to light authentic Confucianism that could be used in political edification.[80] While he spoke of the importance of maintaining the *kongjiao*, he did not support the Confucian religion and criticized the ritualization of Confucianism.[81] For him, *jiao* signified education and transformation (*jiaohua*) rather than religion (*zongjiao*). He believed that this process of *jiaohua* should be embarked upon from the bottom up, without the intervention of the state which could take advantage of Confucianism for authoritarian rule, as in the case of Yuan Shikai.[82] But how could this project be concretized without the state's support? In fact, when Cai Yuanpei was nominated as the Minister of Education of the Beiyang government in 1912, one of the first educational reforms that he undertook was the suppression of the study of Confucian classics.

Furthermore, liberal conservatives could not agree on which traditional learning should be privileged, and much less its concretization at the political level. Their efforts to politicize traditional culture failed to acquire a new dimension in comparison with their rigorous political projects in the 1900s and seemed to become just a vague comparison between Chinese culture and Western politics. Members of the Southern Society were heavily invested in *Minguo ribao*. This newspaper published a literary supplement which reported findings on tra-

---

78 Ibid., 524.
79 Gao Xie, "Lun xueshu san" (On Learning, 3), in *Gao Xie ji*, 18.
80 Gao Xie, "Da Yang Qiuxin shu yi" (Letter to Yang Qiuxin, 1), in *Gao Xie ji*, 389.
81 Ibid.
82 Gao Xie, "Da Wang Jingpan shu zhiliu" (Letter to Wang Jingpan, 6), in *Gao Xie ji*, 371.

ditional elements corresponding to modern Western political values, such as equality, liberty, democracy, and people's rights. However, it remained unclear how they intended to materialize these cultural elements as political reforms. This was especially problematic since the cultural language in which their political aspirations were expressed had already been deemed to be against what they claimed to defend and dismissed by the New Culturalists as reactionary, retrograde, elitist and, with the rise of socialism in China, "feudal". In other words, liberal conservatives failed to explain why tradition still mattered if Western experiences of political modernization could be directly appropriated, as suggested by the New Culturalists. G.A. Cohen's perception of conservatism may help us to understand liberal conservatives' prepossession with traditional culture:

> The conservative impulse is to conserve what is valuable, that is, the particular things that are valuable... If an existing thing has intrinsic value, then we have reason to regret its destruction *as such*, a reason that we would not have if we cared only about the value that thing carries or instantiates... "Conservation of what has value" is, indeed, the canonical phrase here, and not "conservation of value." For if we take the phrase "conservation of value" in the standard way, on the model, that is, of "conservation of energy," or "of matter," a conservation of value policy is entirely unconservative... Conservation of what has value is not in that sense, in the conservation of energy sense, conservation of value, for you lose no value, value itself is conserved, when you destroy something valuable and replace it by a thing of the same value. The conservative disposition is not to keep the value rating high but to keep the things that now contribute to that rating.[83]

In other words, conservatives were pragmatic, but they were not utilitarian. Therefore, they refused to sacrifice traditional culture to achieve a liberal regime, even if totally abandoning traditional culture could hypothetically be a better guide to a less authoritarian future for China. However, incapable of convincingly explaining how traditional culture could politically institutionalize the nation-state, liberal conservatives gradually lost their intellectual vigor. They became marooned amid the dilemma of pursuing the same political aspiration as the progressive liberals, but within a cultural language rejected by the New Culturalists as incompatible with liberty. Liberal conservatives who occupied a place in the state bureaucracy could not escape the fact that their careers were at the mercy of the real political power holders. Without political vigor, conservatives failed to propose concrete measures to politically institutionalize the nation through its sociocultural heritage and concentrated increasingly on proving the enduring value of traditional culture, which had been overwhelmed by the New Culture in the media and in academia.

---

83 G.A. Cohen, *Finding Oneself in the Other* (Princeton: Princeton University Press, 2013), 153.

Also for this reason, their reaction against the excessiveness of the second revolution of social liberation became increasingly dogmatic, without a social space in which their real concern could be well-explained and heard. In 1929, Hu Shi harshly interrogated Ye Chucang: "If ancient China was really a golden world constructed upon virtue, why are we even attempting to create new culture? Why do we not advocate a return to the past?"[84] In the famous debate between science and metaphysics (*xuanxue*) that I will return to later, Hu Shi, Ding Wenjiang (1887–1936), and Chen Duxiu reasoned that science itself was capable of producing a new "conception of life" based upon love, empathy, and social attachment. With this rhetoric, New Culturalists saw conservatives' determination to look back to selected traditional moral teaching as evidence of their backwardness.

Antimodern or Antimodern Conservatism?

Unlike liberal conservatives, antimodern conservatives maintained that China should reject modern Western political and economic systems and return to the sociopolitical arrangements consistent with her agrarian and communitarian cultural tradition. In the Chinese ideological context, modernity was usually associated with Enlightenment ideals, especially with its belief in progress and in reason, which translated into a veneration of "newness," as well as industrialization, economic development, urbanization, democracy, and rational law.[85] In this regard, modernity is a Eurocentric concept, in which the non-West, as the otherness, is created as "the boundary of the space of modernity."[86] Nevertheless, the otherness of modernity was not necessarily invented by the West: it could also be voluntarily established by the modernists in the non-West. During the May Fourth period, Chinese traditional culture, reduced by the New Culturalists to *fengjian* with its pejorative connotation, was deemed to be the otherness

---

84 Hu Shi, "Xinwenhua yundong yu Guomindang" (The New Culture Movement and the KMT), in *Zhongguo xiandai sixiangshi ziliao jianbian* (Sources of Modern Chinese Intellectual history), Vol. 3, ed. Cai Shangsi (Hangzhou: Zhejiang renmin chubanshe, 1982), 148.
85 Edmund S.K. Fung, *The Intellectual Foundations of Chinese Modernity, Cultural and Political Thought in the Republican Era* (New York: Cambridge University Press, 2010), 22–5.
86 Timothy Mitchell, "The Stage of Modernity," in *Questions of Modernity*, ed. T. Mitchell (Minneapolis: University of Minnesota Press, 2000), 16.

of modernity.⁸⁷ This dichotomy between tradition and modernity in the May Fourth narrative led to the construction of an idealized image of Western modern values and institutions to criticize the Chinese past, the history of which should be re-evaluated and rewritten according to the Darwinist concept of progress that was identified with the West.⁸⁸

Antimodern conservatives called into question the modernist idea of linear evolution and progress, as well as industrialization, capitalism, and political institutions associated with modernity. For them, Chinese political-cultural tradition was more capable of offering an adequate solution to the flaws which they perceived in the Western model. This suspicion of the West had much to do with the First World War, which signaled to many the bankruptcy of Western civilization. The drive behind antimodern conservatism was a deep concern for ethics, localism, customs, and habits, from which they developed their own version of progress.

Since it was not only conservatives who cherished these values, it is necessary to distinguish conservatism from anti-modernism. Indeed, in eighteenth and nineteenth century Europe, a branch of conservatism was closely intertwined with anti-modernism. As a criticism of modernism and industrialization, anti-modernism was not clearly distinguishable from conservatism, the core of which expressed the melancholy of the peripheral social group who were excluded from the process of industrialization and modernization and felt incapable and frustrated with the loss of their symbolic and economic stature.⁸⁹ This was one of the reasons that conservative ideology enjoyed popularity, particularly within the marginalized social milieu.⁹⁰

At first glance, it may seem anachronistic to apply the terms of modern and antimodern to the Chinese historical background of the early twentieth century. At the time, progressive intellectuals did not generally use the word modern (*xiandai*) to describe the type of the society that they wished to construct by adopting Western modern technologies and political systems. The dissemination of the word was progressively expanded with the publication of *New Youth*. However, there is little doubt that the reaction of certain intellectuals, such as Zhou

---

87 Viren Murthy, "The Politics of *Fengjian* in Late Qing and Early Republican China," in *Beyond the May Fourth Paradigm: In Search of Chinese Modernity*, ed. Kai-Wing Chow et al. (Lanham: Lexington Books, 2008), 171.
88 Murthy, "The Politics of *Fengjian* in Late Qing and Early Republican China," 172–5.
89 Urs Altermatt, "Conservatism in Switzerland: A Study in Antimodernism", *Journal of Contemporary History* 14, no. 4 (1979): 583–84.
90 Ibid. For an empirical study, see Shulamit Volkov, *The Rise of Popular Antimodernism in Germany: The Urban Master Artisans, 1837–1896* (Princeton: Princeton University Press, 1978).

Zuoren, Lu Xun, and Zhang Taiyan, to the social and cultural effects of the radical changes of China resonated with Western antimoderns. Zhou Zuoren's conception of the utopia of the new village (*xincun*) incarnated the antimodern spirit. His idea of the new village aimed to contest capitalism and urbanization and drew from the Movement of the New Village (*Atarashiki-mura undō*) launched by Mushanokōji Saneatsu (1885–1976) in Moroyama, a village in Saitama, Japan. In 1919, Zhou made a trip to Moroyama. In the same year, he published an article in *New Youth* to introduce the Japanese experience of what he praised as socialist utopia.[91]

Lu Xun's attitude towards modernity was also ambiguous and intellectually much more complex than his colleagues from *New Youth*. Guarding a prudent distance from science and democracy, Lu Xun attacked the dictatorship of the majority and the materialism that could be brought about by democracy and science. His antimodern inclination was comparable to that of Nietzsche (1844–1900). This consisted of affirming the heterogenization of human spirit and cultivating the spontaneity of individuals via cultural critique; both demonstrated a suspicion towards progressivist optimism.[92]

These intellectual orientations that pointed to the defects of modernism were particularly evident in the writings of antimodern conservatives. The abandonment of the city to rectify the human alienation caused by industrialization and urbanization was not Zhou Zuoren's own idea. This was more or less the consensus shared by antimodern conservatives, such as Liang Shuming (1893–1988), Zhang Shizhao, and Liu Yizheng (1880–1956), who were persuaded that a spiritually regenerated rural area could become the basis of the Chinese nation. As to Zhou Zuoren's brother Lu Xun, he was deeply influenced by Zhang Taiyan. Lu Xun attended Zhang's courses in Tokyo during the revolutionary decade and many of his arguments on the individual, nation, and freedom echoed those of his former professor.[93] These intellectual similarities and intersections provided space for a more general reflection on Chinese conservatism. Why were Zhou Zuoren and his brother considered to be progressive New Culturalists, while others were considered conservatives?

The most obvious answer has been provided in the last chapter and has much to do with intellectual competition. That is, it is sometimes not so much the intellectual orientation as the camp that one chooses that determines wheth-

---

91 Zhou Zuoren, "Riben de xincun" (New Village in Japan), *New Youth*, 6, no. 3 (1916).
92 Cheung Chiu-yee, "Lu Xun yu Nicai fan 'xiandaixing' de qihe" (The Similarities in the Anti-Modernism of Lu Xun and Nietzsche), *Twenty-First Century* 29 (1995): 93–4.
93 Viren Murthy, *The Political Philosophy of Zhang Taiyan: The Resistance of Consciousness* (Leiden and Boston: Brill, 2011), 222–242; Veg, "Lu Xun and Zhang Binglin."

er one is progressive or conservative. But here, the philosophical foundations inherent to the intellectual orientation do have a role to play in differentiating antimoderns from antimodern conservatives. Antimodern conservatives sought to bring together politics and traditional culture to tackle problems of modernity, which for the Zhou brothers was illusionary and useless from the very beginning. Even though both had expressed nostalgia for elements of tradition,[94] they rejected tradition as dreadful and in no way useful as an instrument to model the nation or to respond to the challenges of modernity. Deprived of a veritable contemporaneity and marked by a particular malignity, traditional culture and practices should give away to "New Culture" to facilitate the advent of revamped national characters. Hence, antimoderns thought beyond the dichotomy between traditional culture and modernity; they criticized both. In contrast, antimodern conservatives were always oriented by a localist inclination, accompanied by a questioning of modern Western civilization in terms of science, political, and economic systems, and advocated tradition as the solution to the repositioning of the Western model. Antimodern conservatives refused to judge the value of traditional culture according to the Western standard. This cultural ethos was translated into a rediscovery of certain political forms of the past that they praised as the system most adaptable to the Chinese context and capable of redeeming and rectifying the defects of Western modernity.

Antimodern Conservative Theorizers

One of the most eminent antimodern conservatives was Zhang Taiyan. There were always two faces to Zhang Taiyan: a political man and a philological erudite. As a result, his ideas and his activism were not always coherent. A committed revolutionary that fought for the foundation of the Republic of China, he had written several pieces in *Minbao* which condemned nationalism and saluted the advent of an anarchist world. This incoherence might be explained by his awareness of the difficulty to concretize his ideal politics, or by the conviction – not unlike Kang Youwei – that utopianism could only be realized after the dissolution of nationalism.

After the Wuchang Uprising, Zhang returned to China and national interests remained a priority to him. He called for a dissolution of the Tongmenghui so

---

94 In Zhou's case, see Susan Daruvala, *Zhou Zuoren and an Alternative Chinese Response to Modernity* (Cambridge, Mass.: Harvard University Asia Center, 2000). For Lu Xun, see Veg, "Lu Xin and Zhang Binglin."

that national politics could go beyond the revolutionaries' hegemony in order to prevent the Republic from turning into a party-state.⁹⁵ Like Liang Qichao, he served the Yuan Shikai government until Yuan's ambition of restoration was revealed. But his disappointment with politics was not totally due to Yuan Shikai. He was also troubled by the revolutionaries' arbitrariness, which translated into threats or even violence to violate legislative orders.⁹⁶

At the intellectual level, he was influenced by Buddhism and Daoism. He criticized social evolutionism for neglecting the bad in progress and asserted the simultaneous progression of the good, the bad, pleasure, and misfortune.⁹⁷ For him, "what we call progress is nothing but a result of illusion; progress does not exist [since] the quantity of the quality of a certain thing does not change: if it progresses in one aspect, it regresses in another."⁹⁸ Zhang pointed out that the danger of social evolutionism lay in its implication concerning a universal law of progress (portrayed as objective and associated with the West) which internalized the subordination of weak nations to the strong. In this sense, evolution, far from a moral obligation for China, emerged as the hegemony and arrogance of the West that had invaded China, with claims to be the universal ideal.

Zhang remained a lifelong ardent nationalist. At the international level, he called for solidarity among weak nations to resist imperialism.⁹⁹ At the national level, he argued against both the political and the cultural impositions of the central government and the mechanical replication of Western models. Instead, he proposed a return to traditional political practices which he regarded as part of the national essence and concluded that, "to unify the nation... customs and habits of each region should be surveyed and let evolve spontaneously."¹⁰⁰ In such a way, the political nation and the cultural nation again became one.

Zhang insisted upon the preservation of the marriage customs and the institution of family as it existed then. The rules of a family, implemented with a view to maintaining the stability and reinforcing the morality of each member of the family, should be granted the monopoly to prevent immoral behaviors that need

---

95 Tang, *Zhang Taiyan nianpu changbian*, 367.
96 Ma Fei, "Geming wenhua yu minchu xianzheng de bengkui" (Revolutionary Culture and the Collapse of the Constitutionalism at the Beginning of the Republic), *Twenty-First Century* 137 (2012): 44–58.
97 Zhang Taiyan, "Jufen jinhualun" (On the Bilateral Evolution), *Minbao* 7 (1906).
98 Zhang Taiyan, "Sihuolun" (On the Four Illusions), *Minbao* 22 (1908).
99 Zhang Taiyan, "Guojialun" (On the State), *Minbao* 17 (1907).
100 Zhang Taiyan, "Xianzonghe houtongyi lun" (Research before Unification), *Zhang Taiyan zhenglunji*, 552.

to be punished. These rules should be executed without the intervention of the state. He also suggested that public morality should be part of public order. Gambling should be banned by law and police should have the right to arrest all those who participated in "immoral activities," such as dancing and kissing in public.[101]

But this concern for sound morality was not informed by the Confucian religion. Zhang supported laicity and opposed the establishment of Confucianism as the state religion. As an Old Text scholar, his opposition to the Confucian religion was explained not only by the possibility of the state turning the Confucian religion into a potent ideological weapon to justify authoritarian rule, but also because of the lack of religious elements in Confucianism. His rejection of the universal law sanctioning the whole country was informed by the conviction that rational law could not be adapted to each region of the nation. At the core of this concept of rational law, the role of the people occupied a primary place: it was the people who maintained the law and not the reverse, as he said:

> It is illusionary to oblige people to obey a policy that self-proclaimed to be universally applicable without a real research on different habits of different regions…The Qing government ignored local situation and national tradition by establishing laws inspired by Japan and the Tang dynasty, but the new law enacted in this manner was a total failure…This is why reform should always be informed by tradition and the national unification should be based on the respect of regional differences…The concrete politics is never fixed but evolves in accordance with tradition…In order to unify the nation, the establishment of law is not sufficient. Instead, local customs should first be studied and be left evolute spontaneously.[102]

At the political level, the accent put on the preservation of local particularities was translated into his support for local self-government and federalism which he claimed could be traced back to *fanzhen*, a system of administration through regional governors of the Tang Dynasty.[103] Zhang went so far as to oppose the Northern Expedition guided by Chiang Kai-chek to unify the nation. His conception of local self-government was different from that of liberal conservatives. While they all sought resistance to the eventual arbitrariness of the central government through local self-government, Zhang criticized representative democracy for exacerbating social inequality by setting up a parliament to which

---

[101] Zhang Taiyan, "Zhonghua minguo lianhehui diyici dahui yanshuoci" (Speech Given at the First Assembly of the Union of the Republic of China), *Zhang Taiyan zhenglun xuanji*, Vol. 2, 532–5.
[102] Zhang Taiyan, "Xianzonghe houtongyi lun," 550–2.
[103] Zhang Taiyan, "Fanzhen lun", 102.

only the rich and the powerful could be elected. By the same token, he rejected partisan politics, arguing that political parties only served to entrench corruption and vested interests.[104] To prevent the president from obtaining monopolist rule, he proposed the nomination of *fali* and *xueguan* – social instructors who would enjoy the same level of power as the president.[105] He further suggested the re-establishment of the ancient official positions of *jizhongshi* and *yushi* to supervise the government and check on public servants' morality, something that could neither be legislated for nor sanctioned by vote.[106] All these policies were fruits of his indignation towards the corruption of the Republican politics. Several political parties were created after the Revolution of 1911. However, few of them really cared about popular support and the struggles between parties had become a normalcy.[107]

Another example is that of Liu Yizheng (1880–1956). As a historian with the Critical Review Group, he had been a *protégé* of Miao Quansun (1844–1919) and earned his living by teaching during the last few imperial years. In 1901, he found a job in the Publishing House of Translation of Jiangchu (*Jiangchu bianyiju*) set up by Liu Kunyi and Zhang Zhidong in Nanjing, where he was responsible for drafting textbooks on history.[108] In 1902, he went to Japan with Miao to study the Japanese educational system.[109] His textbook *Lidai Shilue* (*A Brief History of Chinese Dynasties*) was in practice a copy of Naka Michiyo's (1851–1908) *Shina tsūshi* (*A General History of China*), in which Liu minimized the struggle between the Han Chinese and the Manchu.[110] After the foundation of the Republic, he was offered a post at the Superior Normal School of Nanjing (*Nanjing gaodeng shifan xuexiao*) which later became the National Southeast University (*Guoli dongnan Daxue*), in which he organized the Historical and Geographical Society (*Shidi yanjiuhui*). The association federated many members of the Critical Review Group, including Chang Chi-yun (1901–1985), Chen Xunci (1901–1991), Jing Changji (1903–1982), and Miao Fenglin.

---

104 Wang Fan-sen, *Zhang Taiyan de sixiang (1868–1919) jiqi dui ruxuechuantong de chongji* (Zhang Taiyan's Thought and Its Percussion on Confucianism) (Taipei: Shibao, 1985), 133–5.
105 Wang, *Zhang Taiyan de sixiang jiqi dui ruxuechuantong de chongji*, 137.
106 Zhang Taiyan, "Yu Zhang Xingyan lun gaige guohui shu" (Lettre on the Reform of Parliament to Zhang Xingyan), in *Zhang Taiyan zhenglunji*, 789–90.
107 Wang, *Zhang Taiyan de sixiang jiqi dui ruxuechuantong de chongji*, 132–6.
108 Wang Xintian, "Liu Yizheng xiansheng nianpu jianbian" (A Short Biographical Chronology of Liu Yizheng), *Zhongguo wenzhe yanjiu tongxun* 9, no. 4 (1999): 148–9.
109 Hon Tze-Ki, "Cultural Identity and Local Self-Government: A Study of Liu Yizheng's 'History of Chinese Culture'," *Modern China* 30, no. 4 (2004): 511.
110 Ibid., 512–3.

His early involvement in the struggle between old culture and new culture concerned the rise of the so-called "Doubting Antiquity School" (*yigupai*) of Peking University. Its proponents applied a critical historiographical approach to Chinese historical sources and claimed that most of them were not authentic. According to Gu Jiegang (1893–1980), the main problem with Chinese history was the disproportionate accent that was put upon the concept of *dao* (Way), which in fact, did not exist. His historiographical approach aimed to meet two objectives: verify the authenticity of ancient history and discern the political usage of the main themes of Chinese ancient history, such as *dao*.

Traditional historians believed in the total concretization of *dao* during the Three Dynasties. They judged the posterior history to be based upon the idealized Three Dynasties. Gu found this absurd and claimed that the Three Dynasties never existed. To prove his point, he argued that Yu, the founder of the first dynasty of the Three Dynasties, the Xia, was an invented figure. *Shangshu* (*Book of Documents*), *Shijing* (*Classic of Poetry*), and *Lunyu* (*Analects*) never linked Yu to the Xia dynasty. He argued that it was common in *Shijing* to add a word indicating the place of origin before the name. In this case, if Yu were really from the Xia dynasty, *Shijing* would have recorded him as Xia Yu, which was not the case. He concluded: "Yu the Great was originally a worm. In the Shang and Zhou eras, people regarded this great worm as a revered sage or the earliest human king. By the time of the Warring States period, Yu had become a figure of remote antiquity, after the Xia dynasty."[111]

Not only did Liu Yizheng criticize Gu Jiegang's "voluntarist" historical approach that undertook research with a preconceived conclusion, but he also questioned whether this type of research had any utility whatsoever for the society:

> Ancient books are not all falsified. Even with a good method to verify the authenticity of history, they will conclude that whatever book spread rumors, how can this type of research contribute to the nation and the society? Many events that had really taken place were also recorded in ancient books, what to do with these incontestable truths then? So, those who are interested by this historical approach refuse to really undertake research and avoid being involved in politics. They do the research for their own satisfaction and retrieve themselves deliberately from useful learning.[112]

---

111 Yang Peng, "Analysis of Japanese Influence on the Three Major Historiographical Trends in Early Modern China," *Chinese Studies in History* 49, no. 1 (2016): 33.
112 Liu Yizheng, "Zhongguoshi yanjiu lunwen jixu" (Preface to the Collection of Articles on Chinese History), in *Liu Yizheng Qutang tiba* (Inscriptions and Postfaces of Liu Yizheng), ed. Liu Dingsheng and Liu Nianzeng (Taipei: Huashi, 1996), 33.

What was the social utility of historical research to Liu Yizheng? One reason why traditional culture could not be denied its value or be judged according to Western standards was that the value that shaped the cultural nation should be distinct and specific. However, this did not mean that Liu rejected foreign cultural influences. He observed in *Zhongguo wenhuashi* (*A History of Chinese Culture*, 1919) three steps of development that Chinese history had gone through: ancient, middle age, and modern. The ancient period went from antiquity to the end of the Han Dynasty, when Buddhism began to flourish in China and led her to the middle age. The contact with Western civilization at the end of the Ming Dynasty brought China into the modern age. Although Chinese culture was influenced periodically by foreign cultures and never remained homogeneous, Liu maintained that Chinese culture was extraordinary in its power to assimilate external cultures. Hence, in the same way that Buddhism was Sinicized and had instilled in Chinese culture the religious spirit that it lacked, a modern China would be capable of assimilating Western culture without sacrificing her roots.[113]

Liu considered this type of cultural progress, which had accrued throughout China's interference from foreign influences, to be the result of the collective efforts of the Chinese people. However, Chinese history is not always a story of progress. In common with Zhang Taiyan, Liu asserted that evolution was always accompanied by counteracting disadvantages and recessions. The fact that progress does not represent the only route of evolution is the outcome of the unpredictability of people's mentality which makes progress possible in the first place. Citing *Hong Fan* (*Grand Rules*), Liu explained that human endeavors triggered "the actions of Heaven" (*tianxing*). It should be noted that, in *Hong Fan*, human endeavors refer to the emperor's conduct, which, if inappropriate, results in punishments imposed by Heaven; an excessive heatwave, for example, symbolizes the emperor's inability to distinguish the good from the bad. Liu partially rejected this belief and explained that it was not the emperor's but people's actions that were directly linked to sociopolitical order. Flood and severe drought, for example, are Heaven's punishments for failures in environmental protection, caused by people's negligence and laziness.[114]

Therefore, in order for history to evolve in a positive direction, people's actions should reflect heavenly principles. Nevertheless, the normative order of heavenly principles cannot dictate people's choices. Based on this philosophical

---

[113] Brian Moloughney, "Nation, Narrative and China's New History," in *Asian Nationalism in an Age of Globalization*, edited by Roy Starrs (New York: Routledge, 2013), 216.
[114] Liu Yizheng, *Zhongguo wenhua shi* (A History of Chinese Culture) (Shanghai: Shanghai guji chubanshe, 2001), 99.

conviction, Liu contested the modernist optimism of evolution and progress and believed that China had her own trajectory of progress, which was fraught with uncertainty. To him, the Chinese principle of progress was filial piety, which went beyond the family level: "Everything that promotes moral quality, improves social costumes, scrutinizes politics and defends our territory is the manifestation of filial piety."[115] The social utility of historical research resided in the revelation of these traditional moral teachings through concrete historical figures and events. While the New Culturalists thought of Confucianism as the nemesis of modernity, Liu retorted that filial piety, a moral principle stressing the reciprocal responsibilities of the Five Relationships, could not be held accountable for national decadence. Instead, the nation's affliction was caused by "Manchu, opium addicts, corrupted officials, disreputable soldiers, scoundrels who claim to be revolutionaries, gentries and politicians who pretend to defend democracy and local rascals," none of whom followed Confucian teachings.[116] The problem therefore was not Confucianism, but the lack of Confucianism.

To restore this virtue which particularized China's trajectory of progress, Liu explained that the Chinese people should first realize that China was a Confucian nation (*ruguo*): her moral dispositions of *ren* (benevolence) and *yi* (justice, righteousness) were unknown in the Western nations of commerce (*shangguo*), whose sole interests were profits which led to the First World War.[117] While he believed scientific technologies and industrialization to be the progress of civilization,[118] the outbreak of the Great War seemed like a tragic melodrama of what scientific progress could do in the wrong hands. It revealed the inherent defects in the Western cultural, economic, and political model which Confucian values could overcome.[119]

To rematerialize Confucian virtue, Liu proposed to re-institutionalize village self-government (*xiangzhi*). Practiced across several centuries, this system created a network of mutual aid in which each member's morality was constantly checked by others. Liu explained that, compared to the Western political system, the Chinese village self-government ensured that elected officials were morally impeccable, since each candidate's conduct was under the supervision of the whole community. In addition, each region could enact laws adapted to its tradition and customs, allowing the rule of virtue to take precedence over the rule

---

115 Liu, *Zhongguo wenhua shi*, 90–3.
116 Liu Yizheng, "Lun Zhongguo jinshi zhi bingyuan" (On the origin of the badness of modern China), *Critical Review* 3 (1922).
117 Liu Yizheng, "Huahua jianbeishi" (On the Sinicization), *Critical Review* 16 (1923).
118 Liu, *Zhongguo wenhua shi*, 910.
119 Liu, "Huahua jianbeishi."

of law. To him, the rule of law, while necessary, could easily be turned into an instrument of political manipulation. His suspicion of the rule of law was largely due to the fact that in the decade after the foundation of the Republic, multiple modifications of the Constitution were made for sole benefit of the holders of power.[120]

Village self-government also inspired Zhang Shizhao. A former liberal invested in initiating an English-style parliament in China, Zhang was disillusioned by the West after the First World War and turned to China's cultural tradition for a better political solution. As Liu, he addressed the agricultural character of the Chinese nation and considered rural self-government to be the answer to the nation's crisis. Opposing capitalism and Western democracy, he saluted the local self-government of rural communities as a system built upon solidarity, virtue, and tradition.[121]

Antimodern Conservative Achievers

None of the aforementioned antimodern conservatives had concretized their culturalist political program. Liang Shuming (1893–1988) and James Yen (Yan Yangchu, 1893–1990) were among the few who had in fact materialized their political projects, although neither managed to extrapolate their political programs to the national level. While some research portrayed James Yen and Liang Shuming as social reformists rather than conservatives who glorified the old rural social and economic system,[122] it should be underlined that social reformists were not necessarily contradictory to conservatives, and conservatism translates by no means into a mere glorification of the good old time while resisting reforms.

Liang was a self-taught philosopher and the son of the Qing official Liang Ji (1858–1918) who drowned himself after the collapse of the Qing dynasty. In his famous book *Dongxi wenhua jiqi zhexue* (*Eastern and Western Cultures and Their*

---

120 Liu Yizheng, "Zhongguo xiangzhi zhi shangdezhuyi" (On the Veneration of Morality in Chinese Rural Self-government), *Critical Review* 17 (1923).
121 Zhang Shizhao, "Zai Shanghai Jinan daxue shangke yanjiang ouyou zhi ganxiang" (Speech on the Voyage to Europe at the Department of Business in the Jinan University in Shanghai), in *Zhang Shizhao quanji*, Vol. 4 (Shanghai: Wenhui chubanshe, 2000), 159–60; "Wenhua yundong yu nongcun gailiang" (Cultural Activities and Rural Reforms), in *Zhang Shizhao quanji*, Vol. 4, 144–6; "Nongcun zizhi" (Rural Self-Government), in *Zhang Shizhao quanji*, Vol. 4, 147–50; "Zhuzhong nongcun shenghuo" (On the Importance of Rural Life) in *Zhang Shizhao quanji*, Vol. 4, 151–2.
122 Zheng Dahua, "Guanyu minguo xiangcun jianshe yundong de jige wenti" (Several Issues on the Rural Construction Movement in the Republic Era), *Shixue yuekan* 2 (2006): 52.

Philosophies), Liang Shuming proposed a theory that China, India, and the West had developed three different cultures that oriented three different political paths and attitudes. The Western type revolved around science and democracy, which aimed to conquer and dominate the world of nature. Indians had developed a will to withdraw from the world and eliminate all desires. The Chinese culture was situated between Western and Indian. It was guided by the principle of introspective balance to live harmoniously with the world of nature.[123]

Liang put these three types of culture into an evolutionist scheme, believing that Western culture would one day be "Easternized."[124] But at that time, if China and Chinese culture were to survive in the present world, they had to be reformed. To him, the revival of the nation depended on the return to the political arrangements consistent with China's agrarian cultural tradition. This project would partly depend on Western technologies, which, however, should not lead China to become a capitalist and industrial nation. Liang launched his experiment of rural reconstruction from 1931 to 1937 in the county of Zouping in the province of Shandong.

For him, capitalism, democracy, and science could not be adopted in China due to the cultural differences between China and the West.[125] He did not oppose collaboration between rural areas and capitalist institutions such as modern banks, believing that loans from banks who invested in rural areas would facilitate the rejuvenation of rural China.[126] However, he warned against a total Westernization and urbanization because the latter destroyed the "old ethos" (*lao daoli*) of the rural areas.[127] But it seems that he was not worried that Chinese culture was in danger of extinction in the face of Westernization. After all, more than 80% of the Chinese population still lived in the countryside and there was no real city in China. Shanghai, for example, was to him just a place for leisure, while authentic Western cities prospered in commercial and industrial activities.[128]

To revive the countryside, Liang counted on the self-consciousness (*zijue*) and the rural associations of the peasants. He criticized the Chinese for being

---

123 Liang Shuming, "Dongxi wenhua jiqi zhexue" (Eastern and Western Cultures and Their Philosophies), in *Liang Shuming quanji* (Jinan: Shandong renmin chubanshe, 2005), 321–548.
124 Fung, *The Intellectual Foundation of Chinese Modernity*, 74–5.
125 Liang Shuming, *Zhongguo minzu zijiu yundong zhi zuihou juewu* (Final Awakening of the National Self-Salvation Movement of the Chinese Nation) (Shanghai: Shanghai shuju, 1933), 97.
126 Liang Shuming "Xiangcun jianshe dayi" (On the Rural Construction), in *Liang Shuming quanji*, Vol. 1, 603.
127 Ibid., 613.
128 Ibid., 608.

Figure 4.2: Liang Shuming and his colleagues in Zouping. Liang Shuming sits at the center. From Liang Shuming, *Xiangcun jianshe lilun* (Theory on the Rural Construction). Beijing: Zhonghua shuju, 2018.

too egoistical to be involved in a communitarian life; this needed to be changed.[129] Liang further pointed out that Chinese rural reconstruction was entirely different from the Western political system. In the West, universal suffrage and individualism were the major themes of political life. However, both these themes ran counter to China's traditional morality of respecting the elders and promoting public goodness.[130] The Five Relationships which designated those between sovereign and officials, father and son, husband and wife, brothers and friends had to be retained, with the relationship between sovereign and officials being replaced by the relationships between peasants.[131] By reviving the Chinese traditional *xiangyue* (rural contractual arrangement), he reasoned that a rural unity would be composed of four types of people: *xuezhong* (peasants), *xuezhang* (leader), *xuedong* (administrator), and *jiaoyuan* (rural activists).[132] The last three types of people played a "paternalist" role in the sense that they educated the peasants and cared for their well-being.

---

129 Ibid., 634.
130 Ibid., 656–9.
131 Ibid., 665.
132 Ibid., 676.

Furthermore, Liang's idea of peasant education was, to a certain extent, based on the folk high school and the educational theories of the Danish conservative pastor, philosopher, and teacher Nikolaj Frederik Severin Grundtvig (1783–1872), and his students Christian Kold (1816–1870), and Ludvig Schrøder (1836–1908). Grundtvig originally conceived the folk high school to educate the peasantry and other people from the lower social strata. He learned of this system from *The Folk High Schools of Denmark and the Development of a Farming Community*, written in 1927 by Holger Begtrup (1859–1937), Hans Lund (1890–1969), and Peter Manniche (1889–1981). This was translated into Chinese in 1931 by Meng Xiancheng (1894–1967), a former student of the George Washington University and the University of London. Inspired by the advantages of this educational system in which education was combined with the cooperative movement in Denmark,[133] Liang turned to Grundtvig, the ideological father of the folk high school. Liang pointed out that this educational system, which was built upon the decentralization of Danish education, freedom for teachers, and the development of people's faculties,[134] made Danish education an authentic "peasant education"[135] and "popular education", which could be appropriated into his rural reconstruction theory to enlighten the peasants.[136]

James Yen, a Christian, took a different path to rejuvenate rural China.[137] Unlike Liang, he was a graduate of Yale and Princeton and affiliated to the YMCA (Young Men's Christian Association).[138] The intellectuals that he brought into the county of Ding where he launched his political experiment from 1926 onwards were all educated in famous national, American or Japanese universities.[139] His Society for the Mass Education (*Pingmin jiaoyu hui*) aimed to tackle "the stupidity, the poverty, the weakness and the egoism" of Chinese peasants. Prior to this, he believed that "all zealous aims to bring down imperialism or imperialist

---

133 K. E. Bugge, "The International Dissemination of Grundtvig's Educational Ideas: I: Motivation and Interpretation," *Grundtvig-Studier* 1 (2012): 168–177 (176).
134 Jens Bjerg et. al., "Danish Education, Pedagogical Theory in Denmark and in Europe, and Modernity," *Comparative Education* 1 (1995): 31–47.
135 Christian Kold, whom Liang mentioned in his article, "was very close to the peasants he worked with, and taught from their perspective". See Robert Thomas Anderson, *Denmark: Success of a Developing Nation* (Cambridge, Mass.: Cambridge University Press, 1975), 109.
136 Liang, *Zhongguo minzu zijiu yundong zhi zuihou juewu*, 287–288.
137 Charles Hayford, *To the People: James Yen and Village China* (New York: Columbia University Press, 1990), 16.
138 Cheek, *The Intellectual in Modern Chinese History*, 62.
139 Li Jinzheng, "Yan Yangchu yu dingxian pingmin jiaoyu shiyan" (Yan Yangchu and the Experiment of Popular Education in the County of Ding), *Twenty-First Century* 85 (2004): 66–7.

capitalism were nothing but a shallow slogan."[140] With more than 400 schools, the incidence of illiteracy had diminished among the young by 34% in 1934. Furthermore, he introduced modern technologies and medical science to the county, which greatly improved agricultural activities and the health of the peasants.[141]

On the other hand, James Yen put great emphasis on the leisure of the peasants. His encouragement of the performance of *yangge* (rice sprout song) during festivals and work greatly contributed to the research on popular literature, since the local government intended to ban this practice on the grounds that it was offensive to public decency.[142] Although the Second Sino-Japanese War was approaching, James Yen still believed that "a nation built of counties reformed one at a time" was still possible.[143] In 1929, Yen asserted that "the main work of the movement in the field of training citizens is the extension... of the political institutions operating for centuries in the 'village republic,' to the larger units of the district community and the nation."[144] The project now sounded reminiscent of the late Qing slogan of "love for the nation begins with love for one's native place." Yen's project was cut short by the Second Sino-Japanese War.

Even if rural reformers had had sufficient funds and the war had not happened, it seems that their efforts were doomed to fail. One the one hand, Liang Shuming realized that there was never a real connection between peasants and intellectual "enlighteners." Once the peasants could no longer attain full benefit from his reform, they quit.[145] The same thing happened in the county of Ding; the patience to persuade villagers to undertake self-reform wore off with time.[146] On the other hand, as Mou Zongsan (1909–1995) pointed out, without evident political and economic progress and a strong state sponsorship, their projects were bound to fail.[147]

---

140 Ibid., 68–9.
141 Ibid., 70.
142 Zhao Jinli, "Lun Yan'an de xin yangge" (On the New Rice Sprout Song of Yan'an), in *Xindai Zhongguo* (Modern China), Vol. 6, ed. Chen Pingyuan (Beijing: Beijing daxue chubanshe, 2005), 111.
143 Kate Merkel-Hess, *The Rural Modernity: Reconstructing the Self and State in Republican China* (Chicago and London: University of Chicago Press, 2016), 125.
144 Ibid., 133.
145 Guy S. Alitto, *The Last Confucian: Liang Shuming and the Chinese Dilemma of Modernity* (Berkeley: University of California Press, 1986), 276–8.
146 Merkel-Hess, *The Rural Modernity*, 13.
147 Ady Van Den Stock, *The Horizon of Modernity: Subjectivity and Social Structure in New Confucian Philosophy* (Leiden and Boston: Brill, 2016), 174–5.

Different Views, Same Dilemma

Despite their different views with respect to the nation's political future, liberal conservatives and antimodern conservatives faced the same dilemma: the difficulty of bringing their projects to the national level. Liberal conservatives advocated a selective preservation of updated moral tradition and the social structure behind it to inform a less authoritarian regime. But the time had passed for liberal conservatives to take the initiative and to concretize their moral ideals. They had to wait until the Nanjing decade (1927–1938) for the KMT to politicize their cultural ideals. However, as I will show in the next chapter, the authoritarian conservatism that the KMT led further tarnished the name of conservatism and the traditional morality to which liberal conservatives adhered. Antimodern conservatives refused both the KMT's top-down rural reconstruction and the CCP's rural revolution in the framework of class struggle. However, the trouble remained that the educated elite found it difficult to connect with the peasants who sought immediate benefits. When they did connect, as in the case of James Yen, the locality that they chose became more of an exhibition spot for foreigners than a real model for the whole nation to follow. As Merkel-Hess observes, "Foreign visitors... treated the whole of China as a social laboratory; they saw it not as a place where particularly Chinese policies were being developed but instead as a blank canvas of undevelopment where universal solutions could be perfected."[148] Nonetheless, the real question concerns whether their projects were realistic for a country in desperate need of industrial investment and a functioning economy. Furthermore, how could they ensure that the masses would accept their guidance, both politically and morally? Their experiences proved that they could not urge the peasants to effectively participate in their experiments. While they celebrated the fact that most of China was still a virgin land of capitalism and industrialization, they had to accept that cultural elites were no longer leaders of the nation able to influence the political decision-making process, even if they still claimed to represent the people.

---

148 Merkel-Hess, *The Rural Modernity*, 146.

5 Philosophical and Authoritarian Conservatism (1920s–1940s)

China in the 1920s was volatile and nationalist. A major milestone early in this decade was the creation of the CCP in 1924 in Shanghai by Chen Duxiu and Li Dazhao. Three years after its foundation, the CCP allied with KMT and formed the First United Front with the aim to end warlordism and unify China. The cooperation lasted for three years and ended when Chiang Kai-shek (the successor of Sun Yat-sen and the leader of the KMT) bloodily purged the Communists from the United Front while the Northern Expedition was only half completed. Those who survived the massacre escaped to the mountains in Hunan and Jiangxi, where Mao Zedong (1893–1976) founded the Chinese Soviet Republic in 1931. Turning against the KMT and receiving orders from the Kremlin, the CCP had become Republican China's biggest opposition party.

The Northern Expedition ended the warlordism in 1928. The capital of the Republic was relocated from Beijing to Nanjing. From the reunification of the nation to the outbreak of the Second Sino-Japanese War in 1937, the CCP and the KMT government entered into the First Chinese Civil War. The Japanese invasion temporarily suspended the Civil War, which was only followed by the Second Civil War after the defeat of Japan in the Second Sino-Japanese War. In 1949, the CCP took over mainland China and expelled the KMT to Taiwan and several small islands of the Fujian province.

By the 1920s, the New Culturalists' domination of the intellectual sphere became increasingly pronounced and a considerable number of the New Culturalists became sympathizers of the CCP. The *New Youth*, for example, was turned into the official body of the CCP in 1924, with its founder Chen Duxiu converted to socialism. Unlike the socialism advocated during the 1900s, the re-emergence of socialism in the 1920s was greatly informed by the October Revolution, Marxism, and Leninism. This political orientation escalated the iconoclasm of the New Culture Movement. Even more remarkable is the Anti-Christian Movement, which from 1922 to 1927 led to a general attack on all sorts of religions or popular beliefs, including Confucianism and Buddhism. Vernacular language also dominated the major print media, with publications in classical Chinese gradually retreating to the private sphere. *Fengjian*, which designates the "feudalism" of the Zhou dynasty, was now used with a pejorative nuance to depreciate everything from the past, with any attempts to restore it denounced as reactionary.

Conservatives from the 1920s to the 1940s were thus confronted with two entities: the denial of the political usage of traditional culture and the rising force of socialism. The two seemed distinct but were in fact interconnected, since traditional culture was belittled as feudal and inconsequential by socialists. It was in this context that the last two types of conservatism emerged in Republican China – the philosophical conservatism and the authoritarian conservatism. My account of philosophical conservatism draws from Jan-Werner Müller's theory:

> [P]hilosophical (or also anthropological) conservatism... is rooted in a particular philosophical anthropology or perhaps social ontology. This stance implies a commitment to realizing a set of substantive values, irrespective of whether these values are already instantiated in the present. In other words, for philosophical conservatives, the primary question is not about what the past suggests, or how, or by which proven method, these values should be implemented. The question is of course what sets of values we are talking about in this context. I claim that philosophical conservatives are primarily invested in the importance of hierarchical relationships, or some more or less naturalized conception of inequality. They do not simply emphasize the particular and the potential importance of its preservation; they attribute differential value to particular sets of human beings, and they emphasize that certain social arrangements distributing power unequally are unalterable.[1]

Furthermore, unlike the liberal conservatives and antimodern conservatives, philosophical conservatives never conceptualized a political ground in which their preservation of value could take root. As I shall show in the following paragraphs, there is no doubt that philosophical conservatism is a political conservatism, but its proponents were usually so attached to the value itself that they did not manage to conceive a political project or institution to materialize their cultural ideals. Members of the Critical Review Group were typical philosophical conservatives. After the dissolution of the National Essence School and the Southern Society, the Critical Review Group was the only intellectual association that claimed to preserve the national essence – a concept already stigmatized in the New Culture period. The association was founded in 1921 by several students educated in the United States who advocated a gradual reform of Chinese culture by incorporating it into New Humanism, a philosophy developed by the American scholar Irving Babbitt (1865–1933) who had taught several members of the group.

By the time that the Critical Review Group was disbanded in 1933, China's political scenario had drastically changed. One of the most remarkable events of the late 1920s was the victory of the Northern Expedition in 1928. This military campaign was launched by the KMT and aimed to reunify China which had become fragmented after the Revolution of 1911 and the death of Yuan Shikai. China entered "the Golden Decade" of strong economic growth and relative po-

---

1 Müller, "Comprehending conservatism," 363.

litical stability. But the Golden Decade was also marked by conflict and fragility. Intellectuals were disenchanted with the nation's experimentation with republicanism and by its backward economy. Their aspiration was for nationalism together with a strong state. This they considered indispensable in tackling the increasing strength of communism and foreign aggressors.[2] Many, including some former liberals, began to condone what they called a "new type of dictatorship" (*xinshi de ducai*). From this emerged an authoritarian conservatism, which transformed philosophical conservatives' cherished values into an ideological weapon to discipline the nation and to fight the CCP and the Japanese invasion.

Wu Mi and the Critical Review Group

Born in 1894 to a well-off family in Shanxi, Wu Mi did not receive a systematic education in the Confucian classics. Instead, following his family to Shanghai when the Imperial family escaped to the capital of Xi'an, Wu Mi recalled that his "ABC books" at the time were Shanghainese magazines and journals.[3] When he returned to Shanxi, he received a modern education in English, Western history, and natural sciences. He even had the chance to read the *Guocui xuebao* and the *Minbao* at school, although neither interested him and he found many texts boring and useless to revolution.[4]

In 1900, he was selected by the province to study in the United States. For this purpose, Wu arrived in Beijing in 1901 and began his studies in the Tsinghua School (*Qinghua xuetang*) – the preparatory school for students selected to study in the United States. The creation of this scholarship was related to the Boxer Rebellion. The Qing Court paid an indemnity to the Eight Nations to appease the Rebellion. The Chinese ambassador in the United States Liang Cheng (1864–1917) negotiated with the United States government so that the latter paid back the supplementary indemnity to China. This was agreed between Theodore Roosevelt (1858–1919) and the Congress on the condition that the reimbursement would be used to fund a scholarship program to sponsor Chinese students to study in the United States. To facilitate this, the Tsinghua School, which later became the famous Tsinghua University, was created to prepare young Chinese stu-

---

[2] Lloyd E. Eastman, *The Abortive Revolution: China under Nationalist Rule*, 140–80.
[3] Wu Mi, *Wu Mi zibian nianpu* (Wu Mi's Self-Compiled Chronological Biography) (Beijing: Sanlian shudian, 1998), 29–30.
[4] Ibid., 81–2.

dents for their trip to the United States.⁵ At the Tsinghua School, Wu made the acquaintance of two of his future colleagues of the Critical Review Group: poet Wu Fangji (1896–1932) and historian Tang Yongtong (1893–1964). The latter was sent to study philosophy in the Hamline University in 1918 before transferring to Harvard in 1920, where he studied Sanskrit and Pali languages under the supervision of Charles Rockwell Lanman (1850–1941).⁶

Wu Mi went to the United States in 1917. His initial intention was to study journalism so that he could "promote national civilization, provide communication between the West and China, shape popular customs, enhance moral norms and guide the society" through print media.⁷ But he was persuaded by Zhou Yichun (1883–1958), president of the Tsinghua School, to study literature at the University of Virginia – a discipline more suited to his character.⁸ During the summer vacation in 1908, he made a trip to the east coast – a trip that would change his life. Through a common friend, he made the acquaintance of Mei Guangdi. Originally from the province of Anhui, Mei obtained the degree of *shengyuan* at the age of 12. Achieving the same scholarship as Wu, he was enrolled in 1911 at Northwestern University to study literature. When he met Wu Mi, he was a master's student at Harvard and his supervisor was Irving Babbitt. Mei had known his future intellectual rival Hu Shi for a long time. In a letter sent to Hu in 1911, Mei considered Hu to be his intellectual comrade and criticized vehemently the National Essence School and the *Guocui xuebao*, claiming that this magazine was nothing but a collection of repetitions of purposeless commentaries on ancient learning and that the defenders of the national essence were a bunch of "laughable and pathetic" people.⁹ He finally broke with Hu over the question of vernacular language. When he met Wu in 1908, he was attempting to rally students of the same mindset to challenge the New Culture Movement on the other side of the Pacific Ocean.¹⁰ Wu was deeply touched by Mei's enthusiasm and promised to join his league straightaway. Mei subsequently intro-

---

5 Ye Weili, *Seeking Modernity in China's Name: Chinese Students in the United States, 1900–1927* (Stanford: Stanford University Press, 2001), 1–16.
6 "Tang Yongtong nianpu jianbian" (A short biography of Tang Yongtong), in *Tang Yongtong quanji* (Complete Works of Tang Yongtong), Vol. 7 (Shijiazhuang: Hebei renmin chubanshe, 2000), 662–683.
7 Wu, *Wu Mi riji: 1910–1915*, 410.
8 Wu, *Wu Mi zibian nianpu*, 149.
9 Mei Guangdi, "Di Shiliu han" (Letter no. 16), *Mei Guangdi wenlu*, 139.
10 Wu, *Wu Mi zibian nianpu*, 177.

duced Wu to his mentor, and Wu decided to transfer to Harvard in September to pursue his studies, also under the supervision of Babbitt.[11]

At Harvard, Wu befriended Chen Yinke (1890–1969), one of the most eminent historians of twentieth century China. Originally from Jiangxi, he had spent 35 years in different universities in Japan, Europe, and the United States.[12] He could read a dozen modern and ancient languages, including English, French, German, Hebrew, Latin, Turkish, Mongolian, Tibetan, and Manchu.[13] Despite his erudition, he returned to China in 1925 without obtaining a diploma. Only under the recommendation of Wu Mi was he recruited to the Tsinghua Institute of Research on National Learning (*Guoxue yanjiuyuan*).[14]

Wu Mi decided not to pursue a PhD, which to him, was only "vain glory."[15] In 1921, he received a letter from Mei Guangdi, who invited him to teach the English language and literature at the National Southeast University and edit a magazine that would be published by the Chung Hwa Book Company (*Zhonghua shuju*). He accepted the offer and arrived at the university in Nanjing in September. Here he federated a number of colleagues to edit the magazine, including Liu Yizheng, the translator Xu Zhen'e (1901–1986), Xu Zeling (1896–1972), graduate of the Illinois University, Xiao Chunjin (1893–1968), economist trained at the University of California, Li Sichun (1893–1960), historian educated at the University of Paris under the supervision of Charles Seignobos (1854–1942), Hu Xiansu (1894–1968), doctor of botany from Harvard, and Liu Boming, who obtained a PhD degree at Northwestern University with a dissertation on the thoughts of Lao Zi under the supervision of Horace Craig Longwell (1876–1929).[16]

The Critical Review Group and the New Culture Movement

Studies on the Critical Review Group tend to regard it as being in opposition to the New Culture Movement due to its implication in the resurrection of Confu-

---

11 Ibid.
12 Axel Schneider, *Zhenli yu lishi: Fu Sinian, Chen Yinke de shixue sixiang yu minzurentong* (Truth and history: historical thought and national identity of Chen Yinke and Fu Sinian), trans. Guan Shan and Li Maohua (Beijing: Shehui kexue wenxian chubanshe, 2008), 22.
13 Wang Cheng-bon, *Duli yu ziyou: Chen Yinke lunxue* (Independence and Liberty: Chen Yinke's Scholarship) (Taipei: Linking, 2011), 66.
14 Schneider, *Zhenli yu lishi*, 28–30.
15 Wu, *Wu Mi riji, 1917–1924*, 139.
16 Mei Guangdi, "Jiunianhou zhi huiyi" (Nine Years after the Death of Liu Biming), *Guofeng banyuekan* 9 (1933): 73.

cianism that had already been stigmatized by the mainstream intellectuals. However, the intellectual quest of the Group was in fact a part of the New Culture Movement since New Humanism, on which the Group elaborated its conservatism, was also a new foreign philosophy to the Chinese public at the time.[17] Most of its members were trained overseas and only Liu Yizheng and his students were devoted to research national learnings. Wu Mi recalled that the magazine *Critical Review* did not have enough texts to publish, because of the lack of articles on national learnings.[18] In the National Southeast University, the intellectual association which specialized in studies on national learning was the Society for the Research on National Learnings (*Guoxue Yanjiuhui*). This society did not collaborate with the Critical Review Group. In fact, to a certain extent, the two groups were rivals. While the Society for the Research on National Learnings openly opposed the Critical Review Group, the latter published an article suggesting that the Society had a narrow view on national learning.[19]

Nevertheless, the Critical Review Group did attempt to challenge the New Culture Movement as it was led by the New Culturalists. They tried to convince the public that their particularistic cultural views acted in the public interest by appropriating the notion of disinterestedness, a concept introduced in China by Miao Fenglin (cf. Chapter 3). The first issue of the *Critical Review* published Mei Guangdi's provocative statement that the New Culturalists were not disinterested intellectuals, but "sophists," "imitators," "opportunists," and "politicians," who seized every chance they could to sacrifice high culture to please the young and the stupid and to gain financial benefits.[20] Wu Mi also maintained that the only reason the vernacular language replaced the classical language to become the national language was because of the New Culturalists who had infiltrated the government.

The Critical Review Group deplored the "cult of newness" that haunted the Chinese intellectual sphere at the time. For its members, the New Culturalists knew little about authentic Western learning and were content to introduce

---

17 Yü Ying-shi, "Neither Renaissance or Enlightenment," in *The Appropriation of Cultural Capital: China's May Fourth Project*, ed. Milena Dolezelova-Velingerova and Oldřich Kral (Cambridge, Mass.: Harvard University Press, 2002), 299–324.
18 Wu, *Wu Mi riji, 1917–1924*, 248.
19 Chen Zhongfan, "Zizhuan" (Autobiography), in *Zhongguo dangdai shehuikexuejia* (Chinese Modern Researchers on Social Sciences), ed. Beijing tushuguan *Wenxian* congkan bianjibu and Jilin sheng tushuguan xuehui huikan bianjibu, Vol. 1 (Beijing: Shumu wenxian chubanshe, 1982); Sun Deqian, "Ping jin zhi zhiguoxuezhe" (On Researchers on National Learning), *Critical Review* 23 (1923).
20 Mei Guangdi, "Ping tichang xinwenhuazhe" (On the New Culturalists), *Critical Review* 1 (1922).

the Chinese public to the most superficial knowledge of the West. Mei resumed their credo in these words: "When it comes to economics and politics, they only cite Russia and Marx; when they talk about philosophy, they only know pragmatism; as to Western literature, they knew nothing more than novels, plays and the most recent decadent works."[21] Therefore, Mei criticized that the New Culturalists were simply frauds who fooled the Chinese public who were not capable of reading Western books and obtained admiration and fame among the young generation by disseminating insolent opinions and ideas.[22] The New Culture Movement needed to be reorganized. For Wu Mi, the first step towards a new inauguration of the New Culture Movement was to comprehend the concept of culture:

> The *Xinwenhua yundong* is translated by the New Culture Movement. The term *wenhua* is thus considered as the equivalent of culture. According to Matthew Arnold, culture is the best of what has been thought and talked about in the world. Thus, to construct a new culture in China, we should naturally select the essence of Chinese and Western culture, establish a dialogue between the two and finally fusion the two. Chinese culture is based on Confucianism and completed by Buddhism, while the Western culture is the fruit of the coalescence of Greco-Roman culture and Christianity. In order to conceive a new culture, ancient culture should first be studied... The New Culture Movement as it is today rejects and despises the vital essence of Chinese and Western cultural essence... in favor of the newness.[23]

The Critical Review Group was unique in the sense that its seemingly "regressive" cultural orientation was guided by a foreign philosophy. In denouncing its intellectual rivals for submitting to the most recent Western thinking, the group did not escape the "cult of the West" that conditioned the Chinese intellectual realm of the time. Indeed, Wu Mi had expressed his frustration that nobody understood the Western origin of his intellectual trajectory, claiming that he was not influenced by Confucianism, but by the Ancient Greek civilization, Christianism, and New Humanism: "I absorbed the essence of the Western spirit through Greek philosophy and Christianity, thanks to which I came to understand the grandeur of Chinese culture and the greatness of Confucianism."[24]

---

[21] Mei Guangdi, "Ping jinrin tichang xueshu zhi fangfa" (On the Way How People Disseminate Knowledge Today), *Critical Review* 2 (1922).
[22] Wu Mi, "Lun Xinwenhua yundong" (On the New Culture Movement), *Critical Review* 4 (1992).
[23] Ibid.
[24] Wu Mi, "Kongxuan shihua" (Literary Critique of Kongxuan), in *Wu Mi shihua* (Literary Critique of Wu Mi) (Beijing: Commercial Press, 2007), 214–5.

Modernism at the time was translated by *congxin zhuyi* (newnessism).²⁵ To assert that they were the real guides of Chinese society, members tried to convince the Chinese public that they brought the most recent Western thoughts into China. The default position of the "cult of newness" does not reside so much in the admiration of the West per se, but rather in the public's incapacity to distinguish what was really new. Thus, Hu Xiansu attempted to convince readers that the heroes of the New Culture Movement, be it Rousseau, Marx or John Dewey (1859–1952), were all demoded in the West, while the New Humanism represented the most exact, the most informed, but more essentially the newest Western philosophy.²⁶ To this end, the Critical Review Group ignored the fact that Babbitt's philosophy was deeply rooted in American society and highlighted that he was extremely familiar with the Chinese situation – in summary, a sage that had the solution to China's crisis.²⁷

This does not imply that Wu Mi's admiration for Babbitt was artificial. In fact, Babbitt was his idol, to whom he showed great respect. In a letter addressed to his professor in 1925, Wu Mi affirmed that the *Critical Review* was created to disseminate Babbitt's ideas and that he considered the translation of his professor's works to be a sacred duty.²⁸ In another letter, Wu Mi wrote: "After having been your disciples, we are like those who, once having a vision of God, would not live contented a life of the Earth, and would have no other gods!"²⁹ He also constantly asked Babbitt's opinion on the publication of the magazine because Babbitt was to him "the one living wise man to whom [he] can look up for intellectual guidance and the conduct of life."³⁰ Pessimistically, he also recounted the situation in China, which he believed was getting worse every day:

> The conditions in China went from bad to worse in the last two years since my return. The country is just now facing an extremely serious political crisis, both internal and foreign. I cannot but be grieved to think that the Chinese people have decidedly degenerated, so that the observations on our national character drawn from history and our past excellencies do not at all fit with the Chinese of today. And I believe, unless the mind and moral character of the Chinese people be completely reformed (by a miracle or a Herculean effort), there is

---

25 Luo Zhitian, *Quanshi zhuanyi: jindai zhongguo de sixiang, shehui yu xueshu* (Shifting of Power: Ideology, Society, and Scholarship in Modern China) (Wuhan: Hebei renmin chubanshe, 1999), 63.
26 Hu Xiansu, "Baibide zhongxi renwen jiaoyu tan" (Babbitt on the Humanist Education in China and in the West), *Critical Review* 3 (1922), 2.
27 Hon Tze-ki, "From Babbitt to 'Bai Bide'" in *Beyond the May Fourth Paradigm: in Search of Chinese Modernity*, ed. Kai-wing Chow et al. (Lanham: Lexington Books, 2008), 258.
28 Wu Mi, *Wu Mi shuxinji* (Letters of Wu Mi) (Beijing: sanlian shudian, 2011), 31.
29 Ibid., 50.
30 Ibid., 39.

no hope even for a political and financial regeneration in the future. Of course, we must work to make a better China; but if no success, then the history of China since 1890 will remain one of the most instructive and interesting pages in the history of the world, with reference to national decadence.[31]

To Wu Mi, "if there are others in China interested in translating [Babbitt's] books (how poor the translation may be), China would never have fallen into the present abyss of material and spiritual decadence."[32] While Wu told his Chinese readers that his mentor had read all works published in China and made New Humanism the spiritual inspiration of the new culture and social reforms, Babbitt had confessed to him that he "[did] not feel qualified to have an opinion."[33] Nevertheless, not only was Babbitt well aware of Wu's intention to evoke his philosophy to combat the proponents of the New Culture Movement, he even suggested that Wu "publish notices of John Dewey's last two volumes of a kind that will expose his superficiality," because Dewey had been exercising a bad influence in the United States and, as Babbitt suspected, in China too.[34] Hence, to a certain extent, the debate between the Critical Review Group and the New Culturalists constituted a continuation of the intellectual rivalry between Babbitt and Dewey in China. Babbitt himself was confident that his New Humanism could engage in dialogue with Confucianism to challenge Dewey and his pragmatism that prevailed in American society at the time.

Babbitt In and Out of China

Babbitt launched the philosophical movement of New Humanism to challenge America's popular democracy and materialist modernity. He developed this philosophy when he was teaching at Harvard. Although he was a professor of the French language and comparative literature, he deeply detested the French language and despised the so-called intellectuals that emerged in France after the Dreyfus Affair. In a letter addressed to his friend Paul Elmer More (1864–1937) during Babbitt's stay in Paris in 1908, he wrote:

> In all purely intellectual ways, I always find this contact with Paris and with France immensely stimulating. My opportunities are unusually good because a number of the friends

---

31 Ibid., 16.
32 Ibid., 32.
33 Wu Xuezhao, "The Birth of a Chinese Cultural Movement: Letters Between Babbitt and Wu Mi," *Humanitas* XVII, no. 1 and 2 (2004): 12.
34 Ibid., 13.

of my student days are now professors at the Sorbonne and the Collège de France, and through them I see a great many of the so-called *intellectuals*. This whole group strikes me as extremely confident, not to say cocky at the present. The reactionary elements ever since the later stages of the Dreyfus affair have been in a state of demoralization. The Revolution has passed from the acute to the chronic state and nearly everybody I meet has something of the toploftiness that one often finds in people whose ideas encounter no effective opposition. Everybody is proud of *les expériences sociales* that France is making for the benefit of the world. These *expériences* so far as I can see may be reduced to one, the attempt of a great nation to dispense utterly with everything that has been traditionally recognized as religion and to offer social sympathy as a substitute.[35]

Babbitt was not content to teach literature produced in such an environment: "I might be willing to sacrifice myself for Greek but I am not going to turn myself into a teaching automaton in order that Harvard sophomores may read French novels of the decadence in the original."[36] Unfortunately for him, he did not have a choice. His professional life was to a certain extent a deception for him. Always aspiring to a post in the department of classical studies, he had to teach the curriculum that he despised because of the "elective system" that the president of the Harvard put forward which allowed students to freely choose their courses with no obligation to learn classical studies. For him, this manifested the idea that the American society at the time was dominated by what he called romantic humanitarianism and scientific humanitarianism that could be traced back to Jean-Jacques Rousseau and Francis Bacon (1561–1626), respectively.

Babbitt reproached Rousseau for accusing society of corrupting innate human love and piety, leading poor men to believe that they were victims of a social conspiracy, instead of introspecting their own soul and conscience.[37] Sentimental humanitarianism is heavily invested in undermining the foundation of traditional education and civilization as constraining human nature and emancipation, leading to nothing but sentimental anarchism, excessive individualism, and destruction of standards.[38] As for scientific humanitarianism, Babbitt viewed it as a doctrinal tool aimed at maximizing materialistic gains, transforming men into instruments serving that end and subjecting men to what he called

---

35 Thomas R. Nevin, *Irving Babbitt: An Intellectual Study* (Chapel Hill: University of North Carolina Press, 1984), 54.
36 Ibid., 12.
37 Irving Babbitt, *Democracy and Leadership* (Boston and New York: Houghton Mifflin Company, 1934), 77.
38 Ibid., 2.

"law for things".³⁹ Both humanitarianisms misunderstand human nature. While sentimental humanitarianism exaggerates the inner goodness of human beings, scientific humanitarianism magnifies the badness. In the university, these two humanitarianisms were fully expressed through the elective system, in which only students' interests mattered, and students were trained as robots for society's gain. What is missing in these two humanitarianisms is the spirit of moderation that is of paramount importance for New Humanism:

> We have seen that the humanist, as we know him historically, moved between an extreme of sympathy and an extreme of discipline and selection, and became humane in proportion as he mediated between these extremes. To state this truth more generally, the true mark of excellence in a man, as Pascal puts it, is his power to harmonize in himself opposite virtues and to occupy all the space between them (*tout l'entredeux*)⁴⁰. By his ability thus to unite in himself opposite qualities man shows his humanity, his superiority of essence over other animals.⁴¹

Men are born with inner dualism and a propensity to succumb to inferior instincts, but only the humanists understand the law of measure and achieve a higher self. The capacity to "unite in himself opposite qualities" constitutes what Babbitt calls inner check, a spontaneous moral uneasiness preventing a moral man from doing wrong when such an impulse emerges.⁴² Babbitt's philosophy is not a religious one. To him, inner check acquired through religious standards is motivated by an exterior authority, while this power of veto is supposed to be completely spontaneous.⁴³ For this reason, Babbitt proclaims his philosophy as entirely individualist. By the same token, he refused the label of conservative, claiming that his insistence on humanist spontaneity made him truly modern, because "the modern spirit is the positive and critical spirit, the spirit that refus-

---

39 Irving Babbitt, *Literature and American College: Essays in Defense of the Humanities* (Boston and New York: Houghton Mifflin Company, 1908), 38–9.
40 "Tout l'entredeux" here refers to Blaise Pascal's remarks: "I do not admire the excess of a virtue as of valor, if I don't see at the same time the excess of the opposite virtue, as in Epaminondas, who had the greatest valor combined with the greatest kindness, since, otherwise, it is not to rise, but to fall. We do not display our greatness by being at one end of extremity, but by touching two sides of extremity at the same time and filling all the intervening space (*tout l'entredeux*). Blaise Plaise, *Les pensées de Pascal* (Paris: P. Lethielleux, 1896), 437.
41 Babbitt, *Literature and American College*, 22–3.
42 Claes G. Ryn, *Will, Imagination and Reason: Babbitt, Croce and the Problem of Reality* (New Brunswick and London: Transaction Publishers, 1997), 150.
43 Irving Babbitt, *Rousseau and Romanticism* (Boston and New York: Houghton Mifflin Company, 1919), 58.

es to take things on authority",⁴⁴ while the "smart young radicals" were merely modernists, since their veneration of positivism, materialism and rationalism turned their pursuit of modernity into a mere dogmatism.⁴⁵ As he said: "The whole modern experiment is threatened with breakdown simply because it has not been sufficiently modern. One should therefore not rest content until one has, with the aid of the secular experience of both the East and the West, worked out a point of view so modern that, compared with it, that of our smart young radicals will seem antediluvian."⁴⁶

The true modern never stands for a radical break with the past but seeks a critical understanding of tradition as a negotiated and complex notion, useful to guide the present and the future. The inner power of veto can only be acquired through this critical understanding and actualization of tradition. The purpose of higher education is to cultivate this intuition among individuals with rigorous classical training.⁴⁷ Not everyone can become a humanist and a university education should be reserved for the selected few who will become leaders of society in the future.⁴⁸ If sympathy towards those who are not intellectually gifted is necessary, sympathy has to be combined with selection. Thus, Babbitt is highly critical of democracy, holding that it can only lead to anarchism, since the masses are doomed to illusion and momentary impulses. New Humanism, as he maintains, is "aristocratic and not democratic in its implication."⁴⁹

Although he did not understand Chinese, Babbitt believed that he truly understood Chinese philosophy, especially that of Confucius:

> Confucius is less concerned with the other world than with the art of living to the best advantage in this. To live to the best advantage in this world is, he holds, to live proportionately and moderately; so that the Confucian tradition of the Far East has much in common with the Aristotelian tradition of the Occident. In one important respect, however, Confucius recalls not Aristotle, but Christ. Though his kingdom is very much of this world, he puts emphasis not merely on the law of measure, but also on the law of humanity. He was humble both in his "submission to the will of Heaven" and in his attitude towards the sage of old. He aspired at most to be the channel through which the moral experience of his race that had accumulated through long centuries and found living embodiment in these sages should be conveyed to the present and the future; in his own words, he was not a creator but a transmitter. A man who looks up to the great traditional models and imitates them, becomes worthy of imitation in his turn. He must be thus rightly imitative if he is to

---

44 Ibid., xi.
45 Babbitt, *Democracy and Leadership*, 317.
46 Babbitt, *Rousseau and Romanticism*, xxiiii.
47 Babbitt, *Literature and American College*, 79–81.
48 Ibid., 80–1.
49 Ibid., 6.

be a true leader. No one has ever insisted more than Confucius on a right example and the imitation that it inspires as the necessary basis of civilized society.[50]

This portrayal of Confucius and Confucianism might well have been a self-portrayal of New Humanism and Babbitt himself, who was confident that the West could learn a lot from Confucianism and considered Confucius the Oriental Aristotle. He praised Confucianism as the real humanism for its emphasis on introspection, responsibility, social hierarchy, morality, and law of measure. He particularly admired Confucian morality of *ren*, rooted in a hierarchical structure, along with social order and authority. He explained that this was exactly what he meant by "sympathy tempered by selection".[51] He compared the law of measure with the Confucian idea of *ju* (rule).[52] With regard to social hierarchy and the importance of social leaders, Babbitt justified his elitism in the framework of Mencius' distinction between men who worked with their hands and men who worked with their minds. Justice demanded that men who worked with their minds guided those who worked with their hands.[53] Citing *Lunyu*, he said: "The virtue of the leader is like unto wind; that of the people unto grass. For it is the nature of grass to bend when the wind blows upon it."[54]

During one of his courses, a Chinese student told Babbitt that what he was teaching was nothing new; it was what the Chinese had known for thousands of years.[55] Particularly popular among Asian students at Harvard, he was invited to give a speech to a Chinese students' association in Boston in 1921. He warned attendees against throwing out the baby – Chinese culture – with the bath water, in the nation's quest for modernization.[56] Critical of democracy and suspicious of scientific materialism, New Humanism greatly fueled the Critical Review Group's criticism of the New Culture Movement that operated under the slogan of "Mr. Democracy and Mr. Science". Inspired by Babbitt, the Group believed the national essence of China was neither the liberal cultural elements nor the agrarian aspects of her tradition, but Confucian values of elitism, social hierarchy,

---

50 Babbitt, *Democracy and Leadership*, 34.
51 Irving Babbitt, "Humanist Education in China and the West," *Chinese Students' Monthly* 2 (Nov. 1921), 89.
52 Babbitt used the law of measure to translate the *ju* in this phrase: 七十而從心所欲, 不踰矩。 Babbitt's translation is: "At seventy I could follow what my heart desired without transgressing the law of measure." See Babbitt, *Rousseau and Romanticism*, 386.
53 Babbitt, *Democracy and Leadership*, 193.
54 Ibid., 35. The original text reads: "君子之德風, 小人之德草。草上之風, 必偃。"
55 A. Owen Aldridge, "Irving Babbitt in and about China," *Modern Age* 4 (1993): 332–339 (333).
56 Babbitt, "Humanistic Education in China and the West," 89.

and doctrine of moderation. These elements had to be reintegrated into society to save the corrupted politics. The *Critical Review* magazine was launched, "striving to performing the following functions:
(1) To interpret the spirit, and to systematize the materials, of Chinese culture.
(2) To introduce and assimilate the standard works and best ideas of Western philosophy and literature.
(3) To discuss the current problems of Chinese life, thought and education, with a sound, intelligent and critical attitude.
(4) To create a modern Chinese prose style, capable of expressing new ideas and sentiments, yet retaining the traditional usage and inherent beauty of the language."[57]

The first issue of the magazine drew parallels between Confucius and Socrates. The Critical Review Group was determined to start a debate around Chinese national essence and the best ideas of Western tradition. Since it emerged as the opponents of the New Culturalists, my account of the Critical Review Group will be structured around its criticism of the two cornerstones of the New Culturalist ideals: democracy and science.

Philosophical Conservatism and Democracy

Disappointed by the political chaos of the Republic, Liu Boming called for a national education that instilled the Confucian teaching of *zhengxin chengyi* (rectifying one's mind and making one's will sincere) – a concept of *Daxue* (*Great Learning*).[58] In promoting political reforms through Confucianism, the Critical Review Group was cautious to distance themselves from reactionaries who also resorted to Confucianism for a different political end. Liu Boming affirmed that the realization of *zhengxin chengyi* could only be achieved in a republican regime. To a certain extent, this point of view echoed Sun Yat-sen's three-staged nation-building strategy, outlined in his *Nationalist Government's Plan for National Construction* (*Guomin zhengfu jianguo dagang*, 1924). Here it was stipulated that constitutional politics (*xianzheng*) would be realized only after the dissolution of the periods of military (*junzheng*) and party rules (*xunzheng*). Nevertheless, Liu refused to legitimize KMT's self-proclaimed exclusive authority of polit-

---

57 Preface of the *Critical Review*.
58 Liu Boming, "Gonghe guomin zhi jingshen" (Spirit of the Republican Citizens), *Critical Review* 10 (1922).

Figure 5.1: Portraits of Confucius and Socrates. From *Critical Review* 1 (1922).

ical education. He insisted upon the primacy of individual moral character (*renge*), the lack of which was the root of all the problems that confronted the Republic. This distinguished him from liberal conservatives, who felt that the morality question should always be combined with the implementation of certain political institutions.

For Wu Mi, the reason that all reforms from the Self-Strengthening Movement onwards had failed was because of the decadence of citizens' moral quality. During his stay at Harvard, Wu Mi was inspired by the Protestant Reform and Renaissance. After graduating, he decided to publish a magazine named *Renais-*

sance to revitalize national spirit.⁵⁹ But when he returned to China, this name had already been taken as the English subtitle of the New Culturalist magazine *Xinchao* (*New Vague*). He was quite disgusted by the iconoclasm of this magazine.⁶⁰ On his return to China, he had to give up *Renaissance* in favor of *Critical Review*, in which he argued for a renewal of Neo-Confucianism. He evoked the concept of *keji fuli* (conquer the self and return to ritual) to translate Babbitt's inner check.⁶¹ Although he reasoned that this traditional spirit should be the cornerstone of the Chinese nation, he failed to answer how Confucian teaching could be instilled in the political nation.

Despite their inability to propose a concrete political initiative to materialize their cultural ideal, the Critical Review Group was highly critical of the major political ideologies of the 1920s, notably democracy, equality, liberty, and socialism. The very danger of democracy is the illusion of equality that leads to eventual destruction of freedom and regression of civilization. Hu Xiansu explained that the reason the French Revolution and the Glorious Revolution produced widely different results was due to differences in their pursuits. Like his mentor Babbitt, he described the French Revolution as a romantic revolution based on an unrealistic ideology of equality. Equality essentially implied a comparison with others: "People desirous of equality want to live the same way and by the same means as others even to detriment of liberty."⁶² Inevitably, the Terror that followed this revolution promised nothing but illusion. The Glorious Revolution, on the other hand, was Britain's fight for liberty. Unlike equality prescribing a general rule on the outside world, liberty, as Hu explained, requires introspection and structures entirely around an individual's mentality and choice. Hence, equality is rooted in jealousy, and liberty is rooted in the virtue of spiritual independence.⁶³ Similarly, the group conceived of socialism as a doctrine of rancor, anchored in a sort of extreme interventionism glorifying the average man's greed.⁶⁴

In the literary field, democracy translated into the abandonment of classical Chinese and the expansion of the commoners' literature, written in the vernacular language. Mei Guangdi held that this type of cultural democracy remained

---

59 Wu, *Wu Mi riji: 1910–1915*, 381
60 Wu, *Wu Mi riji: 1917–1924*, 90–91.
61 Wu Mi, "Wo zhi renshengguan" (My Conception of Life), *Critical Review* 16 (1923).
62 Hu Xiansu, "Wenxue zhi biaozhun" (Standard of Literature), *Critical Review* 31 (1924).
63 Ibid.
64 Ibid; Xiao Chunjin, "Zhongguo tichang shehuizhuyi zhi shangque" (On the Promotion of Socialism in China), *Critical Review* 1 (1922); Xiao Chunjin, "Pingdeng zhenquan" (The Real Equality), *Critical Review* 5 (1922).

nothing more than a means for opportunist intellectuals to earn an exalted reputation among the younger generation who were easily mesmerized and manipulated by their radical and iconoclastic cultural standpoints against high culture.[65] This high culture had much to do with Babbitt's ideal of an American university curriculum, in which Greek, Roman, and Eastern classics were rigorously taught to the selected few who passed beyond an animalistic nature and reached the upper standards with a view to becoming one of society's leaders. Few people have the intellectual competence to understand high culture. Trivializing the high culture that sustained Chinese civilization to feed banality and the mindlessness of the masses in the name of equality is in no circumstance acceptable. To Chang Chi-yun (1901–1985), elitism, with historical origins in Plato's *Republic* and the Zhou dynasty, had the same objective as democracy, i.e., the well-being of the people. Nevertheless, by citing Gustave Le Bon (1841–1931), he explained that only elitism could attain the goal because democracy was the government of the masses condemned to irrationality and extremity.[66] Hu Xiansu pointed out:

> Today's socialism and radicalism are all forms of extreme interventionism. In a democracy and in the framework of law, each one enjoyed the absolute freedom of thought, action, expression and religion. The most significant freedom is related to the private properties. If the means of production and the properties of material life are acquired in the framework of law, the state and the society have absolutely no right to interfere. Socialism and its diverse branches have the sole aim to stealing from others… Under a communist regime, the productivity cannot maintain the civilization, and even if it could, is it moral to steal from others? Economic inequality always exists, just like men are born unequal. As said Mencius: "difference is the natural condition… equalization perturbs the world order."… If equality is imposed… the strong will become weak, the healthy will become ill and the intelligent will be forced to become stupid… Violence of killing people of different qualities will be comparable to that of the French Revolution.[67]

The only legitimate equality is equal chance of self-development and satisfaction to the extent of one's competency and in line with one's intelligence. Xiao Chunjin pointed out that "the nature of democracy is elitism (*xianxian zhuyi*), without recognizing any claim of equality other than equality of opportunity".[68] Further-

---

[65] Mei, "Ping tichang xinwenhuazhe."
[66] Chang Chi-yun, "Bolatu lixiang guo yu Zhouguan" (Plato's Ideal country and Rites of Zhou), *Shidi xuebao* 1 (1920): 1.
[67] Hu, "Wenxue zhi biaozhun."
[68] Xiao, "Pingdeng zhenquan".

more, it was simply nonsense to talk about socialism in a country where capitalism was so poorly developed.[69]

Although the Group did not manage to turn the iconoclast intellectual trend around, it had the merit of insisting on the idea of innate evil in human nature. Western conservatism is, in general, suspicious of the notion of human perfectibility; the prevalence of evil is not due to bad political arrangements, but inherent to human nature.[70] In contrast, Confucianism generally believes in human perfectibility through cultivation.[71] The Critical Review Group surely learnt about this human imperfectability through Babbitt's "inner dualism." Therefore, their Confucian conservatism was shaped by Western conservative intellectual tradition. This idea about human imperfectability could have contributed to the New Culturalists' optimistic pursuit of political reforms during a period where ascension to power was often accompanied by corruption, but the Critical Review Group had inherited Babbitt's suspicion of the manipulation of political machinery.[72] In 1922, for example, Hu Shi and 17 other intellectuals issued a manifesto declaring that a good government should be formed by good people (*haoren*).[73] Although this manifesto did mention the constitutionalism, the Group's suspicion of human perfectibility might have persuaded them that it was not good people who built good government, but that the priority should be the appropriate arrangements of political institutions to impede the badness.

Nevertheless, their conservatism was still highly political. Wu Mi was reluctant to publish articles without political utility. Like Babbitt, he believed that without humanist utility, knowledge was useless. The twenty-fourth and the thirty-first issue of the *Critical Review* published three articles by Ye Yusen (1880–1933), a specialist in oracle bone scripts (*jiaguwen*). Wu Mi recalled several years later that he had no intention of publishing these useless research articles.[74]

This view echoed to some extent the so-called New Confucianism (*Xinrujia*) that emerged during the 1920s. Xiong Shili (1885–1968) and Liang Shuming were

---

69 Xiao, "Zhongguo tichang shehuizhuyi zhi shangque".
70 Kekes, *A Case for Conservatism*, 44–45.
71 Thomas A. Metzger, "Continuities between Modern and Premodern China: Some Neglected Methodological and Substantive Issues," in *Ideas Across Cultures: Essays on Chinese Thought in Honor of Benjamin I. Schwartz*, ed. Paul A. Cohen and Merle Goldman (Cambridge, Mass.: Council on East Asian Studies, Harvard University, 1990), 285–8.
72 Babbitt, *Democracy and Leadership*, 66.
73 Xu Jilin, "Hu Shi: Haoren zhengfu zhuyi de wutuobang" (Hu Shi: The Utopia of Good People's Government), in *Jingshen de lianyu: wenhua bianqian zhong de zhongguo zhishifenzi* (Spiritual Prison: Chinese Intellectuals in Cultural Transitions) (Taipei: Shulin, 1994), 117.
74 Wu, *Wu Mi shihua*, 239.

the first generation of the New Confucians. The second generation, which I will elaborate on in the conclusion, were all students of Xiong.[75] To Xiong, the *ti-yong* relationship was non-existent. As Edmund Fung points out: "If Chinese learning is the *ti*, it must be useful and functional; otherwise, it must seek help from Western learning. If *ti* is totally devoid of function, it is a dead body."[76] During the 1920s and 1930s, only Liang Shuming found a way to integrate *ti* and *yong*, as I have shown in the precedent chapter. New Confucianism had to wait until the 1960s to be fully developed outside mainland China.

Philosophical Conservatism and Science

The Group's insistence on elitism also translated into their distrust of science and its material application – industrialization. Conservatives' wariness of science could certainly be placed at a philosophical level, as was clearly shown by the debate on "science versus metaphysics" in the early 1920s.[77] In 1919, Liang Qichao and Zhang Junmai (1886–1969) traveled to Europe and witnessed the disastrous consequences of the First World War. Shocked by the atrocious effects of the War on the once splendid Europe, they rediscovered Confucianism as the solution to Western materialistic science, which they believed had plunged Europe into warfare. After his exchanges with Henri Bergson (1859–1941), Liang questioned the all-powerful science eulogized by the New Culturalists and appraised the values of Chinese culture in remediating Europe's dream of the mighty power of science. Zhang stayed in Europe to study with Rudolf Eucken (1846–1926), Henri Bergson, and Hans Driesch (1867–1941). For him, a modern society could not be sustained solely by scientific rationalism without regard for what Bergeson called *élan vital*.[78] Within the framework of the Confucian optimism regarding an individual's internal spirit, Zhang explained *élan vital* as the foundation of human endeavors.[79] In 1922, Hans Driesch went to China to lecture, and Zhang served as his interpreter. One year later, his rejection of science

---

75 Fung, *The Intellectual Foundations of Chinese Modernity*, 90.
76 Ibid., 91.
77 Zhang Junmai et al., *Kexue yu renshengguan* (Science and the Conception of Life) (Jinan: Shandong renmin chubanshe, 1997).
78 Henri Bergson, *L'évolution créatrice* (1907; repr. Paris: 1959), 59.
79 Peng Hsiao-yen, "'Renshengguan' yu Ouya houqimeng lunshu" ('The Concept of life' and the Post-Enlightenment Narratives in Europe and in Asia) in *Wenhua fanyi yu wenhua mailuo* (Cultural Translation and Cultural Context), ed. Peng Hsiao-yen (Taipei: Institute of Chinese Literature and Philosophy, Academia Sinica, 2013), 264–65.

as a means of understanding the human experience provoked the debate over "science versus metaphysics." The New Culturalists defended science; the conservatives defended metaphysics.[80]

Worship of science being turned into scientism by the New Culturalists,[81] their next target was religion. At the end of 1922, the Anti-Christian Movement was initiated. In a wider sense, all religions, including Buddhism and the Confucian religion, were despised as superstition. There was a particularly harsh attack on Christianity because of its foreign origin. This movement was supported by both the CCP and the KMT, who associated Christianity with cultural imperialism and capitalism, respectively.[82]

The Critical Review Group reacted against this movement. It was out of the question that science could replace religion and explain the conception of life which was entirely intuitive. As Liu Boming argued, "religion is founded upon emotion and imagination... and is not satisfied with the reality by searching for a way out and building a noble kingdom in which reason can absolutely not intervene."[83] Jing Changji supported this argument, stating that it was unscientific for science to negate the religious transcendent experience. That is to say, science had the vocation to study the material reality of human life, but "mysterious" experience offered by religion did not belong to the scientific realm. Although science was based on rationality, Jing professed that human beings were created to search for the ultimate truth that science could not respond to.[84] But Jing pointed out that there were several mysterious experiences in religion. While the lower experience was superstition, the highest one was the perception of Way in the Chinese philosophy or Buddhist Acinteyya (imponderable) that could only be perceived through concrete engagement in the real world. This reflects Wang Yangming's (1472–1529) concept of *zhixing heyi* (the unity of knowledge and action) that truth could only be revealed through individuals' real action in this profane world. What the Group eulogized was therefore an atheistic religion, without god or superstitious practices, but an authentic engagement in the well-being of the real world.

---

80 Zarrow, *China in War and Revolution*, 178–80.
81 D.W.Y. Kwok, *Scientism in Chinese Thought 1900–1950* (New York: Biblo and Tannen, 1971), 3.
82 Guo Ruoping, "Guo Gong hezuo yu Feijidujiao yundong de lishi kaocha" (The United Front and the History of Anti-Christian Movement), *Zhonggong dangshi yanjiu* 2 (2008): 51.
83 Liu Boming, "Fei zongjiao yundong pinyi" (On the Anti-Religion Movement), *Critical Review* 6 (1922).
84 Jing Changji, "Lun Xuesheng yonghu zongjiao zhi biyao" (On the Necessity for Students to Defend Religion), *Critical Review* 6 (1922).

This contribution to social well-being had little to do with scientific progress. Although Babbitt saw in the contrast between utilitarian science and morality a confrontation between barbarianism and civilization, he recognized the need for China to be industrialized in the context of an increasingly internationalized world economy.[85] His Chinese disciples were much more hostile towards utilitarian science, so far as to ignore or downplay the threats that industrially superior nations posed to an underdeveloped China. Even Hu Xiansu, himself a scientist specialized in botany, warned of the crisis that scientific materialism had brought into China and saw a clear correlation between scientific progress and moral decadence.[86] University professor of literature Hu Jixian (1899–1968) even went so far as to suggest that those who believed China had to be industrialized to compete with the West were only concerned with national interests, but "if the directions and the value of life for humanity of tomorrow are taken into account, these people's opinion are not even worth a laugh."[87]

Liu Boming also issued a book review of Liang Shuming's *Dongxi wenhua jiqi zhexue*, maintaining that Liang had mistakenly classified Western culture as scientific; the authentic science of ancient Greece was not the same thing as the science of the modern West. While ancient Greek science was theoretical and limited to the nature of the universe, in modern times science had become utilitarian and had induced the Industrial Revolution.[88] He was apparently influenced by the same argument put forward by Babbitt in the *Literature and American College*.

What further concerned the Critical Review Group was the change of social order caused by industrialization and a capitalist economy. On the one hand, Chang Chi-yun linked industrialization to the banalization of divorce – an indirect sign of a society fractioned by a disproportionate individualism brought about by industrialization.[89] On the other hand, during his trip to Japan in the 1920s, Hu Xiansu was aggrieved to learn that in Tokyo a rickshaw runner's income could be double that of a teacher. He observed, "This overturn of natural order results from the material civilization of the West."[90] To him, natural order

---

[85] Babbitt, "Humanist Education in China and the West," 86.
[86] Hu, "Shuo jinri jiaoyu zhi weiji" (On the Crisis of Education Today), *Critical Review* 4 (1922).
[87] Hu Jixian, "Jinggao woguo xueshujie" (Respectful Advice to the Academia), *Critical Review* 23 (1923).
[88] Liu Boming, "Ping Liang Shumin zhi Dongxifang wenhua jiqi zhexue" (On Liang Shuming's *Eastern and Western Cultures and Their Philosophies*), *Critical Review* 3 (1922).
[89] Chang Chi-yun, "Bolatu lixiang guo yu Zhou guan".
[90] Hu Xiansu, "Lücheng zaji" (Essay Written During the trip to Europe), *Critical Review* 28 (1923).

intends that those who work with their mind should enjoy a privileged economic and social status compared to those who work with their hands. However, in an industrialized and capitalist society, the value of productivity outweighs that of intelligence, often with little impact on society's economic growth. It is this old social status that cultural elites enjoyed and that the group wished to conserve against the rising newly rich, or in Babbitt's term, natural aristocracy against the aristocracy of money.

Mei Guangdi explained this concern more precisely. He maintained that despite the increasing influence of Western civilization since the end of the Qing dynasty, Chinese intellectuals still retained the prominent place in the system of "four groups of people" (simin) – scholars, peasants, artisans, and merchants – but the expansion of industrialization and capitalism marked the end of Chinese intellectuals' leadership. Hence, Chinese students' resentment against capitalism sprang from the loss of their traditional social status, jeopardized by greedy merchants.[91] However, the Critical Review Group was not antimodern conservative. They did not deny the validity of reason, nor did they aspire to a return to the agrarian communitarian political arrangements. Their conservatism was a purely philosophical one that insisted upon the preservation of what they saw as immutable values.

The Rise of Authoritarian Conservatism

In February 1925, Wu Mi left Nanjing and took a post at the Tsinghua University. One year later, the Northern Expedition began, but the political impact was felt before the initiation of the campaign. In January 1925, Kuo Ping-wen (1879–1969), the president of the National Southeast University, was suddenly dismissed from his post and replaced by Hu Dunfu (1886–1978). The crucial reason for this was Kuo's hostility towards the KMT and its activities on the campus, while the KMT accused him of colluding with local warlords and politicizing education.[92] The Committee of Education of the University refused to recognize this order and rejected the proposal of the government. Under these circumstances, Kuo went to the United States for "investigation," and officially resigned in July. Many emi-

---

[91] Mei Guangdi, "Xifang zai juexing ma?" (Is the West Awakening?), in *Mei Guangdi wenlu*, 210–3.
[92] Lu Fang-sang, "'Xuefa' hu? 'danghua' hu? – 1925 nian de Dongnandaxue xuechao" ('Education-lord'? 'Party' education? – The Student Movement of the Southeast University in 1925), in *Minguo shilun* (On the History of the Republic of China) (Tapei: Commercial Press, 2013), 840–87.

nent professors were outraged and left the National Southeast University. In March 1927, the Northern Expedition army conquered Nanjing and deepened the "party education" (*danghua jiaoyu*) of the University under the pretext that the University was a cluster of "counterrevolutionaries."[93] In 1928, the National Southeast University changed its name to National Central University (*Guoli zhongyang daxue*).

On the eve of the entry of the Northern Expedition army into Beijing in 1928, Wu Mi and Chen Yinke decided that they would never join the KMT and would renounce their posts at the Tsinghua University if the KMT imposed the party education.[94] This was also the reason for which Wu Mi declined the offer from the National Northeast University (*Guoli xibei daxue*), which was dominated by the ideological control of the KMT.[95] He saw himself as truly patriotic. For this reason, he denounced the cult of personality of Chiang Kai-check and the disastrous effects on traditional culture caused by the KMT appropriating Confucianism to the revolutionary rhetoric.[96]

Wu Mi might have not imagined that his philosophical conservatism would have indirectly paved the way for the KMT's authoritarian conservatism. Since 1924, members of the Critical Review Group had been dispersed in various universities, with Wu Mi editing the magazine at the Tsinghua. In 1933, members who remained in Nanjing launched a *"coup d'état"*, requesting Wu Mi to end the collaboration with the Chung Hwa Book Company and have the magazine published by the Zhongshan shuju (Library of Zhongshan), which was founded by Miao Fenglin at the National Central University. Wu Mi refused the proposition, but under pressure from other members, he resigned as the chief editor in favor of Miao Fenglin. However, no more issues were ever published. Instead, the *Critical Review* was replaced by a new magazine named *Guofeng banyuekan* (*National Soul Bi-Monthly*), whose intellectual orientation was linked to authoritarian conservatism – the *enfant terrible* of the late Qing culturalist nationalism.

Whereas the May Fourth New Culture Movement held on to the slogan of "Mr. Democracy" and "Mr. Science," only Mr. Science prevailed in the following decade. This is not to say that the enthusiasm for democracy entirely vanished. Rather, the disenchantment with the experimentation of liberal democracy was accompanied by a growing aspiration for a nationalism that identified with a

---

[93] "Dongnan xuefa chongju zuigao xuefu" (The Education-Lords Reoccupied the Highest Education Institution), *Minguo ribao* (September 9, 1932).
[94] Wu Xuezhao, *Wu Mi yu Chen Yinke* (Wu Mi and Chen Yinke) (Beijing: Qinghua daxue chubanshe, 1996), 49.
[95] Wu, *Wu Mi riji 1939–1940*, 140.
[96] Wu, *Wu Mi yu Chen Yinke*, 91.

strong state to confront foreign aggressors and tackle China's material backwardness. The democracy pursued during the New Culture Movement met with growing opposition and mistrust from many intellectuals at the time, even from intellectuals who were not aligned with the KMT's authoritarianism.[97] They argued that democracy was neither culturally nor politically rooted in China and the redundant administrative process of democracy was simply inappropriate when bold leadership was called for to handle the imminent threats that confronted the nation.[98] More and more intellectuals favored an "enlightened dictatorship", linked with a strong and coordinated government, capable of naturalizing an interventionist policy that stood China in good stead.[99] Another threat that affronted the KMT and its sympathizers was the rise of the CCP. Standing against the spread of the communist "red infection" (*chihua*) in China, authoritarian conservatism also offered an illusion of social solidarity by downplaying class struggles.

One of the major differences between authoritarian conservatism and the other three currents of conservatism discussed previously is the state's endorsement. The cultural nation for authoritarian conservatives was in general a product of the most despotic and repressive elements of Confucian political culture, notably the supremacy of the chief, absolute obedience to the superior and political tutelage. Authoritarian conservatives advocated modernization, but only partially and instrumentally, in the sense that technological modernity and rational bureaucracy served as instruments to nourish an anti-liberal political end. Their conservatism was thus tainted with a cultural and antimodern character, since their openness to technological modernity was combined with a cyclic vision of history and the conviction that spiritual forces of ancient traditions of the glorious past could be restored in order to discipline the population and the nation. In a speech given in Nanchang in 1934, Chiang cited the upturn of Germany after the First World War to illustrate the importance of morality to national recovery. To renew the national spirit and retaliate communist ideology, he explained that China's morality was determined by the "four social bonds" of Confucianism: propriety (*li*), justice (*yi*), integrity (*lian*), and honor (*chi*).[100]

Given Chiang's admiration for Nazi Germany and Confucianism morality, in which he found ideas justifying the maintenance of state power and disciplined

---

97 Tsui, *China's Conservative Revolution*, 16.
98 Lloyd E. Eastman, *The Abortive Revolution: China under Nationalist Rule*, 140–180.
99 Fung, *The Intellectual Foundations of Chinese Modernity*, 151 and 188–189.
100 Chiang Kai-shek "Xinshenghuo yundong zhi yaoyi" (Principles of the New Life Movement), in *Xian Zongtong Jianggong sixiang yanlun zongji* (Works of the Former President Chiang), ed. Chin Hisao-yi, Vol. 12 (Taipei: Zhongyang dangshi weiyuanhui, 1984), 73–4.

society, Frederic Wakeman describes Chiang's politics as "Confucian fascism."[101] It is beyond the scope of this book to discuss whether the KMT's regime could be compared to fascism, but there is indeed a common agreement on its fascist-like rule. In 1934, having already argued with his supporters for the need to "nazify" (*nazuihua*) the country, Chiang started the New Life Movement.[102] Chiang firmly led a secret group known as Lanyishe (the Blueshirts) under the KMT. It could be said as almost in honor of the European fascist regimes of Italy (the Blackshirts) and England (the Brownshirts).[103] Furthermore, in late 1931, the Central Executive Committee of the Kuomintang decreed that "the Kuomintang should follow the example of the organization of Mussolini's Blackshirts in Italy, completely obeying the orders of the leader and creating members who will use blue shirts as the symbol of their will."[104] Many of the Lanyishe went to study military tactics in Japan under Chiang's orders. He extolled the very concept of fascism, quickly assuming the role of *lingxiu* (leader, chief or Führer) amongst the group made up of young officers from the Party's Whampoa Military Academy, as well as impressionable students. Fascism was promoted by one high official as "a very progressive means of resurrecting the nation,"[105] exemplified by their slogan "One doctrine, one party and one leader."[106] Just like the emergent fascist and Nazi ideologies, the Blueshirts glorified violence, taking their aim against communists and dissidents, those they deemed as being "morally deficient" and the Japanese troops.

KMT's fascist-like rule was conservative, but not because of "its resolute opposition to insurrectionary attacks on social hierarchies and inequal treaties" – these were only elements, not the root, of its conservatism. By the same token, the "'thrust towards a *new* type of society,' building 'rhetorically' on the cultural achievements attributed to the former, more 'glorious' or 'healthy' eras rather than suggesting a desire to return to the dynastic past as such" did not make it anticonservative.[107] It was its insistence upon the reintegration of the cultural nation and the political nation that made it conservative in the Chinese context.

---

101 Frederic Jr. Wakeman, "A Revisionist View of the Nanjing Decade: Confucian Fascism," *The China Quarterly* 150 (1997): 395–432.
102 Wakeman, "A Revisionist View of the Nanjing Decade: Confucian Fascism," 396.
103 Xuan Jianxi, "Lanyishe de lailong qumai" (Origins and Ramifications of the Blue Shirts) in *Lanyishe, Fuxingshe, Lixingshe* (The Blue Shirts, The Society for Vigorous Practice, The Renaissance Society), ed. Gan Guoxun et. al. (Beijing: Zhonghua shuju, 2014), 16–45.
104 Wakeman, "A Revisionist View of the Nanjing Decade," 402.
105 Zarrow, *China in War and Revolution*, 255.
106 Yu Zidao and Xu Youwei, "Lixingshe shulun" (On the Society for Vigorous Practice), *Jindaishi yanjiu* 6 (1989): 220.
107 Clinton, *Revolutionary Nativism*, 5.

Authoritarian conservatism represents a distorted continuation of the pre-1912 culturalist nationalism, since liberty was completely trumped by order, discipline, and nationalist collectivism.

Between Order and Liberty: The Case of Zhang Junmai

As previously evoked, liberal conservatives believed in the integration of personal liberty and institutional liberty. Men are born with a moral responsibility towards the communities of which they are part. Thus, individual liberty is inducted without individualist emancipation that discards the authority of the communities. However, authority cannot be turned into an authoritarianism that impedes personal liberty and individual emancipation, because authority sets itself apart from domination and exploitation. Therefore, liberal conservatives usually did not hold a positive view with respect to the KMT and the CCP. But what is the cut-off point between liberty and authority in an era of political turmoil? Were the detriment of liberty, unity, and sacrifice absolutely necessary in order to resist the Japanese invasion and communism?

Liberal conservatives had chosen different political paths to tackle these problems. Some joined the KMT and regarded authoritarianism as the most appropriate policy for China in the 1930s. One example is Ye Chucang, who became the director of the department of propaganda of the central committee of the KMT and an active actor in the New Life Movement, which was inspired ideologically, by fascism, nationalism, militarism, and Confucianism.

Zhang Junmai had chosen "a third way." His conservatism was tainted with liberalism. To him, the nation obtained its independence not only by political autonomy, but also by traditional culture that informed national policy, which should be liberal, constitutional, and democratic. He declared that his ideas were fashioned by German nationalism and English liberalism.[108] Fervently outspoken against Chiang Kai-chek, he was sequestrated by the KMT in 1929 and went to Germany to teach at the University of Jena. He returned to China three years later and assumed a post at the Yenching University, from which he was later forced to resign due to his anti-KMT political standpoint.[109] While he was greatly influenced by Rudolf Eucken during his first trip to Europe, this time

---

[108] Wang Yangwen, "Ziyou yu quanli zhijian – Zhang Junmai xianzheng sixiang de yanbian" (Between Liberty and Power – Evolution of the Constitutional Thought of Zhang Junmai), *Lanzhou xuekan* 11 (2007): 112.
[109] Roger B. Jean, Jr., *Democracy and Socialism in Republican China: The Politics of Zhang Junmai (Carsun Chang), 1906–1941* (Lanham: Rowman & Littlefield Publishers, Inc., 1997), 202.

he returned to China with thoughts of the *Konservative Revolution*, a prominent national conservative revolutionary movement in Germany.[110]

Zhang manifested a great interest in German conservative economists and politicians of the *Konservative Revolution*, including Adolph Wagner (1835–1917), Gustav von Schmoller (1838–1917), both of whom accorded the state the decisive role in economic affairs and the unification of social interests,[111] as well as Erich Ludendorff (1865–1937), an antisemitic military general and politician, who was deeply attached to the politics of *völkisch*.[112] However, Zhang was well aware of the authoritarian seeds in this conservative revolution of the Weimar Republic and wary of the rule of the Soviet Union. Thus, he insisted that without popular consent, no political regime would last long.[113] He proposed what he called the "centralist line" that consisted of implementing a government of a coalition of political parties. This was to reinforce the executive power, to establish a system of civil servants that did not adhere to any political parties and to promote technocracy.[114] In this system, the executive power of the technocratic government would impose itself on the parliament and the illiterate would not have the right to vote.[115] He professed that this political theory was not rooted in English democratic liberalism, nor in the German authoritarianism, nor in the dictatorship of the proletariat, but was situated between the policies of the KMT and the CCP.

The defect of this system is obvious. Liberty risked being impeded by the authority of the government. Without a parliament capable of counterbalancing the executive power, the integrity of the government depended solely upon the omnipotence of the head of the state and the ethical impartiality of the political parties. In fact, Zhang did not hide his willingness to see the appearance of a head of state, who incarnated the spirit of Confucian *junzi* (superior person, gentleman), and was capable of saving China overnight.[116] However, such a political arrangement would produce, unsurprisingly, the reverse effect: an excessive centralization of the authority of the government and an arbitrary head of state. What is more, it seems paradoxical that he was reluctant to grant the illiterate

---

110 Ibid., 215.
111 Suvanto, *Conservatism from the French Revolution to the 1990s*, 82.
112 Karia O. Poewe, *New Religions and the Nazis* (New York and London: Routledge, 2006): 82
113 Wang, "Ziyou yu quanli zhijian," 227.
114 Lin Hongming and Xu Jian'gang, "Zhang Junmai de minzhu zhengzhi he zhongjianluxian sixiang pouxi" (On Zhang Junmai's Democratic Politics and Centralist Line), *Tianzhong xuekan* 22, no. 1 (2007): 100.
115 Jean, *Democracy and Socialism*, 238.
116 Ibid., 230–1.

the right to vote, while denouncing the KMT's rhetoric according to which the intelligence of the Chinese people would not allow democracy to function properly. Despite his opposition to the political tutelage of the KMT, his political project engaged curiously in a dialogue with the KMT who proclaimed themselves to be the educators and governors of the people. As a result, this liberal conservatism would likely produce the same effects as authoritarian conservatism.

Guofeng banyuekan and Authoritarian Conservatism

After the dissolution of the Critical Review Group, several former members launched the *Guofeng banyuekan* in Nanjing, which was published from 1932 to 1936. Although the contributors continued to defend Confucianism and blame the New Culture Movement and Western materialism for the national moral and political decay, their intellectual mentor was no longer Irving Babbitt. Most writers of the journal sympathized with the New Life Movement. Jing Changji, for one, saw Chiang's emphasis on the four bonds as a positive thing, claiming that equality is a hypocritical illusion that should be abandoned in favor of the establishment of social hierarchy deemed necessary to guarantee the well-being of an organic society.[117] While criticizing what he called "unreasonable dictatorship" (*buheli de zhuanzhi*), he showed ambivalence about the political tutelage launched by the KMT, asking rhetorically if in a democratic nation all citizens understand perfectly all government policies.[118]

During the KMT's rule, Chinese intellectuals began to take a serious interest in German conservative thoughts. Zhang's interest in the *Konservative Revolution* was shared by certain members of *Guofeng banyuekan*. For example, the economist and sociologist Werner Sombart (1863–1941) was introduced to the Chinese public through *Guofeng banyuekan*. His *Studiengesellschaft für Geld- und Kreditwirtschaft* (*The Study Society of Currency and Credit*), in which he argued for the policy of *Reagrarisierung* (re-agrarization) to give the state strict control of the economy and to promote the first sector to contend with what he saw as failing capitalism, was translated under the title of *Ziben zhuyi zhi jianglai* (*The Future of Capitalism*) in *Guofeng banyuekan*. The translators highlighted two terms in particular: national character (*minzu de texing*) and leader (*lingxiu*). They underlined that, in certain circumstances, the leader's personal willingness could be transformed into the nation's character and collective willingness, as in the

---
117 Jing Changji, "Shuo siwei" (On the Four Bonds), *Guofeng* 3 (1936): 62.
118 Jing Changji, "Kongzi de zhenmianmu" (Real Faces of Confucius), *Guofeng* 3 (1932): 62–3.

cases of Lenin, Atatürk (1881–1938), and Mussolini (1883–1945). The translators emphasized: "May our fatherland be blessed with the direction of such an individual willingness, otherwise... we would surely become bogged down in a world of disorder."[119] While the *Konservative Revolution* cannot be confused with Nazism,[120] pro-Kuomintang intellectuals found in *Konservative Revolution* useful ideological justifications for the anti-liberal, anti-communist, and authoritarian policies promoted by the state.

Guofeng was more pragmatic than the *Critical Review*. Its contributors no longer conceived of moral rejuvenation as sufficient to ensure national revitalization. Instead, moral reconstruction should be combined with the exigencies of science, technologies, and industrialization. Unlike the previous generation who promoted "Chinese learning as essence and Western learning for application," the main issue here was not the coexistence of Confucian virtue and Western science, but rather the appropriation of the Confucian morality into the spirit of the latter. The change of mentality from scientific and technological development was held in tandem with a deepening of cooperation and ardent devotion to work between social components to achieve national well-being and a planetary social order. One article explains that scientific spirit is composed of Confucian moral qualities of justice, loyalty, honesty, diligence, and persistence, exploiting Confucian doctrines of social hierarchy and order in such a way as to achieve an absolute and mechanical discipline.[121] Such rhetoric agrees with the state's policy of using Confucianism as a necessary ideological basis for industrial development.[122]

Like many antimodern and liberal conservatives, Chang Chi-yun, who maintained a close relation with Chiang Kai-shek, also supported local self-government. Nevertheless, his purpose was neither to form an aggregate of intermediate associations nor to grant each region the possibility of following their own customs. Rather, he wanted to ensure a complete immersion of Confucian virtues of honor, justice, and loyalty into local communities, disciplined like an army, dedicated to the state's mobilization.[123] With the steadily increasing campaign

---

119 Xu Chongyan and Qian Desheng, "Ziben zhuyi zhi jianglai (xia)" (The Future of Capitalism, Second Part), *Guofeng banyuekan* 8 (1934).
120 For the differences between *Konservative Revolution* and Nazism, see Louis Dupeux (ed.), *La "Révolution conservatrice" dans l'Allemagne de Weimar* (Paris: Kimé, 1992), part II.
121 Lu Yudao, "Wei shuli kexue wenhua gao guoren shu" (Letter to the Nation on the Construction of a Scientific Culture), *Guofeng* 7 (1936): 286.
122 Clinton, *Revolutionary Nativism*, 64–97.
123 Chang Chi-yun, "Guofang jiaoyu sijiang" (Four Lessons on National Defense), *Guofeng* 3 (1933): 9.

of Japanese aggression, the admiration for a militarized society also translated into the veneration of war. The belief that peace was only the outcome of equal military, economic, and political power between nations led finally to an aestheticization of violence. Hu Jixian's article in the last issue of the journal, published just before the outbreak of the Second Sino-Japanese War, sublimates the war in such a way as to link it to a world of truth (*zhen*), goodness (*shan*), and beauty (*mei*).[124]

Authoritarian Conservatism during the Second Sino-Japanese War

On July 7, 1937, the Marco Polo Bridge Incident served as a pretext for Japan to declare war on China. Beijing and Tianjin quickly fell into Japanese's hands and became enemy-occupied areas. The three famous universities in the two cities – the Peking University, the Tsinghua University, and the University of Nankai – merged to become the National Southwestern Associated University (*Guoli xinan lianhe daxue*). They retreated first to Changsha and were finally installed in Kunming.[125] Wu Mi also escaped to Kunming in 1938.[126] During the Second Sino-Japanese War, editors of two magazines – *Zhanguoce* (*Strategies of the Warring States*) and *Sixiang yu shidai* (*Thought and Epoch*) – approached Wu Mi, but he declined both offers.

Not all authoritarian conservatives of the 1930s and the 1940s took Confucianism as the Chinese canonic culture. The ephemeral journal *Zhanguoce*, published in Kunming in 1940 by a total of 26 intellectuals (most of whom had studied in the United States and Europe), blamed Confucianism for the nation's weakness against the Japanese invasion. This ultranationalist conservative journal was obsessed with what the founder Lin Tongji (1906–1980) called *li* (fight, force)[127] and that his colleague Tao Yunkui (1904–1944) translated into German as *Kraft*, *Vitalität*, *Energie*, and *Macht*.[128]

---

124 Hu Jixian, "Zhanzheng zhi zhexueguan" (Philosophy of the War), *Guofeng* 12 (1936): 14–18.
125 Xie Yong, "Xinan lianda zhishi fenziqun de xingcheng yu shuailuo" (The Formation and the Decline of the Intellectuals of the Southwestern Associated University), *Twenty-First Century* 38 (1996): 55.
126 Wu, *Wu Mi riji 1936–1938*, 315.
127 Lin Tongji, "Li!" (Force!), in *Shidai zhi bo – Zhanguoce pai wenhua lunzhu jiyao* (Vague of the Epoch – Cultural Comments of the Group of *Zhanguoce*), ed. Wen Rumin and Ding Xiaoping (Beijing: Zhongguo guangbo dianshi chubanshe, 1995), 176–9.

Lin regretted the gradual decline of *li* that he observed throughout Chinese history; the lack of *li*, which resulted in the weakness of the nation in the face of the Japanese invasion, was caused by Confucianism that venerated ethics and pacificism to the detriment of *li*. Accusing Confucianism of tarnishing this vigorous spirit in profit of a refined, mild-mannered but cowardly mindset, Lin Tongji urged that the nation to turn back the clock to the Warring States period, when this consciousness of fight and bravery was fully concretized into military actions.[129] Instead of the Confucian four bonds, his motto was resumed to loyalty (*zhong*), honor (*jing*) bravery (*yong*), and, ultimately, death (*si*).[130]

Another journal worth mentioning is the *Sixiang yu shidai* which appeared in 1941. Less radical than the *Zhanguoce*, it received a monthly subvention from the government, which enabled its publication even during war time. Hu Shi once observed that the journal was anchored in "reactionary spirit, evident conservatism and support for dictatorship."[131] The KMT did make the journal a tool of propaganda. As recalled by Chen Xunci, brother of Chen Bulei (1890–1948) who was Chiang Kai-shek's trusted follower, Chiang was responsible for the funding of the journal: "The articles of the journal... were heavily invested in the explanation and dissemination of the excellent cultural traditional of our nation [...] to hide the role of the KMT in this journal and attenuate the influence of progressist and left-wing publications within the cultural sphere."[132]

Indeed, not all of the authoritarian conservatives expressed their support for the KMT in a straightforward manner. Another example was the 10 professors who issued *Zhongguo benwei de wenhua jianshe xuanyan* (*Declaration of cultural construction on a Chinese base*), in which they highlighted the necessity of adopting "a critical attitude and the scientific method" to "review China's past and grasp its present."[133] It was however published in a 1935 issue of the magazine *Wenhua jianshe* (*Cultural Construction*), a magazine founded by the Association for the Construction of Chinese Culture (*Zhongguo wenhua jianshe xiehui*), controlled by Chen Lifu and Chen Guofu (1892–1951).[134]

---

128 Tao Yunkui, "Liren – yige rengexin de taolun" (Men of Force – A Discussion on the Individual Character), in *Shidai zhi bo*, 184–194.
129 Lin Tongji, "Zhanguo shidai de chongyan" (The Renewal of the Warring State Period), in *Shidai zhi bo*, 60.
130 Lin Tongji, "Dafushi yu shidafu – guoshi shang de liangzhong rengexing" (Soldier and Literati: Two Individual Characters in the National History), in *Shidai zhi bo*, 65.
131 Sanmu, *Xiandai xueren mi'an* (Mystery of Modern Intellectuals) (Taipei: Showwe, 2011), 175.
132 He, *Kexue shidai de renwen zhuyi*, 57.
133 Fung, *The Intellectual Foundations of Chinese Modernity*, 52.
134 Ibid., 114.

The Renewal of Confucianism

During the Republican era, authoritarian conservatives led a last-ditch attempt to revive the culturalist nationalism that unified the political and the cultural nation. It is worth emphasizing that the line between philosophical conservatism and authoritarian conservatism is sometimes blurred. Philosophical conservatism implies a moral commitment to a series of Confucian values, albeit without a clear political program to instantiate these values. Philosophical conservatives primarily underscored the importance of social hierarchy, social order, the law of measure, elitism, and social responsibilities. All these values could easily be appropriated and taken to extremes by authoritarian conservatives in a political arrangement which could turn authority into authoritarianism.

In such a political situation, personal freedoms were disposed invisibly by liberal conservatism to make way for the national causes. Li Zehou proposed that the national crisis of saving the nation (*jiuwang*) had the effect of stifling the May Fourth enlightenment (*qimeng*) project of liberty, equality, and human rights, while Edmund Fung commented that liberal thoughts in the 1920s and 1930s concentrated heavily on the concept of institutional liberty.[135] The motto of liberal conservatism was developed from the proposal that collective aims should not create a vicious circle. Nonetheless, after it was recognized that the Chinese experimentation with the liberal regime had failed, China lost its effective and essential political institution and mechanism, hindering thus the ability to underwrite the legitimacy of the decisions made by the state, as well as the ability to distinguish between personal and institutional liberty. Accordingly, as liberty within the political sphere reduced, so did liberty end in the private environment. All the political measures made in the national interest would be tolerated, regardless of the concept of personal liberty that were ignored somewhat.

From such a backdrop, the defiance of philosophical conservatism and the conversion of certain philosophical conservatives to authoritarian conservatism could be interpreted as a result of the failure of political liberty, which should have gone hand in hand with upholding personal liberty. The Critical Review Group's conservatism, shaped by the New Humanism, was modelled on the assumption of the innate evilness of humanity, which led to suspicion towards

---

135 Li Zehou, "Qimeng yu jiuwang de shuangchong jiezou" (Double Variation of Enlightenment and National Salvation), in *Zhongguo xiandai sixiangshilun* (On Modern Chinese Intellectual History) (Beijing: Dongfang chubanshe, 1987), 7–49. Edmund S. K. Fung, "Were Chinese Liberals Liberal? Reflections on the Understanding of Liberalism in Modern China," *Political Affairs* 81, no. 4 (2008/2009): 559.

free-market capitalism, democracy, and utilitarian science. The Group's members failed to grasp that the values they held in such high esteem could not be realized without taking advantage of appropriate political arrangements, while under the auspices of the New Humanism. Accordingly, the values could be applied forthwith to become an ideological tool of the "new dictatorship," as soon as the moral code of the philosophical conservatism took root in a party-state, facing up squarely to the Japanese invasion and the CCP. This resulted not only in a reduction in personal liberty, but more seriously a complete rejection of conservatism by both the radicals and progressives.

Conclusion

The intellectual history of Chinese conservatism that I have recounted is premised on the assumption that from the late 1890s to the 1940s, a conservative intellectual and political movement did in fact take place. Whereas the existence of this idea is generally recognized, some contemporary political scientists have questioned this presumption. Liu Junning, one the of signatories to the Charter 08 that called for an independent legal system, freedom of speech, and the elimination of one-party rule, professed that China had never "benefited" from conservatism, which in itself did not exist in China. He reasoned that the intellectual and political forces that opposed liberalism or radicalism in the Chinese history of the last century were traditionalism or reactionaryism, which dismissed any reform in order to conserve a tradition that encapsulated few liberties.[1] To him, conservatism's distinguishing feature was the conservation of liberty and as such opposed immobilism, political authoritarianism, anti-liberalism, as well as the oppression of individual liberty, democracy, and human rights.[2] As I have shown throughout this book, this view on Chinese conservatism is too limited.

This book presents a novel interpretation of the development of conservatism in Republican China. As conservatism needs to be understood pluralistically and contextually, the book avoids giving a categorical definition to this intellectual and political movement. Indeed, the question "what is conservatism" appears to be somewhat meaningless, since the answer to this question tends to define conservatism by a list of constituent elements that are not universally applicable, while conservatism adapts and changes in every historical period and contextual background. Instead, this book departs from Disraeli's question "what will you conserve?"

My answer to this question draws from Eugen Weber's theory and Jan-Werner Müller's approach. Conservatism sets out to preserve the present which is in a state of constant change. During the 1890s and the 1900s, the present was the Qing Court, while after 1912, the Republic of China came to be the new reality. Understood in this way, conservatism and its emblematic figures' thoughts were constantly evolving. Today's conservative might be tomorrow's reactionary, while today's conservative might be born from yesterday's radical. Also, it must be stressed that conservatism needed to be understood both culturally and po-

---

[1] Liu Junning, *Baoshou zhuyi* (Conservatism) (Beijing: Zhongguo shehui kexue chubanshe, 1998), 262.
[2] Ibid., 262–3.

litically. Cultural radicalism sometimes acted as the backbone of political conservatism, as in the case of monarchist reformers during the revolutionary decade.

Reacting differently to historical circumstances, Chinese conservatives of the Republican era were all devoted to conserving the late Qing culturalist nationalism. In the imperial context, culturalist nationalism entailed the homogenization of the political nation and the cultural nation. Membership of a nation was acquired, not only by one's implications in various social forms through which nationalist expectations were expressed, but also by individuals' common cultural and ethnic origin. Following this logic, revolutionaries excluded the Manchu population from the Chinese nation and claimed their rule over China as illegitimate. While reformers condemned revolutionaries' exclusion of the Manchu from the Chinese nation, they differentiated outsiders from insiders in a way that pulled reformers and revolutionaries into the same ideological pattern, albeit with diametrically opposite political objectives.

The encounter between Chinese culture and Western culture at the end of the Qing dynasty did not lead to a "disastrous" clash of civilizations, not only because living tradition never ceased to grow, but also because a culture, both traditional and modern, as well as capable of fashioning the nation's political reforms, had taken shape at the behest of progressive intellectuals of this period. Late Qing progressives managed to reconcile modern Western politics and tradition through national essence, and the incompatibility between modernity and tradition was not totally imposed from the outside, but more essentially created by the May Fourth paradigm.[3] Even then, efforts to reintegrate culture into politics never ceased.

Culturalist nationalism requested reformers and revolutionaries to institutionalize the nation-state in the framework of traditional culture. The Japan-imported neologism of national essence flourished in China in this ideological context. As such, cultural preservation cannot be automatically reduced to conservatism. The key issue is the political order that the preservation of certain cultural elements gave rise to. It is only through this lens that the meanings attributed to conservatism can be fully appreciated.

Also, conservatism should be strictly differentiated from traditionalism, reactionaryism, royalism, and loyalism. Conservatism aims to present the present, even to the detriment of the past. The "present" here is the Republic of China. Notwithstanding their different political-cultural standpoints, conservatives in Republican China were devoted to preserving and regenerating this present. The coherence of the Chinese conservatism of the Republican era is manifest

---

3 Chevrier, "Antitradition et démocratie dans la Chine du premier XXe siècle," 392.

in the commitment to defend pre-1912 culturalist nationalism. The four typologies of the conservative intellectual force of this period revolved around this same underlying principle, but the cultural tradition that they set out to preserve diversified. Thus, they differed with respect to the political and economic systems in which the Republic was to function. Placed in the same intellectual camp for the preservation of the same politics of culturalist nationalism, Chinese conservatism is therefore a polysemic term, attributing differential values and judgments to liberalism, capitalism, industrialization, statism, social hierarchy, democracy, and Westernization.

When the Republic was proclaimed in 1912, the racial question no longer held a dominating position in political activists' deliberation about the future. It was still a major theme of modern Chinese history, but attention was focused mainly on the institutionalization of the nation-state. New Culturalists' denial of the political usage of traditional culture and the First World War triggered the proponents of the late Qing culturalist nationalists to react and diversify the meanings of traditional culture. Consequently, notwithstanding their political standpoints during the revolutionary decade, all defenders of the culturalist nationalists came to be dismissed by New Culturalists as conservative. As a result of the New Culturalists' domination of the intellectual field, conservatism became increasingly an intellectual disposition centered on the defense of cultural heritage. Liberal conservatives became marooned amid the dilemma of pursuing the same political aspiration as the progressive liberals, but in a cultural language rejected by the latter as incompatible with liberty. Antimodern conservatives never managed to expand their political programs to the national level. While antimodern conservatives' rural reconstruction might sound like the prelude to the CCP's rural revolution, theirs was based on a revival of the Confucian tradition of community instead of class struggle. This partly explains Liang Shuming's later conflict with Mao Zedong after the foundation of the People's Republic of China. Philosophical conservatism did not even conceive of a concrete political system to implement their cultural ideals.

These efforts to bring back late Qing culturalist nationalism produced a limited effect and faced a major roadblock. During the Republican era, in contrast to the 1910s, culture progressively lost its power to nourish politics.[4] Authoritarian conservatives were the only ones who managed to effectively politicize traditional culture. However, the successful politicization of authoritarian conservatism did not mark a victory for Chinese conservatism or traditional culture. On the one hand, this state-orchestrated conservatism further tarnished traditional cul-

---

4 Chevrier, "La démocratie introuvable (1915–1937)," 474–81.

ture blamed for impeding the construction of a less authoritarian regime in China. On the other hand, as Chevrier reminds us, instead of ensuring the institutionalization of the nation-state, the selective preservation of traditional culture by political parties was merely used to prolong revolutionary agendas that continued after the foundation of the Republic. The failure of the Republic thus voided the substance of cultural conservatism. This line of argument leads us to ask how traditional culture, defended by conservative intellectuals, confronted the rhetoric of revolution in which culture was evoked by political parties, not as a means to build a modern state, but to rationalize revolution. Was the depoliticization of the conservative culture also a consequence of the politicization of the revolutionary traditional culture? Such questions are worth asking to explore the culturalization and depoliticization of the Chinese conservative intellectual movement from a different perspective.

An Impossible Conservatism?

In 1928, Wu Mi wrote: "all the ideal things that I once admired all ebbed away throughout the time."[5] The poem is called *Wilting* (*Luohua*), an appropriate metaphor for something that conservatives pursued but which was discarded by the mainstream intellectuals. Edmund Fung observes that conservatism was neither weak nor negative but constituted a non-negligible force in Republican China.[6] Nevertheless, the continual and permanent existence of conservatism does not suggest that conservatives had exercised a real influence in the political scene. When the government endorsed conservatism, it was usually for the purpose of launching autocratic politics, as in the cases of Liang Qichao's Progressive Party and authoritarian conservatism. Such political usage could only portray conservatism with an extremely negative image. Nevertheless, conservatives failed to propose political alternatives, not only because of the vagueness or even the unrealistic characteristic of their project, but also because of the increasingly restrictive public sphere in which their voice could be heard.

But why did Chinese conservatism fail to secure the popular support achieved by their Western counterparts? Western conservatives' traditionally hostile attitude towards universal suffrage, for example, began to soften by the end of the nineteenth century when they realized that the majority identified with their political visions. As a result, European and American conservatives became

---

5 Wu Mi, "Luohua shi" (Wilting), in *Wu Mi and Chen Yinke*, 69.
6 Fung, *The Intellectual Foundations of Chinese Modernity*, 21–2.

increasingly disposed to extend the right to vote.[7] As Charles Maurras once said: "We have *never* thought about the suppression of universal suffrage, because this suffrage, between many virtues and many defects, possesses a fundamental holding that is inherent to itself: *the universal suffrage is conservative.*"[8]

Suspicious of capitalism and industrialization, philosophical and antimodern conservatives, for example, could have engaged in a dialogue with their Western counterparts and gained public support. Indeed, the concern about the loss of privileged status and the decline of traditional values was a trait of European cultural conservatives on the eve of the Industrial Revolution.[9] On the other side of the Atlantic, a particularly important anti-capitalist battle was unleashed before the American Civil War. For conservatives in the South, the condition of life and work in modern factories was simply appalling for the slaves, whose well-being was always of paramount concern to their owners. The modified version of this highly debatable argument was presented in the manifesto *I'll Take My Stand*, in which a collection of 13 American poets instilled theoretically this anti-capitalism and anti-industrialization of the Southern Agrarians. For them, the expansion of industrial capitalism of the North detached people from their land and resulted in a sentiment of rootlessness. They condoned a social disposition structured around the hierarchy and racial segregation that they believed to be much more compassionate than that of capitalist industrialization, in which the reciprocity between the commanders and the commanded that maintained the human relation in the South simply did not exist in the North. Theodore Roosevelt and Irving Kristol (1920–2009) had both expressed enmity towards industrial capitalism, which altered the morals and the direction of the traditional elite.[10] By the end of the nineteenth century, European conservatism sometimes developed in an antimodern direction. Each individual did not benefit in the same manner from the industrialization and the political revolution of democratization and secularization. The conservative movement was thus supported by those who felt forgotten and marginalized in the process of modernization. Conservative parties in Switzerland, for example, won popular support from the "discriminated" and frustrated peasants, artisans, and traditional merchants in the French and Italian parts of the nation.[11]

---

7 Aughey, *The Conservative Political Tradition in Britain and the United States*, 129–30.
8 François-Guy Trébulle, "Démocratie," in *Le dictionnaire du conservatisme*, 294.
9 Roger Scruton, *Conservatism: Ideas in Profile* (London: Profile Books, 2017), chapter 4.
10 Peter Kolozi, *Conservatives Against Capitalism: From the Industrial Revolution to Globalization* (New York: Columbia University Press, 2017).
11 Altermatt, "Conservatism in Switzerland: A Study in Antimodernism," 586.

In contrast, Chinese conservatives were mostly urban intellectuals detached from the social basis and, except antimodern conservatives like Liang Shuming, did not really care to "go to the people." Even Chinese conservative parties did not care so much about popular support.[12] After all, only 10% of the population had the right to vote and the participation rate fell between 50% and 70% during the first Republican years.[13] Chinese conservative intellectuals' main audience was urban readers, with the intellectual capacity of reading classical Chinese, but conservatives failed to come up with a strategy to attract the urban readership that became more and more attracted to market-oriented magazines of leisure and May Fourth literature. While antimodern conservatives had launched the rural reconstruction in the countryside, they failed to effectively convince the peasants to be sincerely committed to their project, let alone bringing it to the national level. Eighty percent of the Chinese population still lived in rural areas at the time. Since capitalism and industrialization were still under development in China, it seems plausible that the majority of peasants, who did not enjoy the right to vote, did not feel left out in the course of modernization of the nation either. European antimodern conservatives' success by the end of the nineteenth century was thus produced in a historical context totally different from that of twentieth-century China, in which Chinese conservatives were simply not able to emulate their European counterparts' success.

Other factors contributing to the loss of political vigor in conservatives' culturalist discourses were their detachment from political power and, perhaps most importantly, the radicalization of the Chinese intellectual sphere from the late nineteenth century onwards. As Yü Ying-shih observes, the marginalization of the educated elite in the political sphere resulted in the accelerating radicalization of the intellectual sphere.[14] The abolition of the Imperial examination in 1905 deprived the cultural elite of the path to state bureaucracy. They were no longer traditional literati capable of affecting the political process of decision-making, but intellectuals remote from political power. The marginalization of the educated elite in the political sphere accelerated this radicalization of the intellectual sphere.[15] As a result, conservative political agendas were dismissed as

---

12 Li Jiannong, *Zhongguo jinbainian zhengzhishi (1840–1926)* (Political History of the Last 100 Years of China) (Shanghai: Fudan daxue chubanshe, 2002), 328.
13 Chang Peng-yuan, "Cong Minguo chuqi guohui kan zhengzhi canyu – jianlun tuibian zhong de zhengzhi youyi fenzi" (Political Participation in the Early Republican Parliament Election – On the Political Elites in the Transition Period), *Guoli Taiwan shifan daxue lishi xuebao* 7 (1976): 373.
14 Yü, "Zhongguo zhishifenzi de bianyunhua."
15 Ibid.

inconsequential. What is interesting and ironic is that this radicalization of the intellectual field began with revolutionary intellectuals such as those from the National Essence School and the Southern Society, whose version of culturalist nationalism, which marked the debut of intellectual radicalization during the last few years of the Qing dynasty, came to be seen as conservative in Republican China due to the intensification of a radical movement that they themselves initiated.

Conservatism as a political movement in its modern sense appeared in China with the rise of the late Qing revolutionaries, who saw no gain for China in the search for wealth and strength to maintain the monarchy through progressive reforms supported by the constitutionalists. For the radicals, the Manchu Court was neither sincerely committed to nor capable of launching the social and political projects of modernization of which China was in desperate need. Since revolutionaries considered the Manchu to be a foreign nation, they further argued against the Court's legitimacy in leading China's political reforms. Nevertheless, it should be stressed that the 1901 New Policies initiated by the Manchu government was showing significant outcomes in terms of concrete results in institutional systems, economics, education, and military affairs in a short period of time. It was scheduled that a parliament would be established, and political parties would be allowed to operate freely in 1913. The Revolution of 1911 cut short the New Policies and brought an end to the Qing dynasty. However, the foundation of the Republic hardly signaled a victory for revolutionaries, since Liang Qichao's prediction that revolution would undoubtfully lead to political chaos became a reality. In this regard, one might wonder what China's future would look like if reformers' conservative political projects were carried through in the 1900s.

In a study on the relation between conservative parties and democratization in Europe, Daniel Ziblatt brings to the fore two paths to democracy that he distinguishes according to the different consequences that democratic revolution brought to several European nations. Between 1848 and 1950, Britain, Belgium, the Netherlands, Norway, Sweden, and Denmark followed the path of "settled democratization," while Spain, Portugal, Italy, Germany, and France before 1879 followed an "unsettled" path to democratization.[16] Ziblatt argues that the support of conservative elites, rather than the socio-political changes, the demand of the middle class, and the urban bourgeoisie or the movement of the working class, determines whether democratization will bring about perturba-

---

[16] Daniel Ziblatt, *Conservative Parties and the Birth of Democracy* (New York: Cambridge University Press, 2017), 10.

tions or not; if conservatives are convinced that their own fortune, social status, and political power can be maintained after democratization, they are more likely to support political change and thus facilitate the process.[17] It is beyond the scope of this book to reflect on the relevance of Ziblatt's thesis in the Chinese historical context, but his observation does invite us to consider the role that played late Qing conservatives in the abortive political modernization of the Qing dynasty.

In regard to the institutional reforms, the Qing government announced in 1908 that the absolute monarchy would be transformed into constitutional monarchy within 10 years. This decision was not well received among reformers. In the following two years, the Association for Preparation of Constitutional Government (*Yubei lixian gonghui*), founded in 1907 in Shanghai, federated other associations of the same kind in other regions to petition the government to accelerate the political transition, issue a constitution, and establish a parliament and a responsible government. The appeal resonated with many reformist-minded Qing officials.[18] The Association organized three petitions in total. Only after the third petition did the government agree to reduce the transition period to five years, create a cabinet, and draft the constitution.[19] In May 1911, the Prince Qing Cabinet (*Qing Qinwang neige*) was established. However, the composition of the Cabinet disappointed constitutional reformers and enraged revolutionaries. The Cabinet consisted of 13 members, of which nine were Manchus (seven of whom were from the imperial clan) and only four were Han Chinese. Disillusioned and frustrated, many reformers converted to the revolutionary camp.[20] Six months later, the Revolution of 1911 broke out.

By delaying the political transition and setting up the Prince Qing Cabinet, the Qing government seemingly aspired to preserve the prerogatives of the imperial clan. But such a political calculation was simply unacceptable, even to the reformers who defended the Qing's reign. In a word, what Ziblatt terms as the "settled pattern" was never a choice in China. The rising anti-Manchu radicalism, along with the conservative Court's reluctance to usher in a sincere operation of self-reform that provoked the radicalization of political activists, made Chinese conservative thoughts impossible to be materialized from the very beginning. However, its fate reveals the wider evolution of Chinese intellectual and political fields of the first half of the twentieth century.

---

17 Ibid., 16, 18–21.
18 Ching Chih-jen, *Zhongguo lixianshi* (History of Constitutionalization in China) (Taipei: Linking, 1984), 128–30.
19 Kuo, *Jindai Zhongguo shigang*, 378.
20 Ibid., 379.

Afterlives of Conservatism of Republican China

After the government of the Republic of China retreated to Taiwan following the KMT's loss in the Second Chinese Civil War, the New Confucianism, whose origin dated back to the 1920s, began to develop in the peripheries of China. In 1958, the year that marked the beginning of the economic politics of the Great Leap Forward (*dayeujin*), Tang Junyi (1909–1978), Mou Zongsan, Xu Fuguan (1904–1982), and Zhang Junmai cosigned a manifesto published in *Minzhu pinglun* (*Democratic comments*) in Taiwan. In this manifesto, they stated that China's future depended on its capacity to implement democracy and generalize science. Acknowledging that Chinese history had demonstrated a lack of democracy and science in China, they nevertheless refused to admit that China did not possess the democratic and scientific seeds. In order to have these seeds take roots, China should learn from the West, without sacrificing what they called Confucian "spiritual learning" (*xinxing zhixue*):

> Through the development of an ideal culture in line with the nature of Chinese culture we hope that all Chinese, via this spiritual learning, transform themselves in individuals of morality, politics... knowledge and activity of utilitarian technologies... In other words, China needs to be founded on democracy, but also on science and utilitarian technologies... [In this way], the moral quality of Chinese people can be achieved at a higher level and the objective spiritual life of the Chinese nation can [also] be developed at a higher level.[21]

This paragraph recalls Wang Yangming's *zhixing heyi* that insists upon the moral willingness of each individual to become an active actor of industrialization and democratization. Such an achievement could not only be based on the legislation and the institutionalization of the nation-state.

New Confucianism responded to KMT's political need to claim that their government was still culturally and political legitimate after losing the Civil War. Despite the fact that New Confucians were critical of the KMT's authoritarianism, the party appropriated this philosophy to differentiate itself from Communist China. In the mid-1960s, the government actively promoted the Movement of the Restoration of Chinese Culture (*Zhonghua wenhua fuxing yundong*) and Confucianism to resist the Cultural Revolution of mainland China. In this way, it

---

[21] Tang Junyi et al., "Wei Zhongguo wenhua jinggao shijie renshi xuanyan – women dui zhongguo xueshu yanjiu ji zhongguo wenhua yu shijie wenhua qiantu zhi gongtong renshi" (A Manifesto on the Reappraisal of Chinese Culture: Our Joint Understanding of the Sinological Study Relating to World Cultural Outlook), in Zhang Junmai, *Xinrujia sixiangshi* (*History of the Neo-Confucianism*) (Beijing: Zhongguo renmin daxue chubanshe, 2006), 576.

hoped to demonstrate that the Republic of China was the legitimate China, although its real sovereignty only covered Taiwan and several small islands.[22]

While New Confucianism as a philosophical doctrine did not really influence the Taiwanese political decision-making process, it allowed the KMT to pursue a conservative politics that rejected the "total Westernization" of the May Fourth radicalism. Moreover, despite its openness to science and democracy, the New Confucianism was based on an anti-Communist nationalism that the KMT believed to be useful to unify the people of an island that had been governed by Japan for decades, and also to resist, to a reasonable extent, the entry of foreign capital.[23]

When the New Confucianism was flourishing in the periphery of China, China embraced Marxism-Leninism and harshly oppressed tradition, notably Confucianism. Only in the post-Mao era did New Confucianism begin to develop in China. However, the Chinese New Confucians were highly critical of *xinxing zhexue*, which, according to them, voided the political implication of Confucianism. The liberal democracy advocated by the New Confucianism in Hong Kong and Taiwan symbolized to them a submission to the Western political system, which did not uphold any legitimacy in China: values are relative. By the same token, Marxism should also cede its place to Confucianism, which they were determined to re-politicize.

Kang Xiaoguang believed the foundation of liberal democracy was vicious. Not only did it degrade morality and de-structuralize society by inciting extreme individualist liberalism, but the democratic system also revealed a dilemma of the bourgeoisie who wished to prove the legitimacy of their domination through democracy and simultaneously deprive the majority of the democratic rights. The history of democracy is written by blood and combat of the poor, women, and racial minorities. To tackle this dilemma, the bourgeoisie put forward a constitutional regime, which was a conspiration to perpetuate the "oligarchy" and the economic hegemony of the bourgeoisie to control the media, education, and the civil society.[24]

---

22 Li Shu-chen, *Anshen liming: xiandai Huaren gongsi lingyu de tansuo yu chongjian* (Settling and Upholding Oneself: Discussion and Reconstruction of the Private and Public Sphere of Modern Chinese) (Taipei: Linking, 2013), 222.
23 Chan Yao-chi, "Qishi niandai de 'xiandai' lailu: jizhang sumiao" (Origins of the Modernity of the 1970s: Some Sketchs), *Sixiang* 4 (2007): 129–33.
24 Kang Xiaoguang, "Wo weishenme zhuzhang "ruhua" – guanyu Zhongguo weilai zhengzhi fazhan de baozhou zhuyi sikao" (Why do I advocate "Confucianization"? – A Conservative Thought on the Future Chinese Political System), in *Zhongguo bixu zairuhua – "dalu xinrujia"*

Given the fierce opposition against liberal democracy, some scholars see in the Chinese New Confucianism the devotion to a type of fundamentalist Confucianism.[25] Indeed, not only did they conceive political systems that they proclaimed to be totally Confucian, but they aimed also to revive the cosmological order of Confucianism. Jiang Qing is a case in point. His 2003 political project begins by criticizing Yü Ying-shih's point of view that Confucianism no longer holds any political use.[26] Unlike *xinxing zhexue*, his Confucianism is a political one (*zhengzhi ruxue*) that renews Confucianism by an insistence upon the vicious nature of human beings and accentuates the importance of instruction and education. Jiang Qing professes that all reforms from the late Qing period onwards contributed to the degeneration of the Chinese nation. China never had nationalism because the nation, modeled only as an instrument to rationalize the Revolution of 1911, eradicated the cultural roots that should have been used to orient and inform political institutionalization.[27] He finds many similarities between Confucianism and conservatism, but argues that Confucianism is superior in the sense that it judges and shapes the secular political system from a transcendent ideal. The legitimacy of the government came from three levels: Heaven, the consent of the people, and culture.[28] Rejecting liberal democracy, he believes that the consent of the people can be obtained by the morality of leaders, who are obliged to behave themselves according to the order of Heaven and to create political institutions that conform to Confucian political culture.[29]

In 2016, Jiang exhorted politicians to become aware of the responsibility towards the people and assume the task of perpetuating the Way of the Heaven.[30] This confidence on the eventual perfectibility of politicians seems contradictory to his earlier vision on the inherent viciousness of human beings. With regard to the concrete political institution and in line with Kang Youwei, he called for a return to the *Gongyang* learning to institutionalize a parliament. He suggests par-

---

xinzhuzhang (China Has to Be Re-Confucianized – Appeals of Chinese New Confucians), ed. Jiang Qing et al. (Singapore: Global Publishing, 2016), 148–9.
25 Lee Ming-huei, "Dangdai xinrujia "ruxue kaichu minzhu lun" de lilun yihan yu xianshi yiyi" (The Theoretical Meanings and the Pragmatic Significations of the "Development of Democracy from Confucianism" of the Contemporary New Confucianism), *Asian Studies* 2, no. 1 (2014): 15.
26 Jiang Qing, *Zhengzhi ruxue – dangdai ruxue de zhuanxiang, tezhi yu fazhan* (Political Confucianism: Transition, Particularities and Development of Contemporary Confucianism) (Beijing: Sanlian shudian, 2003), 7.
27 Ibid., 399.
28 Ibid., 56.
29 Ibid., 202–10.
30 Jiang Qing, "Wangdao zhengzhi shi rujiao xianzheng de yili juchu" (The Heavenly Way Is the Theoretical Foundation of Constitutionalism), in *Zhongguo bixu zairuhua*, 18.

liament with three houses: House of Confucians (*Tongru yuan*), that symbolizes the transcendent legitimacy of the regime, House of the Commons (*Shumin yuan*), for which members are elected, and House of National Body (*Guoti yuan*), that defends the cultural legitimacy of the nation. The House of Confucius is composed of members selected from the "Confucian wise men" of the society and the Confucian Academy is also to be established. The president of the House of National Body has hereditary status of a descendent of Confucius and nominates the members of the House from descendants of famous scholars, martyrs, literati, and ancient emperors, as well as university professors of national history, retired civil servants, important personages of the society, and the religious realm.[31]

Another Chinese New Confucian, Yao Zhongqiu, remains wary of this system, because parliamentarism, despite its Chinese characteristics in Jiang's plan, does not exist in Confucian classics. He professes that history and the Confucian classics have attested the aboriginal roots of constitutionalist consciousness, decentralization, equality, republican regime, social self-government, and economic freedom. While he admits that he has yet to come up with a project to materialize these cultural elements in a Confucian Republic,[32] he highlights two points. First, democracy is redundant, since the result of an election does not necessarily reflect the morality of the successful candidates. The punctual reformation of government stagnates political advancement, all the more so if the power of a government is sufficiently legitimate, it is not necessary to organize elections regularly to prove its legitimacy.[33] Second, since the legitimacy of the government is also based upon culture, the government should assume the responsibility of cultural inheritor. In this regard, he salutes the efforts of the CCP to preserve Chinese culture during the post-Mao era. In the 1982 constitution, the CCP, which he has accused of having vulgarized the Chinese language for decades to profit from a revolutionary language, finally affirmed that the Chinese political nation is, in nature, a cultural entity. The reintegration of the political nation and cultural nation symbolizes for him the first step towards a Confucian constitutionalism.

However, it is curious that he does not reflect on the reasons that Confucianism regained its place in the Communist vocabulary, after the destructive expression that CCP had left on this way of thinking. Yuan Shikai and Chiang Kai-check had both appropriated Confucianism to rationalize their authoritarianism. It

---

[31] Ibid., 21–22.
[32] Yao Zhongqiu, *Rujia xianzheng lun* (On the Confucian Constitutionalism) (Hong Kong: City University of Hong Kong Press, 2016), 247.
[33] Ibid., 49.

seems difficult to believe that the appropriation of Confucianism by the CCP is not another rhetoric strategy to consolidate its dictatorship in an era where class struggle and Marxism-Leninism no longer had resonance in the population.

If Chinese New Confucianism initially presented itself as the opposite of Marxism-Leninism in the 1990s, certain intellectuals of this movement began to reconcile the New Confucianism with the current political regime. Chen Ming, the president of the Center of Research on Confucianism at the Normal University of the Capital, has explicated, in the framework of *Gongyang*, the slogan of "China's dream" (*Zhongguo meng*), which was postponed by the president Xi in 2012.[34] Xi himself was further described by some as a "red Confucian."[35]

Continual Influence of Western Conservatism in China

While conservatism often rests on particularistic skepticism, where politics concerns individuals forming a community in light of their own traditions, the study of conservatism from a transnational perspective is possible when historically contingent ideas converged in China in the early twentieth century. Through travel and study aboard, foreign thoughts were imported, adopted, and mixed with indigenous elements to tackle the national crisis.

During this period, Chinese intellectuals studied Western political philosophy intensely, searching for catalytic ideas and practices that might hasten national revival. Traditional learning was re-fashioned to accommodate different Western conservative ideas and to support different future policies. This illustrates the plasticity of tradition, which, as this book has shown, was interpreted in numerous ways in accordance with Western thoughts in order to rationalize different or even opposite political movements and projects. Although Western conservatives would probably disagree that their political visions could be deployed in twentieth-century China, some Chinese and Western conservatives shared a suspicion of scientific rationalism and liberal democracy. Their desire to build an alternative modern society made it possible to appropriate Western conservative ideas in a different spatial, political, and temporal context. Western conservatism continues to influence contemporary Chinese conservatism.

In 1978, Deng Xiaoping (1904–1997) led the Opening of China. During the first decade of the reform, market socialism, decentralization, and a growing

---

34 Chen Ming, "Chao zuoyou, tong santong, xin dangguo" (Surpass the Left-Right Dichotomy, Unify the Political Heritages and Create a New Party-State), in *Zhongguo bixu zairuhua*, 73–6.
35 Ma Haoliang, "Hongse rujia Xi Jinping" (Xi Jinping, the Red Confucian), in *Zhongguo bixu zairuhua*, 323–5.

sense of political liberalization began to undermine the authority of the central government. These factors fostered the New Authoritarianism (*Xinquanwei zhuyi*) in the 1980s. Chinese New Authoritarianism was first inspired by Samuel P. Huntington's *Political Order in Changing Societies*, in which Huntington argues that economic development is unsustainable in the absence of a stable political order.[36] Huntington's Chinese followers were not necessarily anti-democratic. However, taking examples from the Four Asian Dragons, whose economy had undergone rapid development under non-democratic or even authoritarian regimes, they regarded a steadfast and effective government as an important prerequisite for China's economic development.[37] Thus, contrary to the common belief that economic liberty would bring about political liberty, Chinese authoritarians supported the CCP as a guarantee of economic reforms in the foreseeable future.[38] This type of conservatism was denounced by Liu Xiaobo (1955–2017) more than two decades previous. For him, economic liberalism was nothing but a tactic for the CCP to perpetuate its dictatorship and political conservatism.[39]

Arguably, the most representative new authoritarian today is Xiao Gongqin. Xiao considers the French Revolution to be a negative development, based solely upon abstract ideas, that brought nothing but terror.[40] Although Xiao urges China to bid farewell to this type of "romantic" revolution,[41] he welcomes Xi's rule as representing an important step on China's path to the so-called "socialist democracy."[42]

Moreover, Chinese conservatism has vested in recent decades a growing interest in Carl Schmitt's (1888–1985) critiques of liberal democracy and parlia-

---

36 Samuel P. Huntington, *Political Order in Changing Societies* (New Haven: Yale University Press, 1973).
37 Peter Moody, *Conservative Thought in Contemporary China* (Lanham: Lexington Books, 2007), 152.
38 See, for example Zheng Yongnian, "Development and Democracy: Are They Compatible in China?" *Political Sciences Quarterly* 2 (1994): 250–52.
39 Liu Xiaobo, "Jiushi niandai Zhongguo de zhengzhi baoshou zhuyi" (Chinese Conservatism of the mid-1990s), *Beijing zhichun* (*Beijing Spring*) 43 (1996), http://bjzc.org/bjs/bc/43/55, accessed February 15, 2018.
40 Xiao Gongqin, "Dangdai Zhongguo xinbaoshouzhuyi de sixiang yuanyuan" (The intellectual Foundation of Modern Chinese Conservatism), *Twenty-First Century* 40 (1997): 126–37.
41 Xiao Gongqin, *Yu zhengzhi langman zhuyi gaobie* (Farewell to Political Romanticism) (Wuhan: Hubei jiaoyu chubanshe, 2001).
42 Zhang Boshu, *Gaibian Zhongguo: Liusi yilai de Zhongguo zhengzhi sichao* (Changing China: Political Trends from June Fourth Onwards) (Hong Kong: Suyuan Books, 2015), 49.

mentarism.⁴³ Nevertheless, it bears stressing that the Weimar Republic and contemporary China are situated in totally different political circumstances. What are the political implications of such an over-exaggeration of the crisis of liberal democracy and parliamentarism in a party-state that has neither?

Indeed, conservatism, as it stands in China today, may not be an adequate political strategy for China. As the English conservative John Kekes remarks, conservatism may be inappropriate during "the times in which the prevailing conditions are so wretched, the enjoyment of life is restricted to so few, and the prospects of a better future are so poor for so many people that drastic change is called for. At these times, it is reasonable to take risks, as it was in Nazi Germany and Stalinist Russia."⁴⁴ But Kekes also reminds us that sometimes "drastic changes can make a bad state of affairs even worse, as they have done in Russia in 1917 and in Iran in 1979."⁴⁵ The same thing might also can be said about the Revolution of 1911; the promise of democracy and constitutionalism was only fully achieved in Taiwan more than half a century later. What does China's political future look like today? If history is any indication, a positive result following a drastic event is never guaranteed, or at least cannot be achieved after a long period of turmoil or "terror."

In summary, more than half a century after the New Life Movement, Chinese conservatism is still divided into two camps. Just like in the Republican era, one camp proposes political propositions that seem difficult or even illusory to concretize, while the other camp supports and rationalizes the authoritarianism of the party-state. In this regard, Chinese conservatism, driven by intellectuals of independent and free spirit, still remains impossible.

---

43 Xu Ben, "Zhongguo buxuyao zheyang de 'zhengzhi' he 'zhuquanzhe jueduan' – 'Shimite re' he guojiazhuyi" (China Does Not Need Such "Politics" and "Decisionism": The Schmitt-Fever and Statism), *Twenty-First Century* 94 (2005): 26, 31.
44 Kekes, *A Case for Conservatism*, 9.
45 Ibid.

Annex

The following charts present divergent opinions of traditionalists, reactionaries, and four types of conservatives on the political, economic, and cultural issues discussed in this book.

The letters mean:

S: support
D: disapproval
C: conditionally support
V: varying view
N: not applicable/non-discussed

Chart I: Social organization

	Agrarianism	Capitalism and industrialization	Equality	Elitism	Family	Social hierarchy	Private property
Traditionalism	S	D	D	S	S	S	C
Reactionaryism	N	N	D	D	D	D	N
Antimodern Conservatism	S	C	S	D	S	D	S
Liberal Conservatism	D	S	C	C	S	C	S
Philosophical Conservatism	N	D	D	S	S	S	S
Authoritarian Conservatism	C	C	D	S	S	S	S

Chart II: Political Movements

	Late Qing Reforms	Revolution of 1911	Republic	Parliament	Social organizations
Traditionalism	D	D	D	D	D
Reactionaryism	N	N	D	D	D
Antimodern Conservatism	S	C	S	D	S
Liberal Conservatism	D	S	C	C	S
Philosophical Conservatism	N	D	D	S	S

https://doi.org/10.1515/9783110740189-010

Chart II: Political Movements *(Continued)*

	Late Qing Reforms	Revolution of 1911	Republic	Parliament	Social organizations
Authoritarian Conservatism	C	C	D	S	S

Chart III: Political Ideologies

	Centralization	Communism/ Socialism	Rule by Men	Rule by Law	Monarchy
Traditionalism	S	N	S	D	D
Reactionaryism	S	V	S	S	D
Antimodern Conservatism	D	D	C	C	D
Liberal Conservatism	C	V	C	S	D
Philosophical Conservatism	N	D	N	N	D
Authoritarian Conservatism	S	D	S	S	D

Chart IV: Cultural Issues

	Inner Evilness	Westernization	Religion	Confucianism	Universality of Chinese Culture
Traditionalism	D	D	D	S	S
Reactionaryism	D	C	S	S	S
Antimodern Conservatism	D	C	D	S	V
Liberal Conservatism	D	C	C	S	V
Philosophical Conservatism	S	C	S	S	S
Authoritarian Conservatism	N	C	N	S	N

Bibliography

Newspapers, Magazines, and Journals

Chinese Students' Monthly
Dagongbao 大公報 (L'Impartial)
Diguo ribao 帝國日報 (Imperial Daily)
Dongfang zazhi 東方雜誌 (The Eastern Miscellany)
Ershi shiji dawutai 二十世紀大舞台 (20th-Century Grand Stage)
Ershi shiji zhi zhina 二十世紀之支那 (20th-Century China)
Guofeng banyuekan 國風半月刊 (National Soul Bi-Monthly)
Guogu yuekan 國故月刊 (National Heritage Monthly)
Guowenbao 國聞報 (National Information)
Guxue huikan 古學彙刊 (Journal of Ancient Learning)
Hubei Xueshengjie 湖北學生界 (Hubei Students)
Jiangsu 江蘇
Jiayin 甲寅 (Tiger)
Jingbao 晶報 (The Crystal)
Minbao 民報 (People's Tribune)
Minguo ribao 民國日報 (Republican Daily)
Minlibao 民立報 (People's Stand)
Minquanbao 民權報 (People's Rights)
New York Times
Shidi xuebao 史地學報 (Journal of Historical and Geographical Society)
Wenxue xunkan 文學彙刊 (Literary Proceedings)
Xinchao 新潮 (New Tides)
Xingshi 醒獅 (Awakening the Lion)
Xinmin congbao 新民叢報 (Journal of New People)
Xinqingnian 新青年 (New Youth)
Xueheng 學衡 (Critical Review)
Zhejiang chao 浙江潮 (The Zhejiang Trend)

Reprinted Periodicals

Guocui xuebao 國粹學報 (Journal of National Essence). Yangzhou: Guangliang shushe, 2006.
Nanshe congke 南社叢刻 (Literary Collection of the Southern Society). Yangzhou: Guangling shushe, 1996.
Nihonjin 日本人 (Japanese). Tokyo: Nihon tosho senta, 1984.
Wan Shiguo 萬世國 and Liu He 劉禾, eds. Tiany-Hengbao 天義•衡報 (Natural Justice & Equity). Beijing: Zhongguo renmin daxue chubanshe, 2016.

Reprinted Sources, Diaries, and Autobiographies

Academia Sinica, ed. *Jindai Zhongguo dui xifang ji lieqiang renshi ziliao huibian* 近代中國對西方及列強認識資料彙編 (Collected Material on Knowledge of the West and the Western Powers in China) Vol. 5. Taipei: Modern History Research Institute, Academia Sinica, 1990.

Cai, Shangsi 蔡尚思, ed. *Zhongguo xiandai sixiangshi ziliao jianbian* 中國現代思想史資料簡編 (Sources of Modern Chinese Intellectual History). Hangzhou: Zhejiang renmin chubanshe, 1982.

Chin, Hisao-yi 秦孝儀, ed. *Xian Zongtong Jianggong sixiang yanlun zongji* 先總統蔣公思想言論總集 (Works of the Former President Chiang). Taipei: Zhongyang dangshi weiyuanhui, 1984.

Chin, Hsiao-yi, ed. *Guofu quanji diyici* 國父全集第一冊 (Complete Works of Sun Yat-sen, Volume 1). Taipei: Guofu quanji bianji weiyuanhui, 1989.

De Bary, Theodore, and Richard Lufrano. *Sources of Chinese Tradition*, Vol. 2. New York: Columbia University Press, 2000.

Feng, Yuxiang 馮玉祥. *Wode shenghuo* 我的生活 (My Life). Shanghai: Jiaoyu shudian, 1947.

Gao, Tian, et al. 高恬, ed. *Gao Xie Ji* 高燮集 (Works of Gao Xie). Beijing: Zhongguo renmindaxue chubanshe, 1999.

Guo, Changhai 郭長海, and Jin Juzhen 金菊貞, eds. *Gao Xu ji* 高旭集 (Works of Gao Xu). Beijing: Shehui kexue wenxian chubanshe, 2003.

Jiang, Yihua 姜義華. *Shehui zhuyi xueshuo zai Zhongguo de chuqi chuanbo* 社會主義學說在中國的初期傳播 (Initial Diffusion of Socialism in China). Shanghai: Fudan daxue chubanshe, 1984.

John, Griffith. *The Cause of the Riots in the Yangtse Valley: A "Complete Picture Gallery."* Hankou: Hankow Mission Press, 1891.

Kang, Youwei 康有為. *Kang Youwei quanji* 康有為全集 (Complete Works of Kang Youwei). Shanghai: Shanghai guji chubanshe, 1990.

Liang, Qichao 梁啟超. *Yinbingshi wenji* 飲冰室文集 (Works of Liang Qichao), Vol. 44. Taipei: Zhonghua shuju, 1960.

Liang, Shuming 梁漱溟. *Liang Shuming quanji* 梁漱溟全集 (Complete Works of Liang Shuming). Jinan: Shandong renmin chubanshe, 2005.

Liang, Shuming. *Xiangcun jianshe lilun* 鄉村建設理論 (Theory on the Rural Construction). Beijing: Zhonghua shuju, 2018.

Liang, Shuming. *Zhongguo minzu zijiu yundong zhi zuihou juewu* 中國民族自救運動之最後覺悟 (Final Awakening of the National Self-Salvation Movement of the Chinese Nation). Shanghai: Shanghai shuju, 1933.

Liu, Dong 劉東, and Tao Wen 文韜, ed. *Shenwen yu mingbian: wanqing minguo de "guoxue" lunzheng* 審問與明辨：晚清民國的「國學」論爭 (Interrogation and Apprehension: The Question of National Learning at the End of the Qing and the Early Republican Period). Beijing: Beijing daxue chubanshe, 2012.

Liu, Shipei 劉師培. *Liu Shenshu xiansheng yishu* 劉師培先生遺書 (Works of Liu Shipei). Edition of Ningwu Nanshi, 1936.

Liu, Sifen 劉斯奮, ed. *Huang Jie shixuan* 黃節詩選 (Selected Poems of Huang Jie). Guangzhou: Guangdong renmin chubanshe, 1993.

Liu, Wu-chi 柳無忌, and Wufei Liu 柳無非, eds. *Zizhuan • Nianpu • Riji* 自傳·年譜·日記 (Autobiography, Chronology and Diaries). Shanghai: Shanghai renmin chubanshe, 1986.

Liu, Yizheng 柳詒徵. *Zhongguo wenhua shi* 中國文化史 (A History of Chinese Culture). Shanghai: Shanghai guji chubanshe, 2001.

Lu, Xun 魯迅. *Lu Xun quanji* 魯迅全集 (Complete Works of Lu Xun). Beijing: Renmin wenxue chubanshe, 2005.

Luo, Gang 羅崗, and Chunyan Chen 陳春豔, eds. *Mei Guangdi wenlu* 梅光迪文錄 (Works of Mei Guangdi). Shenyang: Liaoning jiaoyu chubanshe, 2001.

Ma, Xulun 馬敘倫. *Wo zai liushisui yiqian* 我在六十歲以前 (My Life before the Age of 60). Chongqing: Shenghuo, 1947.

Mo, Shixiang 莫世祥, ed. *Ma Junwu ji* 馬君武集 (Works of Ma Junwu). Wuhan: Huazhong shifandaxue chubanshe, 1991.

Qian, Zhonglian 錢仲聯, ed. *Gu Yanwu wenji* 顧炎武文集 (Works of Gu Yanwu). Suzhou: Suzhou daxue chubanshe, 2001.

Second Historical Archives of China, ed. *Zhonghua minguo shi dang'an shiliao huibian•disanji•jiaoyu* 中華民國史檔案史料彙編·第三輯·教育 (Archives of the Republic of China, Volume 3, Education). Nanjing: Jiangsu guji chubanshe, 1991.

Shen, Yunlong 沈雲龍, ed. *Jindai Zhongguo shiliao congkan xubian di nianqi ji* 近代中國史料叢刊續編第廿七輯 (The 27th Volume of the Second Edition of Sources on Modern Chinese History). Taipei: Wenhai chubanshe, 1976.

Shen, Yunlong, ed. *Jindai Zhongguo shiliao congkan xubian di nianba ji* 近代中國史料叢刊續編第廿八輯 (The 28th Volume of the Second Edition of Sources on Modern Chinese history). Taipei: Wenhai chubanshe, 1976.

Song, Liankui 宋聯奎. *Guanzhong congshu* 關中叢書 (Series of the Central Shaanxi Plain), Vol. 8. Shanxi tongzhiguan edition, 1936.

Sun Yat-sen Research Institute, Department of History, Sun Yat-sen University, ed. *Sun Zhongshan quanji* 孫中山全集 (Complete Works of Sun Yat-sen). Beijing: Zhonghua shuju, 1985.

Tan, Sitong 譚嗣同. *Tan Sitong quanji* 譚嗣同全集 (Complete Works of Tan Sitong), Vol. 1. Beijing: Sanlian shudian, 1954.

Tang, Yongtong 湯用彤. *Tang Yongtong quanji* 湯用彤全集 (Complete Works of Tang Yongtong). Shijiazhuang: Hebei renmin chubanshe, 2000.

Tang, Zhijun 湯志鈞, ed. *Zhang Taiyan zhenglun xuanji* 章太炎政論選集 (Political Comments of Zhang Taiyan). Beijing: Zhonghua shuju, 1977.

Tang, Zhijun, ed. *Kang Youwei zhenglunji* 康有為政論集 (Political Comments of Kang Youwei). Beijing: Zhonghua shuju, 1981.

Terrien de Lacouperie. *Western Origin of the Early Chinese Civilisation, from 2,300 B.C. to 200 A.D.* London: Asher & Co., 1894.

Tomita, Noboru 冨田昇. "Shakai shugi kōshūkai oyobi Ashū washin kai kanren shiryō" 社会主義講習会及び亜洲和親会関連資料" (Sources of the Society for the Study of Socialism and the Asiatic Humanitarian Brotherhood). In *Chūgoku bunka to sono shūhen* 中国文化とその周辺 (Chinese culture and its peripheries), by Hosoya Yoshio 細谷良夫. Sendai: Tōhoku gakuin daigaku chūgokugaku kenkyūkai, 1992.

Wen, Rumin 溫儒敏, and Ding Xiaoping 丁曉萍, eds. *Shidai zhi bo – Zhanguoce pai wenhua lunzhu jiyao* 時代之波——戰國策派文化論著輯要 (Vague of the Epoch – Cultural Comments of the *Zhanguoce* Group). Beijing: Zhongguo guangbo dianshi chubanshe, 1995.

Wu, Guoyi 鄔國義, and Wu Xiuyi 吳修藝, eds. *Liu Shipei shixue lunzhu xuanji* 劉師培史學論著選集 (Selected Historical Articles of Liu Shipei). Shanghai: Shanghai guji chubanshe, 2006.

Wu, Mi 吳宓. *Wu Mi riji* 吳宓日記 (Wu Mi's Diaries). Beijing: Sanlian shudian, 1998.

Wu, Mi. *Wu Mi shihua* 吳宓詩話 (Literary Critique of Wu Mi). Beijing: Shangwu yinshuguan, 2007.

Wu, Mi. *Wu Mi shuxinji* 吳宓書信集 (Letters of Wu Mi). Beijing: sanlian shudian, 2011.

Wu, Mi. *Wu Mi zibian nianpu* 吳宓自編年譜 (Wu Mi's Self-Compiled Chronological Biography). Beijing: Sanlian shudian, 1998.

Xu, Jingbo 徐靜波, ed. *Liang Shiqiu piping wenji* 梁實秋批評文集 (Liang Shiqiu's critiques). Zhuhai: Zhuhai chubanshe, 1998.

Yan, Fu 嚴復. *Shehui tongquan* 社會通詮 (A General Interpretation of Society). Beijing: Shangwu yinshuguan, 1981.

Yan, Fu. *Tianyan Lun* 天演論 (On Evolution). Beijing: Shangwu yinshuguan, 1981.

Yao, Kunqun 姚昆群, et al., eds. *Yao Guang quanji* 姚光全集 (Complete Works of Yao Guang). Beijing: Shehui kexue wenxian chubanshe, 2007.

Zhang, Junmai 張君勱, et al. *Kexue yu renshengguan* 科學與人生觀 (Science and the Conception of Life). Jinan: Shandong renmin chubanshe, 1997.

Zhang, Pinxing 張品興, et al., eds. *Liang Qichao quanji* 梁啟超全集 (Complete Works of Liang Qichao). Beijing: Beijing chubanshe, 1999.

Zhang, Shizhao 章士釗. *Zhang Shizhao quanji* 章士釗全集 (Complete Works of Zhang Shizhao). Shanghai: Wenhui chubanshe, 2000.

Zhang, Taiyan 章太炎. *Zhang Taiyan quanji* 章太炎全集 (Complete Works of Zhang Taiyan). Shanghai: Shanghai renmin chubanshe, 1984.

Zhang, Yi 張夷, ed. *Chen Qubing quanji* 陳去病全集 (Complete Works of Chen Qubing). Shanghai: Shanghai guji chubanshe, 2009.

Zhang, Zhan 張枬, and Renzhi Wang 王忍之, eds. *Xinhai geming qianshinianjian shilun xuanji* 辛亥革命前十年間時論選集 (Anthology of Political Comments in the Decade before the 1911 Revolution). Beijing: Sanlian shudian, 1960.

Zhang, Zhidong 張之洞. *Quanxuepian* 勸學篇 (Exhortation to Learning). Zhengzhou: Zhongzhou guji shubanshe, 1998.

Zhi, Zhi 郅志, ed. *Meng huitou: Chen Tianhua Zou Rong ji* 猛回頭：陳天華鄒容集 (Fierce Back: Works of Chen Tianhua and Zou Rong). Shenyang: Liaoning renmin chubanshe, 1994.

References

Aldridge, A. Owen. "Irving Babbitt in and about China." *Modern Age* 4 (1993): 332–339.

Alitto, Guy S. *The Last Confucian: Liang Shuming and the Chinese Dilemma of Modernity.* Berkeley: University of California Press, 1986.

Altermatt, Urs. "Conservatism in Switzerland: A Study in Antimodernism." *Journal of Contemporary History* 14, no. 4 (1979): 583–584.

Anderson, Benedict. *Imagined Communities: Reflections on the Origins and Spread of Nationalism.* London: Verso, 1991.

Anderson, Robert Thomas. *Denmark: Success of a Developing Nation*. Cambridge, Mass.: Cambridge University Press, 1975.

Arnold, Matthew. *Essays in Criticism*. London and Cambridge: Macmillan and Co., 1865.

Asanuma, Chie 浅沼千恵. "Guanyu Mingzhi moqi riben jiaoxi paiqian beijing de yanjiu" 關於明治末期日本教習派遣背景的研究 (On the Context in Which Japan Sent Consultants to China at the End of the Meiji Era). In *Zhejiang yu Riben* 浙江與東亞 (Zhejiang and Japan), 265–284. Beijing: Zhongguo shehui kexue chubanshe, 2019.

Aughey, Arthur, at al. *The Conservative Political Tradition in Britain and the United States*. London: Pinter Publishers, 1992.

Babbitt, Irving. *Democracy and Leadership*. Boston and New York: Houghton Mifflin Company, 1934.

Babbitt, Irving. *Literature and American College: Essays in Defense of the Humanities*. Boston and New York: Houghton Mifflin Company, 1908.

Babbitt, Irving. *Rousseau and Romanticism*. Boston and New York: Houghton Mifflin Company, 1919.

Bagehot, Walter. *The English Constitution*. Boston: Little, Brown, and Company, 1873.

Bakounine, Michel. *Œuvres*. Paris: P. V. Stock, 1895.

Banerjee, Milinda, et al., eds. *Transnational Histories of the 'Royal Nation'*. London: Palgrave Macmillan, 2017.

Bergson, Henri. *L'évolution créatrice*. 1907; repr. Paris: 1959.

Berlin, Isaiah. "Two Concepts of Liberty." In *Four Essays on Liberty*, 118–172. Oxford: Oxford University Press, 1969.

Bieler, Stacey. *"Patriots or Traitors"? A History of American-Educated Chinese Students*. London and New York: Routledge, 2015.

Billingsley, Phil. *Bandits in Republican China*. Stanford: Stanford University Press, 1988.

Bjerg, Jens, et. al. "Danish Education, Pedagogical Theory in Denmark and in Europe, and Modernity." *Comparative Education* 1 (1995): 31–47.

Blanckert, Claude. *La nature de la société: organicisme et sciences sociales au XIXe siècle*. Paris: L'Harmattan, 2004.

Bonner, Stephan Eric, ed. *Twentieth Century Political Theory: A Reader*. New York: Routledge, 2006.

Bugge, K.E. "The International Dissemination of Grundtvig's Educational Ideas: I: Motivation and Interpretation." *Grundtvig-Studier* 1 (2012): 168–177.

Butler, Eammon, and Madsen Pirie, eds. *Hayek – On the Fabrik of Human Society*. London: Adam Smith Institute, 1987.

Calhoun, Craig. *Nationalism*. Minneapolis: University of Minneapolis Press, 1997.

Calinescu, Matei. *Five Faces of Modernity: Modernism, Avant-garde, Decadence, Kitsch, Postmodernism*. Durham: Duke University Press, 1987.

Chan, Hok-yin 陳學然. "Gexing qishi: 'Dongnan hubao' yu 'Liangguang duli' zhong zhi gefang zhengzhi choumou" 各行其是：「東南互保」與「兩廣獨立」中之各方政治籌謀 (Dynamics of different political policies: From Southern Mutual Protection to Guangdong and Guangxi independence). *Historical Inquiry* 49 (2012): 65–109.

Chan, Yao-chi 詹曜齊. "Qishi niandai de 'xiandai' lailu: jizhang sumiao" 七十年代的「現代」來路：幾張素描 (Origins of the modernity of the 1970s: some sketch). *Sixiang* 4 (2007): 115–140.

Chang, Hao. *Chinese Intellectuals in Crisis*. Berkeley: University of California Press, 1987.

Chang, Peng-yuan 張朋園. "Cong Minguo chuqi guohui kan zhengzhi canyu – jianlun tuibian zhong de zhengzhi youyi fenzi" 從民國初期國會看政治參與——簡論蛻變中的政治優異份子 (Political Participation in the Early Republican Parliament Election – on the Political Elites in the Transition Period). *Guoli Taïwan shifan daxue lishi xuebao* 7 (1976): 363–447.

Chang, Peng-yuan. *Liang Qichao yu Minguo zhengzhi* 梁啟超與民國政治 (Liang Qichao and Republican Politics). Shanghai: Sanlian shudian, 2013.

Chang, Peng-yuan. *Minguo chunian de zhengdang* 民國初年的政黨 (Political Parties of the Early Republic Era). Changsha: Qiulu shushe, 2004.

Chang, Yu-fa 張玉法. *Qingji de geming tuanti* 清季的革命團體 (Revolutionary Associations at the End of the Qing). Taipei: Institute of Modern History, Academia Sinica, 1982.

Charles, Christophe. *Naissance des intellectuels:1880–190*. Paris: Les Éditions de Minuit, 1990.

Cheek, Timothy. *The Intellectual in Modern Chinese History*. Cambridge: Cambridge University Press, 2015.

Chen, Chun-chi 陳俊啟. "Wu Mi yu Xinwenhua yundong" 吳宓與新文化運動 (Wu Mi and the New Culture Movement). *Bulletin of the Institute of Modern History, Academia Sinica* 56 (2007): 45–89.

Chen, Gonglu 陳恭祿. *Zhongguo jindaishi* 中國近代史 (Chinese Modern History). Hong Kong: Open Page, 2017.

Chen, Jianhua 陳建華. "Minguo chuqi xiaoxian zazhi yu nuxing huayu de zhuanxing" 民國初期消閒雜誌與女性話語的轉型 (The Magazines of Leisure and the Transition of the Female Discourses at the Early Republican Era). *Zhongzheng hanxue yanjiu* 22 (2013): 355–386.

Chen, Jianhua. "Minguo chuqi Zhou Shoujuan de xinli xiaoshuo – jianlun 'Libailiu pai' yu 'Yuanyang hudie pai' zhi bie" 民國初期周瘦鵑的心理小說——兼論「禮拜六派」與「鴛鴦蝴蝶派」之別 (Zhou Shoujuan's Psychological Fiction in the Early Republican Era and the Difference Between the *Saturday* School and the Mandarin Ducks and Butterflies School). *Xiandai zhongwen xuekan* 11 (2011): 37–49.

Chen, Kaiyi. "Missionaries and the Early Development of Nursing in China." *Nursing History Review* 4 (1996): 129–49.

Chen, Lai. *Tradition and Modernity: A Humanist View*. Leiden: Brill, 2009.

Chen, Pingyuan 陳平原, ed. *Xindai Zhongguo* 現代中國 (Modern China), Vol. 6. Beijing: Beijing daxue chubanshe, 2005.

Chen, Pingyuan. *Wanqing tuxiang:* Dianshizhai huabao *zhiwai* 晚清圖像：《點石齋畫報》之外 (Illustrating the Late Qing: Besides *Dianshizhai* Illustrated Newspaper). Hong Kong: Open Page, 2015.

Chen, Qi 陳奇. *Liu Shipei nianpu changbian* 劉師培年譜長編 (Chronological Biography of Liu Shipei). Guiyang: Guizhou renmin chubanshe, 2007.

Chen, Xulu 陳旭麓. "Wuxu shiqi weixinpai de sheshuiguan – qunxue" 戊戌時期維新派的社會觀——群學 (The Concept of Society among the Reformers of the Wuxu Era – *Qunxue*). *Jindaishi yanjiu* 近代史研究 2 (1984): 161–175.

Cheng, Anne. *Histoire de la pensée chinoise*. Paris: Seuil, 1997.

Cheung, Chiu-yee 張釗貽. "Lu Xun yu Nicai fan 'xiandaixing' de qihe" 魯迅與尼采反「現代性」的契合 (The Similarities in the Anti-modernism of Lu Xun and Nietzsche). *Twenty-First Century* 29 (1995): 91–96.

Chiang, Yung-chen 江勇振. *Shewo qishe: Hu Shi* 捨我其誰：胡適 (Who Else but Me: Hu Shi), Vol. 3. Taipei: Linking, 2018.
Ching, Chih-jen 荊知仁. *Zhongguo lixianshi* 中國立憲史 (History of Constitutionalization in China). Taipei: Linking, 1984.
Ching, May Bo 程美寶. "You ai xiang er ai guo: Qing mo Guangdong xiangtu jiaocai de guojia huayu" 由愛鄉而愛國：清末廣東鄉土教材的國家話語 (From Loving One's Native Place to Loving the Nation: the National Discourse on Native-Place Textbooks of Guangdong in Late Qing). *Lishi yanjiu* 4 (2003): 68–84.
Chiu, Eugene. *Qimeng, lixing yu xiandaixing: jindai Zhongguo qimeng yundong, 1895–1925* 啟蒙、理想與現代性：近代中國啟蒙運動, 1895–1925 (Enlightenment, Reason and Modernity: Modern Chinese Movement of Enlightenment, 1895–1925). Taipei: National Taiwan University Press, 2019.
Cho, Se-Hyun 曹世鉉. *Qingmo minchu wuzhengfupai de wenhua sixiang* 清末明初無政府派的文化思想 (Cultural Thoughts of Anarchism at the End of the Qing and the Early Republic). Beijing: Kexue wenxian chubanshe, 2003.
Chow, Kai-Wing, et al., eds. *Beyond the May Fourth Paradigm: In Search of Chinese Modernity*. Lanham: Lexington Books, 2008.
Clinton, Maggie. *Revolutionary Nativism: Fascism and Culture in China, 1925–1937*. Durham: Duke University Press, 2017.
Coble, Parks M. Jr. *The Shanghai Capitalists and the Nationalist Government, 1927–1937*. Cambridge, Mass.: Council on East Asian Studies, Harvard University Press, 1986.
Cohen, G.A. *Finding Oneself in the Other*. Princeton: Princeton University Press, 2013.
Cohen, Paul A., and John E. Schrecker, eds. *Reform in 19th Century China*. Cambridge, Mass.: East Asian Research Center, Havard University, 1976.
Cohen, Paul A., and Merle Goldman, eds. *Ideas Across Cultures: Essays on Chinese Thought in Honor of Benjman I. Schwartz*. Cambridge, Mass.: Council on East Asian Studies, Harvard University, 1990.
Compagnon, Antoine. *Les Antimodernes. De Joseph de Maistre à Roland Barthes*. Paris: Gallimard, 2005.
Confino, Michael. "On Intellectuals and Intellectual Traditions in Eighteenth- and Nineteenth-Century Russia." *Daedalus* 2 (1972): 117–149.
Cox, Judy. "An Introduction to Marx's Theory of Alienation." *International Socialism, Quarterly Journal of the Socialist Workers Party (Britain)* 79 (1998). http://pubs.social istreviewindex.org.uk/isj79/cox.htm. Accessed February 8, 2018.
Crossley, Pamela Kyle. "Manzhou Yuanliu Kao and the Formalization of the Manchu Heritage." *The Journal of Asian Studies* 46, no. 4 (1987): 761–790.
Daruvala, Susan. *Zhou Zuoren and an Alternative Chinese Response to Modernity*. Cambridge, Mass.: Harvard University Asia Center, 2000.
Delmas-Marty, Mireille, and Pierre-Étienne Will, eds. *La démocratie et la Chine*. Paris: Fayard, 2007.
Ding, Wenjiang 丁文江, and Zhao Fengtian 趙豐田. *Liang Qichao nianpu changbian* 梁啟超年譜長編 (A Chronological Biography of Liang Qichao). Shanghai: Shanghai renmin chubanshe, 1983.
Dirlik, Arif. *Anarchism in the Chinese Revolution*. Berkeley, Los Angeles and London: University of California Press, 1991.
Disraeli, Benjamin. *Coningsby, or The New Generation*. Leipzig: Bernh. Tauchnitz Jun., 1844.

Disraeli, Benjamin. *Vindication of the English Constitution*. London: Saunders and Otley, 1835.
Doak, Kevin. *A History of Nationalism in Modern Japan: Placing the People*. Leiden: Brill, 2007.
Doleželová-Velingerová, Milena, et al. *The Appropriation of Cultural Capital: China's May Fourth Project*. Cambridge, Mass.: Cambridge University Press, 2001.
Du, Chunmei. *Gu Hongming's Eccentric Chinese Odyssey*. Philadelphia: University of Pennsylvania Press, 2019.
Duan, Lian 段煉. *Shisu shidai de yiyi tanxun – Wusi qimeng sixiang zhong de xindaodeguan yanjiu* 世俗時代的意義探詢——五四啟蒙思想中的新道德觀研究 (Search for Meaning in a Secular Age – Study of the New Moral in the Thought of Enlightenment of the May Fourth Period). Taipei: Xiuwei chuban, 2012.
Dupeux, Louis, ed. *La "Révolution conservatrice" dans l'Allemagne de Weimar*. Paris: Kimé, 1992.
Earl, David Magarey. *Emperor and Nation in Japan, Political Thinkers and the Tokugawa Period*. Seattle: University of Washington Press, 1964.
Eastman, Lloyd. *Seeds of Destruction: Nationalist China in War and Revolution, 1937–1949*. Stanford, CA: Stanford University Press, 1984.
Eastman, Lloyd. *The Abortive Revolution: China under Nationalist Rule, 1927–1937*. Cambridge, MA: Harvard University Press, 1974.
Elias, Norbert. *La civilisation des mœurs*, translated by Pierre Kamnitzer. Paris: Pocket, 2015.
Elman, Benjamin A. *On Their Own Terms: Science in China 1550–1900*. Cambridge, Mass.: Harvard University Press, 2005.
Elwick, James. "Herbert Spencer and the Disunity of the Social Organism". *History of Science* 41 (2003): 35–72.
Esherick, Joseph W. *The Origins of the Boxer Uprising*. Berkeley: University of California Press, 1987.
Fang, Weigui 方維規. "Lun jindai sixiangshi shang de 'minzu,' 'nation' yu 'Zhongguo'" 論近代思想史上的「民族」、「Nation」與「中國」 (On *Minzu*, Nation and China in Modern Intellectual History). *Twenty-First century* 70 (2002): 33–43.
Feng, Youlan 馮友蘭. *Zhongguo zhexueshi* 中國哲學史 (A History of Chinese Philosophy). Beijing: Zhonghua shuju, 1961.
Finkel, Steven E., and Karl-Dieter Opp. "Party Identification and Participation in Collective Political Action." *The Journal of Politics* 53, no. 2 (1991): 339–371.
Fogel, Joshua A., and Peter G. Zarrow, eds. *Imagining the People: Chinese Intellectuals and the Concept of Citizenship*. Armonk: M. E. Sharpe, 1997.
Forster, Elisabeth. *1919 – The Year that Changed China: A New History of the New Culture Movement*. Berlin: De Gruyter, 2018.
Foucault, Michel. *Qu'est-ce que les Lumières?* Rosny: Bréal, 2004.
Frohnen, Bruce. *Virtue and the Promise of Conservatism: The Legacy of Burke and Tocqueville*. Lawrence: University Press of Kansas, 1993.
Fukuzawa, Yukichi 福沢諭吉. *Minjō Isshin* 民情一新 (Transition of People's Way of Thinking). Tokyo: Joshazohan,1879.
Fung, Edmund S.K. "Were Chinese Liberals Liberal? Reflections on the Understanding of Liberalism in Modern China." *Political Affairs* 81, no. 4 (2008/2009): 557–576.

Fung, Edmund S.K. "The Idea of Freedom in Modern China Revisited: Plural Conceptions and Dual Responsibility." *Modern China* 32, no. 4 (2006): 453–482.

Fung, Edmund S.K. *The Intellectual Foundations of Chinese Modernity: Cultural and Political Thought in the Republican Era.* New York: Columbia University Press, 2010.

Furth, Charlotte Furth, ed. *The Limits of Change: Essays on Conservative Alternatives in Republican China.* Cambridge, Mass.: Harvard University Press, 1976.

Gan, Guoxun 干國勛, et. al., eds. *Lanyishe, Fuxingshe, Lixingshe* 藍衣社 復興社 力行社 (The Blue Shirts, The Society for Vigorous Practice, The Renaissance Society). Beijing: Zhonghua shuju, 2014.

Gavin, Masako. *Shiga Shigetaka, 1863–1927: The Forgotten Enlightener.* London: Routledge Curzon, 2001.

Gellner, Ernest. *Nation and Nationalism.* Oxford: Basil Blackwell, 1983.

Gernet, Jacques. *La vie quotidienne en Chine: à la veille de l'invasion mongole (1250–1276).* Paris: Hachette, 1978.

Giddens, Anthony. *The Consequences of Modernity.* Stanford: Stanford University Press, 1990.

Goosaert, Vincent. "1898: The Beginning of the End for Chinese Religion?" *Journal of Asian Studies* 65, no. 2 (2006): 307–336.

Gray, John. *Liberalism.* Milton Keynes: Open University Press, 1995.

Greenfeld, Liah. "Nationalism and Modernity." *Social Research* 63, no. 1 (1996): 3–40.

Grypma, Sonya. *Healing Henan: Canadian Nurses at the North China Mission, 1888–1947.* Vancouver and Toronto: UBC Press, 2008.

Guo, Ruoping 郭若平. "Guo Gong hezuo yu Feijidujiao yundong de lishi kaocha" 國共合作與非基督教運動的歷史考察 (The United Front and the History of Anti-Christian Movement). *Zhonggong dangshi yanjiu* 2 (2008): 49–57.

Habermas, Jürgen. "The European nation-state. Its achievements and its limitations. On the past and future of sovereignty and citizenship." *Ratio Juris* 19, no. 2 (1996): 125–137.

Hall, Stuart, and Bram Gieben, eds. *Formations of Modernity.* Cambridge: Polity Press, 1992.

Hao, Shiyuan 郝時遠. "Zhongwen 'minzu' yici yuanliu kaobian" 中文「民族」一詞源流考辨 (On the Origin of the Word "Minzu" in the Chinese language). *Minzu yanjiu* 6 (2004): 60–69.

Hayek, F.A. "Individualism: True and False." In *Individualism and Economic Order*, by F.A. Hayek, 1–32. Chicago: The University of Chicago Press, 1948.

Hayford, Charles. *To the People: James Yen and Village China.* New York: Columbia University Press, 1990.

He, Fangyu 何方昱. *Kexue shidai de renwen zhuyi: Sixiang yu shidai yuekan (1941–1948) yanjiu* 科學時代的「人文主義」：《思想與時代》月刊 (1941–1948)研究 (Humanism in the Era of Science: Research on the Magazine *Thoughts and Epoch* (1941–1948)). Shanghai: Shanghai shudian, 2008.

He, Xiaoming 何曉明. *Fanben yu kaixin – jindai Zhongguo wenhua baoshou zhuyi xinlun* 返本與開新——近代中國文化保守主義新論 (Reconstruction and Renovation – New Arguments on Chinese Modern Cultural Conservatism). Beijing: Shangwu yinshuguan, 2006.

Herbert, Spencer. *Social Statics, or The Conditions Essential to Human Happiness Specified, and the First of Them Developed.* London: John Chapman, 1851.

Hill, Michael Gibbs. *Lin Shu, Inc.: Translation and the Making of Modern Chinese Culture.* Oxford: Oxford University Press, 2013.

Hobsbawm, Eric, and Terence Ranger, eds. *The Invention of Tradition*. Cambridge: Cambridge University Press, 1983.
Hockx, Michel, ed. *The Literary Field of Twentieth-Century China*. Honolulu: University of Hawai'i Press, 1999.
Hon, Tze-Ki. "Cultural Identity and Local Self-Government: A Study of Liu Yizheng's 'History of Chinese Culture'." *Modern China* 30, no. 4 (2004): 506–542.
Hon, Tze-ki. *Revolution as Restoration: Guocui xuebao and China's Path to Modernity, 1905–1911*. Leiden: Brill, 2013.
Howland, Douglas. "The Dialectics of Chauvinism: Minority Nationalities and Territorial Sovereignty in Mao Zedong's New Democracy." *Modern China* 37, no. 2 (2011): 170–201.
Hsu, Immanuel C.Y. *The Rise of Modern China*. New York: Oxford University Press, 1970.
Hu, Fengxiang 胡逢祥. *Shehui biange yu wenhua chuantong: Zhongguo jindai wenhua baoshouzhuyi sichao yanjiu* 社會變革與文化傳統：中國近代文化保守主義思潮研究 (Social Evolution and Cultural Tradition: Study on Chinese Modern Cultural Conservatism). Shanghai: Shanghai renmin chubanshe, 2000.
Hu, Shi 胡適, et al. *Huainian Fu Sinian* 回憶傅斯年 (In Memory of Fu Sinian). Taipei: Xiuwei chuban, 2014.
Huang, Ray. *China: A Macro History*. Armonk and London: M.E. Sharpe, INC., 1997.
Huang, Xingtao 黃興濤. "'Minzu' yici jiujing heshi zai zhongwen li chuxian?" 「民族」一詞究竟何時在中文裡出現？ (When Did the Word *Minzu* Appear in the Chinese Language?). *Zhejiang xuekan* 1 (2002): 168–170.
Huguenin, François. *Le conservatisme impossible: libéralisme et réactionnaires en France depuis 1789*. Paris: Édition de La Table Ronde, 2006.
Huntington, Samuel P. *Political Order in Changing Societies*. New Haven: Yale University Press, 1973.
Huntington, Samuel P. "Conservatism as an Ideology." *The American Political Science Review* 2 (1957): 454–473.
Hutchinson, John. *Dynamics of Cultural Nationalism: The Gaelic Revival and the Creation of the Irish Nation State*. London: Allen & Unwin, 1987.
Huters, Theodore. *Bringing the World Home: Appropriating the West in Late Qing and Early Republican China*. Honolulu: University of Hawai'i Press, 2005.
Hwang, Jinlin 黃金麟. "Jindai Zhongguo de junshi shenti jian'gou, 1895–1949" 近代中國的軍事身體建構 (The Construction of Military Body in Modern China 1895–1949). *Bulletin of the Institute of Modern History, Academia Sinica* 43 (2004): 173–221.
Ishikawa, Yoshihiro 石川禎浩. "Ershi shiji chunian Zhongguo liuri xuesheng "huangdi" zhi zaizao – paiman, xiaoxiang, xifang qiyuan lun" 20世紀初年中國留日學生「黃帝」之再造——排滿、肖像、西方起源論 (The Rebuilding of "Huang Di" in the 20th Century: Excluding the Manchu, Portraits and Western Originality Theory). *Qingshi yanjiu* 4 (2005): 51–62.
Jean, Roger B. Jr. *Democracy and Socialism in Republican China: The Politics of Zhang Junmai (Carsun Chang): 1906–1941*. Lanham: Rowman & Littlefield Publishers, Inc., 1997.
Jenks, Edward. *A History of Politics*. London: J. M. Dent, 1900.
Ji, Xiao-bin. *Politics and Conservatism in Northern Song China: The Career and Thought of Sima Guang (A.D. 1019–1086)*. Hong Kong: The Chinese University Press, 2005.
Jiang, Guo 姜國. *Nanshe xiaoshuo yanjiu chutan* 南社小說研究初探 (Preliminary Study on the Southern Society's Novels). Changchun: Jilin daxue chubanshe, 2012.

Jiang, Qing 蔣慶 et al., eds. *Zhongguo bixu zairuhua – "dalu xinrujia" xinzhuzhang* 中國必須再儒化——「大陸新儒家」新主張 (China Has to Be Re-Confucianized – Appeals of Chinese New Confucians). Singapore: Global Publishing, 2016.

Jiang, Qing. *Zhengzhi ruxue – dangdai ruxue de zhuanxiang, tezhi yu fazhan* 政治儒學——當代儒學的轉向、特質與發展 (Political Confucianism: Transition, Particularities and Development of Contemporary Confucianism). Beijing: Sanlian shudian, 2003.

Jin, Guantao, and Qingfeng Liu. *Xingsheng yu weiji: lun Zhongguo shehui chaowending jiegou* 興盛與危機：論中國社會超穩定結構 (The Cycle of Growth and Decline: On the Ultra-Stable Structure of Chinese society). Hong Kong: The Chinese University Press, 1992.

Jin, Guantao 金觀濤, and Qinfeng Liu 劉青峰. "Cong 'qun' dao 'shehui', 'shehuizhuyi' – Zhongguo jindai gonggong lingyu bianqian de sixiangshi yanjiu" 從「群」到「社會」、「社會主義」——中國近代公共領域變遷的思想史研究 (From "Grouping" to "Society" and "Socialism" – An Intellectual History on the Evolution of Chinese Modern Public Sphere). *Bulletin of the Institute of Modern History, Academia Sinica* 35 (2011): 1–66.

Jing, Tsu, and Benjamin A. Elman, eds. *Science and Technology in Modern China, 1880s–1940s*. Leiden: Brill, 2014.

Jones, Emily. *Edmund Burke & The Invention of Modern Conservatism, 1830–1914*. Oxford: Oxford University Press, 2017.

Judge, Joan. "Talent, Virtue, and the Nation: Chinese Nationalisms and Female Subjectivities in the Early Twenty Century." *The American Historical Review* 106, no. 3 (2001): 765–803.

Kalberg, Stephan. "Max Weber's Types of Rationality: Cornerstones for the Analysis of Rationalization Processes in History." *The American Journal of Sociology* 85, no. 5 (1980): 1145–1179.

Karl, Rebecca E., and Peter Zarrow, eds. *Rethinking the 1898 Reform Period: Political and Cultural Change in Late Qing China*. Cambridge, Mass.: Harvard University Press, 2002.

Karpik, Lucien. *Les avocats. Entre l'État, le public et le marché, XIIIe–XXe siècle*. Paris: Gallimard, 1995.

Kekes, John. *A Case for Conservatism*. Ithaca and London: Cornell University Press, 1998.

Kirk, Russell. *The Conservative Mind: From Burke to Eliot*. Washington, D.C.: Regnery Publishing, Inc., 2001.

Kolozi, Peter. *Conservatives Against Capitalism: From the Industrial Revolution to Globalization*. New York: Columbia University Press, 2017.

Kuhn, Philip A. "Ideas Behind China's Modern State." *Harvard Journal of Asiatic Studies* 55, no. 2 (1995): 295–337.

Kuo, Ting-yee 郭廷以. *Jindai Zhongguo shigang* 近代中國史綱 (History of Modern China). Hong Kong: The Chinese University Press, 2012.

Kuo, Ya-Pei. "Polarities and the May Fourth Polemical Culture: Provenance of the 'Conservative' Category." *Twentieth-Century China* 2 (2019): 174–189.

Lee, Kuo-chih 李國祁. *Zhongguo zaoqi de tielu jingying* 中國早期的鐵路經營 (Early Railway Management in China). Taipei: Institute of Modern History, Academia Sinica, 2015.

Lee, Ming-huei 李明輝. "Dangdai xinrujia 'ruxue kaichu minzhu lun' de lilun yihan yu xianshi yiyi" 當代新儒家「儒學開出民主論」的理論意涵與現實意義 (The Theoretical Meanings and the Pragmatic Significations of the "Development of Democracy from Confucianism" of the Contemporary New Confucianism). *Asian Studies* 2, no. 1 (2014): 7–18.

Lee, Ou-fan 李歐梵. *Zhongguo xiandai wenxue yu xiandaixing shijiang* 中國現代文學與現代性十講 (Ten lessons on Modern Chinese Literature and Modernity). Shanghai: Fudan daxue chuanshe, 2008.

Leibold, James. "Competing Narratives in Republican China: From the Yellow Emperor to Peking Man." *Modern China* 32, no. 2 (2006): 181–220

Lenin, Viladimir. *L'impérialisme, stade suprême du capitalism.* 1916. http://marxiste.fr/lenine/imp.pdf. Accessed October 21, 2017.

Levenson, Joseph R. *Confucian China and Its Modern Fate: A Trilogy.* Berkeley and Los Angeles: University of California Press, 1968.

Li, Changli 李長莉. *Zhongguo jindai shehui shenghuo shi* 中國近代社會生活史(History of Chinese Modern Social Life). Beijing: Zhongguo shehui kexue chubanshe, 2015.

Li, Feng. "'Feudalism' and Western Zhou China: A Criticism." *Harvard Journal of Asiatic Studies* 63, no. 1 (2003): 115–144.

Li, Hsiao-t'I 李孝悌. *Qingmo de xiacengshehui qimeng yundong* 清末的下層社會啟蒙運動 (Movement of Social Enlightenment of the Inferior Society at the End of the Qing). Taipei: Institute of Modern History, Academia Sinica, 1998.

Li, Jiannong 李劍農. *Zhongguo jinbainian zhengzhishi (1840–1926)* 中國近百年政治史 (Political History of the Last 100 years of China). Shanghai: Fudan daxue chubanshe, 2002.

Li, Jing 李靜. *Xin Qingnian Zazhi huayu yanjiu* 《新青年》雜誌話語研究 (On the Discourse of the *New Youth*). Tianjin: Tianjin daxue chubanshe, 2010.

Li, Jinxi 黎錦熙. *Guoyu yundong shigang* 國語運動史綱 (History of the Movement of National Language). Shanghai: Shangwu yinshuguan, 1935.

Li, Jinzheng 李金錚. "Yan Yangchu yu dingxian pingmin jiaoyu shiyan" 晏陽初與定縣平民教育實驗 (Yan Yangchu and the Experiment of Popular Education in the County of Ding). *Twenty-First Century* 85 (2004): 64–73.

Li, Nan 李楠. *Wanqing, Minguo shiqi Shanghai xiaobao yanjiu* 晚清、民國時期上海小報研究 (Study on Tabloids in Shanghai from the late Qing Period to the Republican Era). Beijing: Renmin wenxue chubanshe, 2005.

Li, Shu-chen 李淑珍. *Anshen liming: xiandai Huaren gongsi lingyu de tansuo yu chongjian* 安身立命：現代華人公私領域的探索與重建 (Settling and Upholding Oneself: Discussion and Reconstruction of the Private and Public Sphere of Modern Chinese). Taipei: Linking, 2013.

Li, Xiangying 李向英. "Seikyōsha no tai shinninshiki: zasshi *Nihonjin* wo chūshin ni" 政教社の対清認識：雑誌「日本人」を中心に (The Qing Dynasty in the Eyes of the Society for Political Education: On the Magazine *Japanese*). *Nihon kenkyū* 日本研究 18 (2015): 92–101.

Li, Xiaobing, and Patrick Fuliang Shan, eds. *Ethnic China: Identity, Assimilation, and Resistance.* Lanham: Lexington Books, 2005.

Li, Xizhu 李細珠. *Wanqing baoshou sixiang de yuanxing: Woren yanjiu* 晚清保守思想的原型：倭仁研究 (The Archetype of Late Qing Conservatism: Research on Woren). Beijing: Shehui kexue wenxian chubanshe, 2000.

Li, Zehou 李澤厚. *Zhongguo xiandai sixiangshilun* 中國現代思想史論 (On Modern Chinese Intellectual History). Beijing: Dongfang chubanshe, 1987.

Li, Zhiting 李治亭. *Qingshi* 清史 (A History of the Qing Dynasty). Shanghai: Shanghai renmin chubanshe, 2002.

Lin, Chih-hung 林志宏. *Minguo nai diguo ye: zhengzhi wenhua zhuanxing xia de Qing yimin* 民國乃敵國也：政治文化轉型下的清遺民 (Republic of China Is the Enemy State: Qing Loyalists under the Transformation of Politics and Culture). Taipei: Linking, 2009.

Lin, Hongming 林紅明, and Jian'gang Xu 許建剛. "Zhang Junmai de minzhu zhengzhi he zhongjianluxian sixiang pouxi" 張君勱的民主政治和中間路線思想剖析 (On Zhang Junmai's Democratic Politics and Centralist Line). *Tianzhong xuekan* 22, no. 1 (2007): 99–101.

Lin, Xiaoqing. "Historicizing Subjective Reality: Rewriting History in Early Republican China." *Modern China* 1 (1999): 3–43.

Lin, Yü-sheng. *The Crisis of Chinese Consciousness: Radical Antitraditionalism in the May Fourth Era.* Madison: University of Wisconsin Press, 1979.

Link, Perry. *Mandarin Ducks and Butterflies. Popular Fiction in Early Twentieth-Century Chinese Cities.* Berkeley: University of California Press, 1991.

Liu, Dingsheng 柳定生, and Nianzeng Liu 柳曾符, eds. *Liu Yizheng Qutang tiba* 柳詒徵劬堂題跋 (Inscriptions and Postfaces of Liu Yizheng). Taipei: Huashi, 1996.

Liu, Junning 劉軍寧. *Baoshou zhuyi* 保守主義 (Conservatism). Beijing: Zhongguo shehui kexue chubanshe, 1998.

Liu, Lydia. *The Clash of Empires: The Invention of China in Modern World Making.* Cambridge, Massachusetts and London: Harvard University Press, 2004.

Liu, Lydia. *Translingual Practice: Literature, National Culture, and Translated Modernity – China, 1900–1937.* Stanford: Stanford University Press, 1995.

Liu, Tong 劉統. *Tangdai jimizhou yanjiu* 唐代羈縻府州研究 (On *Jimizhou* of the Tang Dynasty). Xi'an: Xibei daxue chubanshe, 1998.

Liu, Wu-chi, ed. *Nanshe Jilue* 南社紀略 (History of the Southern Society). Shanghai: Shanghai renmin chubanshe, 1983.

Liu, Wu-chi, and Anru Yin 殷安如. *Nanshe renwuzhuan* 南社人物傳 (Biographies of the Members of the Southern Society). Beijing: Shehui kexue chubanshe, 2002.

Liu, Xiaobo 劉曉波. "Jiushi niandai Zhongguo de zhengzhi baoshou zhuyi" 九十年代中國的政治保守主義 (Chinese Conservatism of the mid-1990s). *Beijing zhichun* 北京之春 (*Beijing Spring*) 43 (1996). http://bjzc.org/bjs/bc/43/55. Accessed February 15, 2018.

Liu, Xiaofeng 劉小楓, and Liwei Lin 林立偉, eds. *Jingji lunlu yu jinxiandai Zhongguo shehui* 經濟倫理與近現代中國社會 (Economic Ethos and Modern Chinese society). Hong Kong: The Chinese University Press, 1998.

Lu, Fang-sang 呂芳上. *Minguo shilun* 民國史論 (On the History of the Republic of China). Tapei: Shangwu yinshuguan, 2013.

Luan, Meijian 欒梅健. *Minjian de wenren yaji: Nanshe yanjiu* 民間的文人雅集：南社研究 (Elegant Assembly of the Men of Letters: A Study on the Southern Society). Shanghai: Dongfang chubanshe zhongxin, 2016.

Luo, Gang. "Lishi zhong de Xueheng" 歷史中的學衡 (The Critical Review Group in History). *Twenty-First Century* 28 (1995): 40–48.

Luo, Zhitian 羅志田. "Qingji baocunguocui de chaoye nuli jiqi guannian yitong" 清季保存國粹的朝野努力及其觀念異同 (Efforts to Preserve the National Essence by Government and Society and Differences and Similarities Between the Two). *Jindaishi yanjiu* 2 (2001): 28–100.

Luo, Zhitian. *Guojia yu xueshu: Qingji minchu guanyu "guoxue" de sixiang lunzheng* 國家與學術：清季民初關於「國學」的思想論證 (The State and Academia: Debates over the

Thought of National Learning at the End of the Qing and the Beginning of the Republic). Beijing: Sanlian shudian, 2003.

Luo, Zhitian. *Liebian zhong de chuancheng – 20 shijiqianqi de Zhongguo wenhua yu xueshu* 裂變中的傳承：20世紀前期的中國文化與學術 (Inheritance Within Rupture: Chinese Culture and Scholarship in the Early 20th Century). Beijing: Zhonghua shuju, 2003.

Luo, Zhitian. *Quanshi zhuanyi: jindai zhongguo de sixiang, shehui yu xueshu* 權勢轉移：近代中國的思想、社會與學術 (Shifting of Power: Ideology, Society, and Scholarship in Modern China). Wuhan: Hebei renmin chubanshe, 1999.

Ma, Fei 馬飛. "Geming wenhua yu minchu xianzheng de bengkui" 革命文化與民初憲政的崩潰 (Revolutionary Culture and Collapse of the Constitutionalism at the Beginning of the Republic). *Twenty-First Century* 137 (2012): 44–58.

Manent, Pierre. *Tocqueville et la nature de la démocratie*. Paris: Gallimard, 2012.

Mannheim, Karl. *La pensée conservatrice*, translated by Jean-Luc Evard. Paris: Édition de la revue Conférence, 2009.

Marx, Karl. *Revolution and Counter-Revolution*. Chicago: Charles H. Kerr & Company, 1907.

Maspero, Henri. "Les origines de la civilisation chinoise." *Annales de géographie* 35, no. 194 (1926): 135–154.

McClelland, J.S., ed. *The French Right from de Maistre to Maurass*. New York, Evanston, San Francisco and London: Harper Torchbooks, 1970.

Merkel-Hess, Kate. *The Rural Modernity: Reconstructing the Self and State in Republican China*. Chicago and London: University of Chicago Press, 2016.

Mesure, Sylvie. "Durkheim et Tönnies: regards croisés sur la société et sur sa connaissance." *Sociologie* 4, no. 2 (2013): 201–211.

Metzger, Thomas. *Escape from Predicament: Neo-Confucianism and China's Evolving Political Culture*. New York: Columbia University Press, 1977.

Milner, Jean-Claude. *Les penchants criminels de l'Europe démocratique*. Lagrasse: Éditions Verdier, 2003.

Mitchell, Timothy, ed. *Questions of Modernity*. Minneapolis: University of Minnesota Press, 2000.

Mittler, Barbara. *A Newspaper for China? Power, Identity, and Change in Shanghai's News Media, 1872–1912*. Cambridge, Mass.: Harvard University Press, 2004.

Moody, Peter. *Conservative Thought in Contemporary China*. Lanham: Lexington Books, 2007.

Mühlhahn, Klaus. *Making China Modern: From the Great Qing to Xi Jingping*. Cambridge, Mass.: The Belknap Press of Harvard University, 2019.

Müller, Jan-Werner. "Comprehending Conservatism: A New Framework for Analysis." *Journal of Political Ideologies* 3 (2006): 359–365.

Murthy, Viren, and Axel Schneider, eds. *The Challenge of Linear Time: Nationhood and the Politics of History in East Asia*. Leiden: Brill, 2014.

Murthy, Viren. *The Political Philosophy of Zhang Taiyan: The Resistance of Consciousness*. Leiden and Boston: Brill, 2011.

Nagai, Kazumi 永井算巳. "Iwayuru Shinkoku ryūgakusei torishimari kisoku jiken no seikaku" 所謂清国留学生取締規則事件の性格 (The Nature of the Rule on the Control of Qing Students). *Shinshū daigaku kiyō* 信州大学紀要 2 (1952): 11–33.

Nakanome, Tōru 中野目徹. *Meiji no seinen to nashonnarizumu: Seikyōsha • Nippon shinbunsha no gunzō* 明治の青年とナショナリズム：政教社. 日本新聞社の群像 (The

Young Generation of the Meiji era and Nationalism: The Society of Political Education and the Society of Japanese Daily). Tokyo: Yoshikawa kōbunkan, 2014.

Nathan, Andrew J. *Peking Politics 1918–1923: Factionalism and the Failure of Constitutionalism*. Ann Arbor: University of Michigan Center for Chinese Studies, 1998.

Nevin, Thomas R. *Irving Babbitt: An Intellectual Study*. Chapel Hill: University of North Carolina Press, 1984.

Niu, Ming-shi 牛銘實. "Wanqing difang zizhi yundong de fansi," 晚晴地方自治運動的反思 (Reflection on the Local Self-Government at the End of the Qing Dynasty). *Twenty-First Century* 98 (2006): 40–47.

Ogiwara, Takashi 萩原隆. "Shiga Shigetaka ni okeru kokusuishugi no kannen" 志賀重昂における国粋主義の観念 (Shiga Shigetaka's Concept of National Essence). *Nagoya gakuindaigaku ronshū* 名古屋学院大学論集 45, no. 2 (2008): 23–37.

Pen, Geoffrey. *The Fist of Righteous Harmony: A History of the Boxer Uprising in China in the Year 1900*. London: Leo Cooper, 1991.

Peng, Chunling 彭春凌. *Rujia zhuanxing yu wenhua xinming: yi Kang Youwei, Zhang Taiyan wei zhongxin (1898–1927)* 儒學轉型與文化新命：以康有為、章太炎為中心（1898–1927）(The Transformation of Confucianism and the New Culture Movement: On Kang Youwei and Zhang Taiyan (1898–1927)). Beijing: Beijing daxue chubanshe, 2014.

Peng, Hsiao-yen 彭小妍, ed. *Wenhua fanyi yu wenhua mailuo* 文化翻譯與文本脈絡 (Cultural Translation and Cultural Context). Taipei: Institute of Chinese Literature and Philosophy, Academia Sinica, 2013.

Plaise, Blaise. *Les pensées de Pascal*. Paris: P. Lethielleux, 1896.

Poewe, Karia O. *New Religions and the Nazis*. New York and London: Routledge, 2006.

Pye, Lucian. "Zhongguo minzuzhuyi yu xiandaihua" (Chinese nationalism and modernization). *Twenty-First Century* 9 (1991): 13–26.

Pyle, Kenneth B. *The New Generation in Meiji Japan: Problems of Cultural Identity, 1885–1895*. Stanford: Stanford University Press, 1969.

Qian, Liqun 錢理群, et al. *Zhongguo xiandai wenxue sanshinian* 中國現代文學三十年 (30 Years of Chinese Literature). Taipei: Wunan, 2002.

Rémond, René. *Les trois droites en France*. Paris: Aubiers, 2014.

Rhoads, Edward J.M. *Manchus and Han: Ethnic Relations and Political Power in Late Qing and Early Republican China, 1861–1928*. Seattle and London: University of Washington Press, 2000.

Rouvillois, Frédéric, et al., eds. *Le dictionnaire du conservatisme*. Paris: Les Édition du Cerf, 2017.

Rutt, Richard. *Zhouyi: A New Translation with Commentary of the Book of Change*. Abingdon and New York: Routledge, 2013.

Ryn, Claes G. *Will, Imagination and Reason: Babbitt, Croce and the Problem of Reality*. New Brunswick and London: Transaction Publishers, 1997.

Sadahira, Motoshiro 定平元四良. "Shiga Shigetaka, hito to shisō" 志賀重昂人と思想 (Shiga Shigetaka, Life and Thought). *Kwanseigakuindaigaku shakaigakubu kiyō* 関西学院大学社会学部紀要 24 (1972): 33–38.

Sanmu 散木. *Xiandai xueren mi'an* 現代學人謎案 (Mystery of Modern Intellectuals). Taipei: Xiuwei chuban, 2011.

Sapiro, Gisèle. "Forms of politicization in the French literary field." *Theory and Society* 32 (2003): 633–652.

Schneider, Axel. *Zhenli yu lishi: Fu Sinian, Chen Yinke de shixue sixiang yu minzurentong* 真理與歷史：傅斯年、陳寅恪的史學思想與民族認同 (Truth and History: Historical Thought and National Identity of Chen Yinke and Fu Sinian). Translated by Guan Shan 關山 and Li Maohua 李貌華. Beijing: Shehui kexue wenxian chubanshe, 2008.

Schoppa, Keith R. "Local Self-Government in Zhejiang, 1909–1927." *Modern China* 2, no. 4 (1976): 503–530.

Scruton, Roger. *Conservatism: Ideas in Profile*. London: Profile Books, 2017.

Scruton, Roger. *Fools, Frauds and Firebrands: Thinkers of the New Left*. London and New York: Bloomsbury Continuum, 2015.

Scruton, Roger. *How to Be a Conservative*. London: Bloomsbury Publishing, 2014.

Shinichi, Anzai. "Unmediated Nationalism: Science and Art in Shigetaka Shiga's *The Japanese Landscape* (1894)." *Journal of the Faculty of Letters, The University of Tokyo, Aesthetics* 34 (2009): 65–81.

Sih, Paul K.T., ed. *The Strenuous Decade: China's Nation-Building Efforts, 1927–1937*. Jamaica, NY: St. John's University Press, 1970.

Smith, Anthony D. *The Ethnic Origins of Nations*. Malden: Blackwell Publishing, 1988.

Starrs, Roy, ed. *Asian Nationalism in an Age of Globalization*. New York: Routledge, 2013.

Stenner, Karen. "Three kinds of 'Conservatism'." *Psychological Inquiry* 20 (2009): 142–159.

Sun, Kuang-the 孫廣德. *Wanqing chuantong yu xihua de zhengduan* 晚清傳統與西化的爭論 (Disputes Between Tradition and Westernization at the End of the Qing). Taipei: Taiwan Shangwu yinshuguan, 1995.

Sun, Yingying. *Cultivation Through Classical Poetry: The Poetry and Poetic Studies of Huang Jie (1973–1935)*. Unpublished PhD dissertation, University of Hong Kong, 2013.

Sun, Zhimei 孫之梅. *Nanshe yanjiu* 南社研究 (A Study on the Southern Society). Beijing: Renmin wenxue, 2003.

Suvanto, Pekka. *Conservatism from the French Revolution to the 1990s*. Translated by Roderick Fletcher. New York: St. Martin's Press, 1997.

Suzuki, Sadami 鈴木貞美, and Liu Jianhui 劉建輝, eds. *Higashi Ajia ni okeru kindai shogainen no seiritsu* 東アジアにおける近代諸概念の成立 (Formation of Modern Concepts in East Asian). Kyoto: International Research Center for Japanese Studies, 2012.

Sven, Saaler, and W.A. Szpilman, eds. *Pan-Asianism: A Documentary History, Volume 1: 1850–1920*. Maryland: Rowman & Littlefield Publishers, 2011.

Tam, Kwok-kan. *Chinese Ibsenism: Reinventions of Women, Class and Nation*. Singapore: Springer, 2019.

Tang, Te-Kong 唐德剛. *Wanqing qishinian* 晚清七十年 (The Last 70 years of the Qing Dynasty). Taipei: Yuanliu, 1998.

Tang, Zhijun. *Kang Youwei zhuan* 康有為傳 (Biography of Kang Youwei). Taipei: Taiwan Shangwu yinshuguan, 1997.

Tang, Zhijun. *Zhang Taiyan nianpu changbian* 章太炎年譜長編 (A Chronological Biography of Zhang Taiyan). Beijing: Zhonghua shuju, 1979.

Thompson, Larry Clinton. *William Scott Ament and the Boxer Rebellion: Heroism, Hubris and the Ideal Missionary*. Jefferson, NC: McFarland, 2009.

Tien, Hung-mao. *Government and Politics in Kuomintang China, 1927–1937*. Stanford: Stanford University Press, 1972.

Touraine, Alain. "Une sociologie sans société." *Revue française de sociologie* 22, no. 1 (1981): 3–13.

Tseng, Chun-hai 曾春海. *Zhongguo jindangdai zhexueshi* 中國近當代哲學史 (History of Modern and Contemporary Chinese Philosophy). Taipei: Wunan, 2018.

Tsui, Brian. *China's Conservative Revolution: The Quest for a New Order, 1927–1949*. Cambridge: Cambridge University Press, 2018.

Van Den Stock, Ady. *The Horizon of Modernity: Subjectivity and Social Structure in New Confucian Philosophy*. Leiden and Boston: Brill, 2016.

Veg, Sebastian. "Lu Xun and Zhang Binglin: New Culture, Conservatism and Local Tradition." *Sixiangshi* 思想史 6 (2016): 151–193.

Vincent, Jean-Phillipe. *Qu'est-ce que le conservatisme? Histoire intellectuelle d'une idée politique*. Paris: Les belles lettres, 2016.

Vogel, Ezra F. *China and Japan: Facing History*. Cambridge, Mass.: The Belknap Press of Harvard University Press, 2019.

Vogelsang, Kai. "Chinese 'Society': History of a Troublesome Concept." *Oriens Extremus* 51 (2012): 155–192.

Volkov, Shulamit. *The Rise of Popular Antimodernism in Germany: The Urban Master Artisans, 1837–1896*. Princeton: Princeton University Press, 1978.

Wakeman, Frederic Jr. "A Revisionist View of the Nanjing Decade: Confucian Fascism." *The China Quarterly* 150 (1997): 395–432.

Wall, Steven. *The Cambridge Companion to Liberalism*. Cambridge: Cambridge University Press, 2015.

Wang, Cheng-bon 王震邦. *Duli yu ziyou: Chen Yinke lunxue* 獨立與自由：陳寅恪論學 (Independence and Liberty: Chen Yinke's Scholarship). Taipei: Linking, 2011.

Wang, Fan-sen 王汎森, ed. *Zhongguo jindai sixiangshi de zhuanxing shidai* 中國近代思想史的轉型時代 (The Age of Transformation of Modern Chinese Intellectual history). Taipei: Linking, 2007.

Wang, Fan-sen. "Tansuo Wusi lishi de liangtiao xiansuo" 探索五四歷史的兩條線索 (Two Paths to Explore the May Fourth History). *Twenty-First Century* 4 (2019): 18–31.

Wang, Fan-sen. *Sikao shi shenghuo de yizhong fangshi: Zhongguo jindai sixiangshi de zaisikao* 思考是生活的一種方式：中國近代思想史的再思考 (Thinking as a Lifestyle: Reconsiderations on Modern Chinese Intellectual History). Taipei: Linking, 2017.

Wang, Fan-sen. *Zhang Taiyan de sixiang (1868–1919) jiqi dui ruxuechuantong de chongji* 章太炎的思想（1868–1919）及其對儒學傳統的衝擊 (Zhang Taiyan's Thought and Its Percussion on Confucianism). Taipei: Shibao, 1985.

Wang, Fan-sen. *Zhongguo jindai sixiang yu xueshu de xipu* 中國近代思想與學術的系譜 (A Genealogy of Modern Chinese Thought and Scholarship). Taipei: Linking, 2003.

Wang, Ke 王柯. "Minzu: yige laizi Riben de wuhui" 「民族」：一個來自日本的誤會 (Nation: A Misunderstanding from Japan). *Twenty-First Century* 77 (2003): 73–83.

Wang, Qisheng 王奇生. *Geming yu fangeming – shehui wenhua shiye xia de Minguo zhengzhi* 革命與反革命——社會文化視野下的民國政治 (Revolution and Counter-Revolution, Republican Politics from a Sociocultural Perspective). Beijing: Shehui kexue wenxian chubanshe, 2010.

Wang, Xiaoling 王曉苓. "Lusuo 'pubian yizhi' gainian zai Zhongguo de yinjie jiqi lishi zuoyong" 盧梭「普遍意志」概念在中國的引介及其歷史作用 (On Rousseau's "general will" in China and its historical function). *Sixiangshi* 3 (2014): 1–66.

Wang, Xiaoling. "Liu Shipei et son concept de *contrat social chinois*." *Études chinoises* XVII, no. 1–2 (1998): 155–190.
Wang, Xintian 王新田. "Liu Yizheng xiansheng nianpu jianbian" 柳詒徵先生年譜簡編 (A Short Biographical Chronology of Liu Yizheng). *Zhongguo wenzhe yanjiu tongxun* 中國文哲研究 9, no. 4 (1999): 147–155.
Wang, Yangwen 王仰文. "Ziyou yu quanli zhijian – Zhang Junmai xianzheng sixiang de yanbian"自由與權利之間——張君勱憲政思想的演變 (Between Liberty and Power – Evolution of the Constitutional Thought of Zhang Junmai). *Lanzhou xuekan* 11 (2007): 111–115.
Waters, Malcolm. *Daniel Bell*. New York: Routledge, 1996.
Weber, Eugen. "Ambiguous Victories." *Journal of Contemporary History* 13 (1978): 819–823.
Wong, Young-tsu 汪榮祖. *Cong chuantong zhong qiubian – Wanqing sixiangshi yanjiu* 從傳統中求變——晚清思想史研究 (Seeing Change in Tradition: Study on Late Qing Intellectual History). Nanchang: Baihuazhou wenyi chubanshe, 2001.
Wong, Young-tsu. *KangZhang helun* 康章合論 (On Kang Youwei and Zhang Taiyan). Taipei: Linking, 1988.
Wong, Young-tsu. *Search for Modern Nationalism: Zhang Binglin and Revolutionary China 1860–1936*. Hong Kong: Oxford University Press, 1989.
Wu, Xuezhao 吳學昭. *Wu Mi yu Chen Yinke* 吳宓與陳寅恪 (Wu Mi and Chen Yinke). Beijing: Qinghua daxue chubanshe, 1996.
Wu, Xuezhao. "The Birth of a Chinese Cultural Movement: Letters Between Babbitt and Wu Mi." *Humanitas* XVII, no. 1 and 2 (2004): 6–25
Wu, Zhongliang 吳忠良. "Deng Shi shixue sixiang lunxi" 鄧實史學思想論析 (Interpretation of Deng Shi's historical thought). *Dongfang luntan* 東方論壇 2 (2003): 111–116.
Xiao, Gongqin 蕭功秦. "Dangdai Zhongguo xinbaoshouzhuyi de sixiang yuanyuan" 當代中國新保守主義的思想淵源 (The Intellectual Foundation of Modern Chinese Conservatism). *Twenty-First Century* 40 (1997): 126–137.
Xiao, Gongqin. *Yu zhengzhi langman zhuyi gaobie* 與政治浪漫主義告別 (Farewell to Political Romanticism). Wuhan: Hubei jiaoyu chubanshe, 2001.
Xiao, Liang 尚 朗. "Fukuzawa Yukichi to Chūgoku no keimou shisou: Liang Qichao to no shisouteki kanren o chūshin ni" 福沢諭吉と中国の啓蒙思想——梁啓超との思想的関連を中心 (Fukuzawa Yukichi and Chinese enlightenment: a study on Liang Qichao's enlightenment ideas). *Nagoya Daigaku Kyōikugakubu Kiyō* 名古屋大学教育学部紀要 40, no. 1 (1993): 63–77.
Xie, Yong 謝泳. "Xinan lianda zhishi fenziqun de xingcheng yu shuailuo" 西南聯大知識分子的形成與衰落 (The formation and the decline of the intellectuals of the Southwestern Associated University). *Twenty-First Century* 38 (1996): 55–66.
Xong, Yuezhi 熊月之. *Xixue dongjian yu wanqing shehui* 西學東漸與晚清社會 (The Dissemination of Western Learning and the Late Qing Society). Shanghai: Shanghai renmin chubanshe, 1995.
Xu, Aymeric. "Mapping Conservatism of the Republican Era: Genesis and Typologies." *Journal of Chinese History* 4 (2020): 135–159.
Xu, Ben 徐賁. "Zhongguo buxuyao zheyang de 'zhengzhi' he 'zhuquanzhe jueduan' – 'Shimite re' he guojiazhuyi" 中國不需要這樣的「政治」和「主權者決斷」——「施米特熱」和國家主義 (China Does Not Need Such "Politics" and "Decisionism": The Schmitt-Fever and Statism). *Twenty-First Century* 94 (2005): 26–39.

Xu, Guoqi. *China and The Great War: China's Pursuit of a New National Identity and Internationalization*. Cambridge: Cambridge University Press, 2005.

Xu, Jilin 許紀霖. "Geren zhuyi de qiyuan – 'Wusi' shiqi de ziwoguan yanjiu" 個人主義的起源——「五四」時期的自我觀研究 (The Origin of Individualism: A Study of the View on the Self in the May Fourth Period). *Tianjin shehui kexue* 天津社會科學 6 (2008): 113–124.

Xu, Jilin. *Jingshen de lianyu: wenhua bianqian zhong de zhongguo zhishifenzi* 精神的煉獄：文化變遷中的中國知識份子 (Spiritual Prison: Chinese Intellectuals in Cultural Transitions). Taipei: Shulin, 1994.

Xu, Xiaoqun. *Chinese Professionals and the Republican State: The Rise of Professional Associations in Shanghai, 1927–1937*. Cambridge: Cambridge University Press, 2004.

Yang, Peng. "Analysis of Japanese Influence on the Three Major Historiographical Trends in Early Modern China." *Chinese Studies in History* 49, no. 1 (2016): 28–38.

Yang, Tianshi 楊天石, and Wang Xuezhuang 王學莊. *Nansheshi changbian* 南社史長編 (Chronological History of the Southern Society). Beijing: Zhongguo renmin daxue chubanshe, 1995.

Yang, Zhiqiang 楊志強. *"Cong 'Miao' dao 'Miaozu' – lun jindai minzu jituan xingcheng de 'tazhexing' wenti"* 從「苗」到「苗族」——論近代民族集團形成的「他者性」問題 (From "Miao" to the "Miao Nation" – On the Question of the "Otherness" in the Formation of Modern Nation). *Xinan minzu daxue xuebao* 6 (2010): 1–7.

Yao, Dianzhong 姚奠中, and Dong Guoyan 董國炎. *Zhang Taiyan xueshu nianpu* 章太炎學術年譜 (Academic Biographical Chronology of Zhang Taiyan). Taiyuan: Jiangxi guji chubanshe, 1996.

Yao, Zhongqiu 姚中秋. *Rujia xianzheng lun* 儒家憲政論 (On the Confucian Constitutionalism). Hong Kong: City University of Hong Kong Press, 2016.

Ye, Weili. *Seeking Modernity in China's Name: Chinese Students in the United States, 1900–1927*. Stanford: Stanford University Press, 2001.

Yü, Yingshi 余英時. *Xiandai ruxue de huigu yu zhanwang* 現代儒學的回顧與展望 (A Retrospective and Prospective View on Contemporary Confucianism). Beijing: Sanlian shudian, 2004.

Yü, Ying-shih. "Zhongguo zhishifenzi de bianyunhua" 中國知識份子的邊緣化 (The Marginalization of Chinese intellectuals). *Twenty-First Century* 15 (2003): 15–25.

Yu, Zidao 余子道, and Xu Youwei 徐有為. "Lixingshe shulun" 力行社述論 (On the Society for Vigorous Practice). *Jindaishi yanjiu* 6 (1989): 217–237.

Zarrow, Peter. "Historical Trauma: Anti-Manchuism and Memories of Atrocity in Late Qing China." *History and Memory* 16, no. 2 (2004): 67–107.

Zarrow, Peter. *After the Empire: The Conceptual Transformation of the Chinese State, 1885–1924*. Stanford: Stanford University Press, 2012.

Zarrow, Peter. *Anarchism and Chinese Political Culture*. New York: Columbia University Press, 1990.

Zarrow, Peter. *China in War and Revolution, 1895–1949*. London: Routledge, 2005.

Zhang, Boshu 張博樹. *Gaibian Zhongguo: Liusi yilai de Zhongguo zhengzhi sichao* 改變中國：六四以來的中國政治思潮 (Changing China: Political Trends from June Fourth Onwards). Hong Kong: Suyuan Books, 2015.

Zhang, Chuntian 張春田. *Geming yu shuqing: Nanshe de wenhua zhengzhi yu Zhongguo xiandaixing (1903-1923)* 革命與抒情：南社的文化政治與中國現代性（1903-1923）

(Revolution and Lyricism: Cultural Politics of the Southern Society and Chinese Modernity, 1903–1923). Shanghai: Shanghai renmin chubanshe, 2015.

Zhang, Junmai 張君勱. *Xinrujia sixiangshi* 新儒家思想史(History of the Neo-Confucianism). Beijing: Zhongguo renmin daxue chubanshe, 2006.

Zhang, Yi 張夷. *Chen Qubing nianpu* 陳去病年譜 (Chronological Biography of Chen Qubing). Shanghai: Shanghai guji chubanshe, 2009.

Zhang, Yongjiu 張永久. *Yuanyang hudie pai wenren* 鴛鴦蝴蝶派文人 (Writers of the Mandarin Ducks and Butterflies School). Taipei: Xiuwei chuban, 2011.

Zhao, Erxun 趙爾巽, et al. *Qingshigao* 清史稿 (Draft History of Qing). Taipei: Dingwen, 1981.

Zhao, Lu. *In Pursuit of the Great Peace: Han Dynasty Classicism and the Making of Early Medieval Literati Culture*. Albany: SUNY, 2019.

Zheng, Dahua 鄭大華, and Zou Xiaozhan 鄒小站, eds. *Zhongguo jindaishi shang de minzuzhuyi* 中國近代史上的民族主義 (Nationalism in Modern Chinese History). Beijing: Shehui kexue wenxian chubanshe, 2007.

Zheng, Dahua. "Guanyu minguo xiangcun jianshe yundong de jige wenti" 關於民國鄉村建設運動的幾個問題 (Several Issues on the Rural Construction Movement in the Republic Era). *Shixue yuekan* 史學月刊 2 (2006): 53–60.

Zheng, Xiaowen. *The Politics of Rights and the 1911 Revolution in China*. Stanford: Stanford University Press, 2018.

Zheng, Yimei 鄭逸梅, *Nanshe congtan* 南社叢談 (On the Southern Society). Beijing: renmin chubanshe, 1981).

Zheng, Yongnian. "Development and Democracy: Are They Compatible in China?" *Political Sciences Quarterly* 2 (1994): 235–259.

Zhongguo wenhua fuxing weiyuanhui 中國文化復興委員會 (Committee for the restoration of Chinese culture), ed. *Zhongguo jindaixiandaishi lunji* 中國近代現代史論集 (Collection of Articles on Modern and Contemporary Chinese History), Vol. 18. Taipei: Taiwan Shangwu yinshuguan, 1985.

Zhou, Huimei 周慧梅. *Minzhong jiaoyuguan yu Zhongguo shehui bianqian* 民眾教育館與中國社會變遷 (Popular Schools and the Evolution of the Chinese society). Taipei: Xiuwei chuban, 2013.

Zhu, Ying 朱英. "Qingmo xinxing shangren ji minjian shehui" 清末新興商人及民間社會 (New Businessmen and the Popular Society at the end of Qing). *Twentieth-First Century* 3 (1990): 37–44.

Ziblatt, Daniel. *Conservative Parties and the Birth of Democracy*. New York: Cambridge University Press, 2017.

Glossary

Acinteyya　不可思議
aiqing xiaoshuo　愛情小說
Ajia　亜細亜
Anlewo　安樂窩
ansha　暗殺
Atarashiki-mura undo　新しき村運動
badao　霸道
bahsian　白相
baihua　白話
Bairi weixin　百日維新
baixing　百姓
Bao Tianxiao　包天笑
Baoguo hui　保國會
baoguo　保國
baohuang　保皇
Baojiao gonghui　保教公會
baojiao　保教
baoshou/hoshu　保守
baozhong　保種
Beifa　北伐
Beijing tiaoyue　北京條約
Beiting　北廷
buheli de zhuanzhi　不合理的專制
bunmei　文明
Cai Yuanpei　蔡元培
Cao Kun　曹錕
Chang Chi-yun　張其昀
Charter 08 08　憲章
Chen Duxiu　陳獨秀
Chen Guofu　陳果夫
Chen Lifu　陳立夫
Chen Qimei　陳其美
Chen Qubing　陳去病
Chen Xunci　陳訓慈
Chenbao　晨報
Cheng Fangwu　成仿吾
chenjie　臣節
chenmin　臣民
chi　恥
chihua　赤化
Chinese Communist Party　中國共產黨
Chongzhen　崇禎
Chouanhui　籌安會

Chunqiu　春秋
Cixi　慈禧
congxin zhuyi　從新主義
da hanzu zhuyi　大漢族主義
da minzu zhuyi　大民族主義
Dagonghe bao　大共和報
Dai Jitao　戴季陶
danghua jiaoyu　黨化教育
dao　道
datong　大同
Datsu-A Ron　脱亜論
Daxue　大學
dayeujin　大躍進
de xiansheng　德先生
Deng Shi　鄧實
difang gongyi shiye　地方公益事業
Diguo ribao　帝國日報
Ding Wenjiang　丁文江
Dongfang zazhi　東方雜誌
Dongnan hubao　東南互保
Dongxi wenhua jiqi zhexue　東西文化及其哲學
Du Yaquan　杜亞權
Duan Qirui　段祺瑞
Duanfang　端方
dushan qishen　獨善其身
Duyin tongyihui　讀音統一會
Erci geming　二次革命
Ershi shiji dawutai　二十世紀大舞台
fali　法吏
Fan Yanqiao　範煙橋
fandong　反動
fangzhen　方鎮
fanwan　飯碗
fanzhen　藩鎮
fazhi　法治
fen　分
fengjian　封建
fengshui　風水
fenshu kengru　焚書坑儒
fugu　復古
Fukuzawa Yukichi　福沢諭吉
fuqing mieyan　扶清滅洋

Fushe 復社
gailiang xinju 改良新劇
gangchang 綱常
Gao Xie 高燮
Gao Xu 高旭
Gao Yihan 高一涵
Gao Zeng 高增
Geming jun 革命軍
geming 革命
geren 個人
Gi aku shu Nihonjin 偽惡醜日本人
gong 工
gong 公
gongde 公德
gonghe 共和
gongmin 公民
gongming 功名
Gongyang 公羊
gongzhu 共主
Gu Hongming 辜鴻銘
Gu Yanwu 顧炎武
Guan Zhong 管仲
guangfu 光復
Guangfuhui 光復會
Guangxu 光緒
Guanhua 官話
guihua 歸化
guizu wenxue 貴族文學
Gujing jingshe 詁經精舍
Guocui pai 國粹派
Guocui xuebao 國粹學報
guocui 國粹
Guofeng banyuekan 國風半月刊
Guogu yuekan 國故月刊
Guoli dongnan daxue 國立東南大學
Guoli xibei daxue 國立西北大學
Guoli xinan lianhe daxue 國立西南聯合大學
Guoli zhongyang daxue 國立中央大學
Guomin zhengfu jianguo dagang 國民政府建國大綱
Guomin 國民
guomin/kokumin 國民/国民
guoti 國體
guowen 國文
Guoxue baocunhui 國學保存會
Guoxue Yanjiuhui 國學研究會

guoxue 國學
Guoyu tongyi choubeihui 國語統一籌備會
Guoyu yanjiuhui 國語研究會
guoyu yundong 國語運動
guwen 古文
Guxue huikan 國學彙刊
haiku 俳句
Han Yu 韓愈
hanguan weiyi 漢官威儀
Hanzhong 漢種
haoren 好人
Hayashi Gonsuke 林權助
He Xiu 何休
He Zhen 何震
Heimin sha 平民社
heimu xiaoshuo 黑幕小說
Hengbao 衡報
Hong Fan 洪範
Hou Hongjian 侯鴻鑒
Hu Dunfu 胡敦復
Hu Xiansu 胡先驌
huachang 花場
Huang Jie 黃節
Huang Xing 黃興
Huang Zongxi 黃宗羲
Huang Zunxian 黃遵憲
Huaxinghui 華新會
huayi zhi bian 華夷之辨
Hua 華
huidang 會黨
Hundred Days' Reform 百日維新/戊戌變法
Huo Qubing 霍去病
huowenxue 活文學
Itō Hirobumi 伊藤博文
Inoue Enryō 井上円了
Inoue Kaoru 井上薫
Isawa Shūji 近衛篤麿
ji 集
Jiading tucheng jilue 嘉定屠城紀略
Jiaguwen 甲骨文
Jian Chaoliang 簡朝亮
Jian'an caotang 簡岸草堂
Jianai 兼愛
Jiang Fangzhen 蔣方震
Jiangchu bianyiju 江楚編譯局
Jiangsu 江蘇
Jiangxi 江西

jiaohua 教化
jiaoyuan 教員
Jiaozhou wan 膠州灣
jie 節
jimizhou 羈縻州
Jin 金
Jing Changji 景昌極
Jing Yaoyue 景耀月
jing 敬
jing 經
Jingbu dang 進步黨
jingshi zhiyong 經世致用
Jingzhong ribao 警鐘日報
jinqian zhishang 金錢至上
jinqu 進取
jinshi 進士
jinwen 今文
Jishe 幾社
Jiushe 酒社
Jiwang 救亡
jiyū 自由
jizhongshi 給中使
Ju'e yiyong dui 拒俄義勇隊
Ju'e yundong 拒俄運動
jubin 舉兵
Juemin she 覺民社
Juluanshi 據亂世
jun guomin 軍國民
jundao 君道
junguo/guojia shehui 軍國/國家社會
junquan 君權
junxian 郡縣
junxue 君學
junzheng 軍政
Kang Youwei 康有為
kaozheng 考證
Kedilun 客帝論
keji fuli 克己復禮
keju 科舉
Kikuchi Kumatarō 菊池熊太郎
Kōtoku Shūsui 幸德秋水
kojin 個人
Kokubu Tanenori 国府種德
kokugaku 国学
kokusui 国粋
Kon Sotosaburō 今外三郎
kongjiao 孔教

Kuaihuo 快活
Kuga Katsunan 陸葛南
Kuo Ping-wen 郭秉文
Kuomintang 國民黨
Lanyishe 藍衣社
lao daoli 老道理
Lao Naixuan 勞乃宣
Li Dazhao 李大釗
Li Hongzhang 李鴻章
Li Shucheng 李書城
Li Sichun 李思純
Li Yuanhong 黎元洪
Li Yuerui 李岳瑞
Li Zicheng 李自成
li 力
li 禮
lian 廉
Liang Cheng 梁誠
Liang Qichao 梁啟超
Liang Shih-chiu 梁實秋
Liang Shuming 梁漱溟
Libailiu 禮拜六
Lidai Shilue 歷代史略
liji 利己
Lin Shu 林紓
Lin Xie 林獬
lingxiu 領袖
lishi minzu 歷史民族
lita 利他
Liu Boming 劉伯明
Liu Kunyi 劉坤一
Liu Mingchuan 劉銘傳
Liu Nianzeng 柳念曾
Liu Shipei 劉師培
Liu Xiaobo 劉曉波
Liu Xin 劉歆
Liu Yazi 柳亞子
Liu Yizheng 柳詒徵
Liu Yu 劉餘
longmai 龍脈
Lu Xun 魯迅
Lu Yin 盧隱
Lunyu 論語
Luo Jialun 羅家倫
Luo Zhenyu 羅振玉
Ma Xulun 馬敘倫
manyi/tuteng shehui 蠻夷/圖騰社會

Manzhou yuanliu kao 滿洲源流考
May Fourth Movement 五四運動
Meiji 明治
meishu 美術
Miao Fenglin 繆鳳林
Miao Quansun 繆荃孫
mimi zhi yin 靡靡之音
min 民
Minbao 民報
minben 民本
mingjiao 名教
mingshi 名士
Minguo xinwen 民國新聞
Minhu ribao 民呼日報
Minjō isshin 民情一新
Minlibao 民立報
minquan zhuyi 民權主義
minquan 民權
Minquanbao 民權報
Minsheng ribao 民生日報
minsheng zhuyi 民生主義
minzu de texing 民族的特性
minzu/minzoku 民族
Miyake Setsurei 三宅雪嶺
Miyazaki Michimasa 宮崎道正
Mou Zongsan 牟宗三
muleer guniang 穆勒爾姑娘
Mushanokōji Saneatsu 武者小路實篤
Nahan 吶喊
Naka Michiyo 那珂通世
Nanbei yihe 南北議和
Nanjing gaodeng shifan xuexiao 南京高等師範學校
Nanshe congke 南社叢刻
Nanshe 南社
nazuihua 納粹化
minyue/minyaku 民約
New Culture Movement 新文化運動
New Life Movement 新生活運動
New Policies 新政
nianhao 年號
nihon bunshi daha shigi 日本分子打破旨義
Nihon fūkeiron 日本風景論
Nihon yobini Nihonjin 日本及日本人
Nihonjin 日本人
Nippon 日本

nong 農
Ou Jujia 歐榘甲
pendang 朋黨
pingmin wenxue 平民文學
Putonghua 普通話
Puyi 溥儀
qi 妻
qi 齊
qimeng 啟蒙
Qian Xuantong 錢玄同
Qiangxuehui 強學會
Qianlong 乾隆
Qimin she 齊民社
Qin Shi Huang 秦始皇
Qinghua xuetang 清華學堂
Qing Qinwang neige 慶親王內閣
Qingshigao 清史稿
Qingyibao 清議報
Qiu Jin 秋瑾
Qiu Jun 丘濬
Qiushu 訄書
Quanguo jiaoyu lianhehui 全國教育聯合會
Quanxuepian 勸學篇
qun 群
qunzhi 群治
ren 仁
rendao zhuyi 人道主義
renge 人格
renmin 人民
renzhi 人治
Riben guozhi 日本國誌
Rokumeikan 鹿鳴館
ruguo 儒國
sai xiansheng 賽先生
Sandai 三代
sangang wuchang 三綱五常
sanmin zhuyi 三民主義
sanminzhuyi wenyi 三民主義文藝
Sanshi 三世
Sapporo nōgakkō 札幌農學校
Seikyōsha 政教社
Self-Strengthening Movement 自強運動
shang 商
shangceng/shangdeng shehui 上層/上等社會
shangding wenhao 商定文豪
Shangguo 商國

Shangshu 尚書
shangxiatong 上下通
shanhaijing 山海經
Shao Lizi 邵力子
Shehui tongquan 社會通詮
shehui 社會
Shehui zhuyi jiangxihui 社會主義講習會
sheji 社稷
Shenbao 申報
shengpingshi 昇平世
shengyuan 生員
Shenjiao she 神交社
Shenzhou ribao 神洲日報
shi 士
shi 史
Shibao 時報
Shidi yanjiuhui 史地研究會
Shiga Shigetaka 志賀重昂
Shiji 詩經
Shin zen bi Nihonjin 真善美日本人
Shina bunkatsu ron 支那分割論
Shina hozen 支那保存
Shina bunmeishi 支那文明史
Shina tsūshi 支那通史
Shinkoku ryūgakusei torishimari kisoku 清国留学生取締規則
Shirakawa Jirō 白川次郎
Shiwubao 時務報
shizhe shengcun 適者生存
shoujiu 守舊
shu er bu zuo 述而不作
Shuowen jiezi 說文解字
shuren 庶人
si 死
si 私
side 私德
Sima Guang 司馬光
simin 四民
siwenxue 死文學
Sixiang yu shidai 思想與時代
Song Jiaoren 宋教仁
Song Shu 宋恕
Song Yuren 宋育仁
Subao 蘇報
Sun Jiansheng 孫諫聲
Sun Yat-sen 孫逸仙/孫中山
suo chuanwen shi 所傳聞世

suojianshi 所見世
suowenshi 所聞世
Tao Xisheng 陶希聖
taipingshi 太平世
Taipingyang bao 太平洋報
Taiwan nichinichi shinbun 台湾日日新聞
Tan Sitong 譚嗣同
Tanahashi Ichirō 棚橋一郎
Tang Caichang 唐才常
Tang Junyi 唐君毅
Tang Yongtong 湯用彤
ti 體
tiaohe lun 調和論
Tianduo bao 天鐸報
tianming 天命
tianran minzu 天然民族
tianxia 天下
tianxing 天行
Tianyan Lun 天演論
Tianyi 天義
Tianzuhui 天足會
tifa yifu 剃髮易服
Tirenge daxueshi 體仁閣大學士
Tōa Dōbunkai 東亜同文会
Tōhō Kyōkai 東邦協會
Tokutomi Sohō 德富蘇峰
tomatsu shigi 塗抹旨義
tonghua 同化
Tongmenghui 同盟會
Tongsu jiaoyu yanjiuhui 通俗教育研究會
Wang Anshi 王安石
Wang Chuanshan 王船山
Wang Dungen 王鈍根
Wang Jingwei 汪精衛
Wang Jingxuan 王敬軒
Wang Kangnian 汪康年
Wang Yangming 王陽明
wangdao 王道
wangtong 王統
Wanmu Caotang 萬木草堂
wenchang 文娼
wengai 文丐
Wenhua jianshe 文化建設
wenhua shoucheng zhuyi 文化守成主義
wenming jiehun 文明結婚
wenren 文人
wenyan 文言

wenzi geming 文字革命
Woren 倭人
Wu Fangji 吳芳吉
Wu Mi 吳宓
Wu Yu 吳虞
Wuchang qiyi 武昌起義
wujing tianze 物競天擇
wulun 五倫
Wusi yundong 五四運動
wuxia xiaoshuo 武俠小說
wuzu gonghe 五族共和
xiaceng/xiadeng shehui 下層/下等社會
xiandai 現代
xiangshan qiqun 相善其群
Xiangtu lishi jiaokeshu 鄉土系列教科書
xiangzhi 鄉治
xianqi liangmu 賢妻良母
xianxian zhuyi 賢賢主義
xianzheng 憲政
Xiao Chunjin 蕭純錦
xiao minzu zhuyi 小民族主義
xiao 孝
xiaobao 小報
Xiaoshuo yuebao 小說月報
Xin Guangdong 新廣東
Xin Hunan 新湖南
Xin quanweizhuyi 新權威主義
Xinchao 新潮
Xinchou tiaoyue 辛醜條約
xincun 新村
Xinhai geming 辛亥革命
Xinshenghuo yundong 新生活運動
Xinwenhua yundong 新文化運動
Xinzheng 新政
Xingshi 醒獅
Xingzhonghui 興中會
Xining bianfa 熙寧變法
Xinmin congbao 新民叢報
Xinmin shuo 新民說
Xinqingnian 新青年
xinrujia 新儒家
xinshi de ducai 新式的獨裁
Xiong Shili 熊十力
Xu Fuguan 徐復觀
Xu Shouwei 許守微
Xu Tong 徐桐
Xu Zeling 徐則陵

Xu Zhenya 徐枕亞
Xu Zhiheng 許之衡
xuanchuan 宣傳
Xuantong 宣統
xuanxue 玄學
Xuechi xuehui 雪恥學會
xuedong 學董
xueguan 學官
Xueheng pai 學衡派
xuezhang 學長
xuezhong 學眾
xunzheng 訓政
yacui 亞粹
yaji 雅集
Yan Yangchu 晏陽初
Yang Zhu 楊朱
yangge 秧歌
Yangzheng shushu 養正書塾
Yangzhou tucheng qinli 揚州屠城親歷
Yansheng gong 衍聖公
Yao Guang 姚光
Yao Hechu 姚鶴雛
Yazhou heqinhui 亞洲和親會
Ye Chucang 葉楚傖
Ye Yusen 葉玉森
Yi Baisha 易白沙
yi 義
yigupai 疑古派
yihetuan 義和團
Yijing 易經
yimin 遺民
yinsi 淫祀
yong 用
yong 勇
Youxi zazhi 遊戲雜誌
Yu Dafu 郁達夫
Yu Lianruan 余聯沅
Yu Yue 俞樾
Yu 禹
Yuan Shikai 袁世凱
Yuan 院
Yuanqiang 原強
Yuanyang hudie pai 鴛鴦蝴蝶派
Yubei lixian gonghui 預備立憲公會
yushi 御使
Zeng Guofan 曾國藩
Zhang Ji 張繼

Zeng Jize 曾紀澤
Zhang Junmai 張君勱
Zhang Shizhao 章士釗
Zhang Taiyan 章太炎
Zhang Xun 張勳
Zhang Zhidong 張之洞
Zhang Ziping 張資平
Zhanguoce 戰國策
Zhangyuan 張園
Zhao Bingjun 趙秉鈞
Zhejiang chao 浙江潮
Zheng Zhenduo 鄭振鐸
zhengke 政客
zhengti 政體
zhengxin chengyi 正心誠意
zhengxue 正學
Zhengyi tongbao 政藝通報
zhentai xiaoshuo 偵探小說
Zhibao 直報
Zhina ren 支那人
Zhina wangguo erbai sishi'er nian jinianhui 支那亡國二百四十二年紀念會
zhiye 職業
zhong 忠
zhongceng/zhongdeng shehui 中層/中等社會
Zhongguo baihuabao 中國白話報
Zhongguo guohui 中國國會
Zhongguo jiaoyuhui 中國教育會
Zhongguo minzhu dang 中國民主黨
Zhongguo wenhuashi 中國文化史
Zhonghua minzu 中華民族
Zhonghua wenhua fuxing yundong 中華文化復興運動

Zhonghua xinbao 中華新報
zhongti xiyong 中體西用
Zhongguo benwei de wenhua jianshe xuanyan 中國本位的文化建設宣言
zhonguo meng 中國夢
Zhongwaitong 中外通
Zhongguo wenhua jianshe xiehui 中國文化建設協會
Zhongxi wenming zhi pingpan 中西文明之評判
zhongzu sixiang 種族思想
Zhou Shi 周實
Zhou Shoujuan 周瘦鵑
Zhou Xiangjun 周祥駿
Zhou Yichun 周詒春
Zhouli 周禮
Zhu Ciqi 朱次琦
Zhu Zhixin 朱執信
Zhuangzi 莊子
zhuyi/shugi 主義
zhuzi baijia 諸子百家
zi 子
Ziben zhuyi zhi jianglai 資本主義之將來
zijue 自覺
Ziqiang yundong 自強運動
ziyou 自由
Ziyoutan 自由談
zizhu zhi quan 自主之權
zongfa shehui 宗法社會
zongjiao 宗教
zongshe dang 宗社黨
zuoyi wenxue 左翼文學

Index of Persons

Arnold, Matthew   126, 186

Babbitt, Irving   181, 183 f., 187–192, 195–197, 200 f., 207
Bagehot, Walter   156 f.
Bakunin, Mikhail   103
Baudelaire, Charles   146
Begtrup, Holger   177
Bergson, Henri   198
Bonaparte, Napoléon   71
Burke, Edmund   3, 11, 144, 156 f.

Cai Yuanpei   53, 55, 63, 162
Cao Kun   127
Chang Chi-yun   170, 196, 200, 208
Chen Duxiu   26, 74 f., 133 f., 137, 139, 150 f., 154, 164, 180
Chen Guofu   210
Chen Lifu   17, 210
Chen Qimei   60, 159
Chen Qubing   55 f., 59–61, 71, 86, 91, 130
Chen Tianhua   113
Chen Xunci   170, 210
Chen Yinke   184, 202, 216
Chiang Kai-shek   27, 180, 203, 208, 210
Chongzhen   34, 56
Cixi   8, 19, 22, 29, 43 f., 53, 57, 75, 96
Constant, Benjamin   144

Dai Jitao   60
de Bonald, Louis   35
de Chateaubriand, François-René   147
de Lacouperie, Terrien   23, 66, 89–91
de Maistre, Joseph   4, 16, 35, 146 f.
de Tocqueville, Alexis   144 f.
Deng Shi   61–63, 66 f., 76, 78, 92 f., 113, 119
Deng Xiaoping   225
Dewey, John   187 f.
Dicey, Albert Venn   156 f.
Disraeli, Benjamin   11, 144 f., 160, 213
Dong Zhongshu   47, 91, 93, 98
Driesch, Hans   198

Du Yaquan   6, 133 f., 137, 153
Duan Qirui   24, 69

Engels, Friedrich   103
Enryō, Inoue   81
Eucken, Rudolf   198, 205

Feng Yuxiang   106
Fukuzawa Yukichi   6 f.

Gao Xie   59, 97, 152 f., 162
Gao Xu   56, 59 f., 71, 112, 114, 127, 161
Gao Yihan   25, 151
Gao Zeng   59
Grundtvig, Nikolaj Frederik Severin   177
Gu Jiegang   171
Gu Yanwu   88, 93, 116
Guan Zhong   96
Guangxu   7, 19, 22, 44, 52 f., 58, 65

Hayek, Friedrich   144
He Xiu   46
He Zhen   63
Hou Hongjian   117
Hu Dunfu   201
Hu Jixian   200, 209
Hu Shi   4, 72, 136 f., 149 f., 159, 164, 183, 197, 210
Hu Xiansu   184, 187, 195 f., 200
Huang Jie   61, 63, 67, 87, 95 f., 105, 111, 115, 119
Huang Xing   53
Huang Zongxi   88, 93, 116 f.
Huang Zunxian   5

Isawa Shūji   85
Itō Hirobumi   84

Jenks, Edwards   99
Jiang Fangzhen   6
Jin Tianhe   103
Jing Changji   170, 199, 207
Jing Yaoyue   110

Kang Youwei 19, 29, 45, 47–53, 56, 58, 61, 65, 67, 69f., 73, 76, 97, 104, 114, 133, 167, 223
Kikuchi Kumatarō 80
Kold, Christian 177
Kon Sotosaburō 80
Konoe Atsumaro 84
Kōtoku Shūsui 63
Kuga Katsunan 21
Kuo Ping-wen 201

Lanman, Charles Rockwell 183
Lao Naixuan 72
Lenin, Vladimir 103, 208
Li Dazhao 150, 152, 180
Li Hongzhang 38, 44, 84
Li Sichun 184
Li Zicheng 34
Liang Ji 174
Liang Qichao 6f., 19, 22, 29, 45, 48–50, 52, 56f., 67, 69f., 73, 84, 94, 97, 106–108, 113, 120, 158, 168, 198, 216, 219
Liang Shuming 166, 174–176, 178, 197f., 200, 215, 218
Lin Shu 5, 120, 124
Lin Tongji 209f.
Liu Bannong 136
Liu Boming 26, 184, 193, 199f.
Liu Kunyi 44, 170
Liu Shipei 22, 63–66, 71, 86, 90f., 98–100, 107, 111, 113–115, 124, 126, 133
Liu Yazi 34, 56f., 59f., 71, 95, 120, 125, 128f., 132, 139, 141
Liu Yizheng 166, 170–174, 184f.
Lu Xun 24f., 121, 129f., 135–137, 166f.
Lu Yin 154
Lund, Hans 177
Luo Zhenyu 72, 84

Ma Xulun 62, 86, 120
Manniche, Peter 177
Mao Zedong 180, 215
Marx, Karl 146, 186f.
Maurras, Charles 147, 217
Mei Guangdi 136–138, 183–186, 195, 201
Meng Xiancheng 177
Miao Fenglin 126, 170, 185, 202

Miao Quansun 170
Miyake Setsurei 80
Miyazaki Michimasa 81
Mou Zongsan 178, 221
Mushanokōji Saneatsu 166

Naitō Konan 82
Naka Michiyo 170
Nietzsche, Friedrich 166

Oakeshott, Michael 4, 8, 37
Ou Jujia 117

Qian Xuantong 25, 120, 134, 136
Qianlong 105
Qiu Jin 56

Rousseau, Jean-Jacques 56, 98, 187, 189–192

Saint-Étienne, Rabaut 144
Schrøder, Ludvig 177
Shao Lizi 159
Shiga Shigetaka 20–22, 79–83, 86
Song Jiaoren 60, 70, 100
Song Shu 85, 89
Song Yuren 72
Stalin, Joseph 103
Sugiura Jūgō 80
Sun Jiansheng 106
Sun Yat-sen 22, 27, 44, 53, 57, 59, 70, 77, 109, 120, 128, 161, 180, 193

Tan Sitong 19, 49
Tanahashi Ichirō 81
Tang Caichang 58
Tang Yongtong 183
Tao Yunkui 209f.
Tian Tong 127
Tokutomi Sohō 83

Wang Chuanshan 88, 93
Wang Jingwei 60, 74, 99, 104f., 108, 114
Wang Kangnian 57
Wang Xiaonong 109
Wang Yangming 199, 221
Weber, Max 146

Woren 12 f.
Wu Fangji 183
Wu Mi 119 f., 136 f., 139, 182–188, 194 f., 197, 201 f., 209, 216
Wu Zhihui 155

Xi Jinping 225
Xiao Chunjin 184, 195 f.
Xiong Shili 197
Xu Shouwei 112
Xu Tong 43
Xu Zeling 184
Xu Zhen'e 184
Xu Zhiheng 111
Xuantong (Puyi) 24, 70, 121

Yan Fu 75 f., 94, 99
Yang Zhu 97
Yao Guang 59, 100, 149
Ye Chucang 60, 159, 164, 205
Yen, James 174, 177–179
Yi Baisha 151
Yu Dafu 137
Yu Yue 39, 57 f., 70

Yuan Shikai 24 f., 44, 70, 106, 120, 127, 130, 158 f., 162, 168, 181, 224

Zaifeng, Prince Chun 53
Zeng Guofan 40
Zeng Jize 40
Zhang Dongsun 159 f.
Zhang Ji 63
Zhang Junmai 198, 205 f., 221
Zhang Shizhao 156–158, 166, 174
Zhang Taiyan 15, 23, 34, 39, 48, 52, 54, 57–59, 62 f., 65, 67, 70, 84, 87, 90 f., 105, 108, 110 f., 117, 120, 123 f., 127, 166–170, 172
Zhang Xun 24, 69
Zhang Zhidong 44 f., 50, 58, 85, 114, 170
Zhang Ziping 137
Zheng Zhenduo 131
Zhou Shi 100
Zhou Xiangjun 100
Zhou Yichun 183
Zhu Zhixin 107
Zou Rong 104 f., 114

Index of Subjects

anarchism   63f., 189, 191
antimodern   4, 64, 146f., 164–167, 201, 203
antimodern conservatism   31f., 144, 155, 159, 164–179, 181, 208, 215, 217f.
Asianism   82f.
Asiatic Humanitarian Brotherhood   64f.
authoritarian conservatism   32, 179–182, 201–203, 205, 207, 209, 211, 215f.

Boxer Rebellion   41–44, 58, 114, 182
– Boxer Protocol   43f.
– Eight-Nation Alliance   43
Britain   38, 141, 146, 195, 217, 219
Buddhism   10, 32, 48, 168, 172, 179–182, 186, 199, 201–205, 207–211

Chinese Association for Education (*Zhongguo jiaoyuhui*)   55, 57, 63
Chinese Communist Party (CCP)   1, 27, 33, 35, 103, 127, 179f., 182, 199, 203, 205f., 212, 215, 224–226
Chinese Democratic Party (*Zhongguo minzhu dang*)   158
Chinese Empire Reform Association (*Baohuanghui*)   52
Chinese Parliament (*Zhongguo guohui*)   58
Chouanhui (Peace Planning Society)   71
Christianity   40, 48, 92, 199
– Anti-Christian Movement   32, 180, 199
classical Chinese (*wenyan*)   22, 122–125, 128, 130, 134, 136, 141, 180, 195, 218
Confucianism   9f., 13, 17, 25, 27, 29f., 32f., 46, 48, 50–52, 57, 65f., 68, 73, 85, 91f., 104, 111, 113–115, 121f., 131, 133, 149f., 153, 161f., 169f., 173, 180, 185f., 188, 192f., 197f., 202f., 205, 207–211, 221–225, 229
– All under Heaven (*tianxia*)   74
– Confucian orthodoxy (*zhengxue*)   45, 50, 52, 67
– Confucian paternalism   158
– Confucian religion (*kongjiao*)   48, 50, 52, 65, 71, 87, 162, 169, 199
– filial piety (*xiao*)   136, 150, 173
– Five Relationships   9, 51, 173, 176
– *Gongyang*   46, 223, 225
– Grand Unity (*datong*)   47, 103
– hegemonic way (*badao*)   161
– Mandate of Heaven (*tianming*)   9, 34, 46
– Neo-Confucianism   95, 150, 161, 195, 221
– New Confucianism   33, 197f., 221–223, 225
– New Text scholarship   50, 67
– *ren* (benevolence)   34, 173, 192
– royal way (*wangdao*)   161
– rule of man   115
– Three Ages (*sanshi*)   46f., 65
– three principles and five virtues   30
– *yi* (justice, righteousness)   20, 44, 48, 55, 60, 67, 69, 75, 84f., 89, 91, 100, 128, 136, 149, 151, 161f., 173, 203
– *zhengxin chengyi* (rectifying one's mind and making one's will sincere)   193
constitutionalism   30, 65, 69, 112, 159, 168, 197, 223f., 227
counterrevolution   35, 143, 202
Critical Review Group   14, 26, 32, 119, 121, 126, 136, 141, 170, 181–188, 192f., 195, 197, 199–202, 207, 211
– National Southeast University   170, 184f., 201f.
– New Humanism   141, 181, 185–192, 211f.

democracy   12, 16, 25, 30–32, 47, 52, 64, 69, 71, 103, 112, 121, 123, 142, 155, 157, 159, 163f., 166, 169, 173–175, 188f., 191–193, 195–197, 202f., 205–207, 212f., 215, 219, 221–227
division of power   112

Education Society for Strength (*Qiangxuehui*)   49
elitism   32, 192, 196, 198, 211, 228
enlightened dictatorship   203

Index of Subjects

Enlightenment 1, 7, 9f., 25, 31, 64, 95f., 121, 130, 146f., 155, 164, 185, 198, 211
equality 30, 64, 98, 108, 114, 144f., 149, 163, 195f., 207, 211, 224, 228

fangzhen (or *fanzhen*) 116f.
fascism 13, 204f.
federalism 30, 87, 112, 116f., 127, 169
fengjian 116f., 159, 164f., 180
feudalism 116, 180
First Civil War 180
First Sino-Japanese War 7, 45, 53, 55, 82–84
First United Front 180
Five People Under One Union (*wuzu gonghe*) 108f.
France 7, 18, 35, 38, 51, 54, 74, 107, 143, 188f., 219
– Dreyfus Affair 54, 188f.
– French Revolution 4, 9, 18, 35, 51, 73, 113f., 144, 160, 195f., 206, 226
Francoist Spain 35

Germany 16, 24, 69, 84, 107, 117, 134, 165, 203, 205f., 219, 227
– *Konservative Revolution* 206–208
gong (public) 116f.
great-Hanism (*da hanzu zhuyi*) 108
Guangfuhui 53

Huaxinghui 53
Hundred Days' Reform 2, 7, 19, 29, 44f., 51, 53, 57, 97, 114
Hundred Schools of Thought 92f., 113

Imperial examination 9f., 14, 45, 55, 61, 63, 111, 125, 218
imperial learning (*junxue*) 92, 110f.
individualism 78, 96, 144, 147–149, 151, 153f., 176, 189, 200
– *geren* (individual) 147, 151, 154
Industrial Revolution 9, 200, 217

Jiangnan Arsenal 61
– Shanghai School for the Diffusion of Languages 61

jimizhou 106f.
junxian 116

kokugaku (Japanese national learning) 110f.
Kuomintang (KMT) 1f., 12–14, 16, 27, 33, 35, 70f., 109, 127, 158f., 179–181, 193, 199, 201, 204–210, 221f.
– Blue Shirt cliques (*Lanyishe*) 13
– Central Club („CC" Club) 13
– constitutional politics (*xianzheng*) 193
– military rules (*junzheng*) 193
– New Life Movement 13, 27, 203–205, 207, 227
– party rules (*xunzheng*) 193
– Tongmenghui 22, 29, 53f., 59, 63, 65, 68, 70, 95, 125, 167

Legalism 92, 115f.
– rule of law 115
liberal conservatism 31, 143–145, 147–164, 207, 211
liberalism 2, 9, 16, 28, 31, 144f., 147, 151, 155, 157, 205f., 211, 213, 215, 222, 226
– *ziyou* (freedom, liberty) 147
local self-government 87, 115–117, 159f., 162, 169, 174, 208
loyalism 16, 28, 34, 72, 214

Manchukuo 72, 109
Mandarin Ducks and Butterflies School 120, 122, 128f., 131, 134–138
May Fourth Movement 1, 17, 24f, 30, 119, 202.
Meiji restoration 7
Minbao (*People's Tribune*) 23, 52, 59, 78, 87, 99, 102, 104f., 107–110, 113f., 167f., 182
minben (people as root) 99
Ming dynasty 34, 54, 56, 61, 106f., 172
Mohism 92, 114
– *jian'ai* (universal love) 114
Movement of the New Village (*Atarashikimura undō*) 166
Mutual Protection of Southeast China (*Dongnan hubao*) 44, 58

national essence (*guocui*)   20, 22 f., 25 f., 29 f., 84–87, 89, 91 f., 105, 110–112, 115, 130, 142, 156, 161, 168, 181, 183, 192 f., 214
National Essence School   12–14, 22, 54, 63, 65, 86, 88, 112, 123, 181, 183, 219
– *Guocui xuebao*   63, 86, 89, 111 f., 119, 182 f.
– Society for the Preservation of National Learnings (*Guoxue baocunhui*)   54 f., 63, 66 f., 72, 119
Nationalism   14–27, 54, 64–66, 73–79, 81, 86, 102 f., 105, 107–109, 116–119, 125, 149, 151, 167, 182, 202, 205, 222 f.
– culturalist nationalism   17, 21–27, 29–32, 73 f., 79, 89, 93, 102, 118–121, 141–144, 156, 202, 205, 211, 214 f., 219
– great nationalism   108, 117
– *guomin*   19, 22, 76, 99
– *minzu*   19 f., 22, 105, 108, 207
– small nationalism   108, 117
– *Zhonghua minzu*   108 f.
national learning (*guoxue*)   86, 92, 110–112, 115, 123, 140, 161, 185
National Southwestern Associated University   209
Nazism   208
New Authoritarianism   33, 226
New Culture Movement   1 f., 4, 17, 23–25, 30, 34, 48, 119, 124, 129, 133, 136–139, 149 f., 155, 164, 180, 183–188, 192, 202 f., 207
New Policies (*Xining bianfa*)   8
New Policies (*Xinzheng*)   2, 7, 28 f., 44, 53, 55, 63, 93, 111, 115, 219
New Youth   25 f., 120, 134, 137, 150 f., 153, 165 f., 180
Northern Expedition   27, 35, 169, 180 f., 201 f.
North-South Conference   120

Opium War   7, 39 f.

parliamentarianism   30 f., 51
Peking University   119, 139, 171, 209

philosophical conservatism   32, 143, 181, 193–202, 211 f., 215
Progressive Party (*Jingbu dang*)   158, 216

qun   49, 75 f., 93 f.

radicalism   17, 23, 29 f., 35 f., 44–68, 73, 78, 108–110, 118 f., 138, 196, 213 f., 220, 222
Railway Protection Movement   103
reactionaryism   8, 17, 28 f., 34 f., 68–72, 133, 135, 213 f., 228 f.
Revolution of 1911   1, 8, 23, 25, 29, 67, 69 f., 108, 118 f., 127, 130, 170, 181, 219 f., 223, 227
– Wuchang Uprising   6, 67, 167
Royalist Party (*Zongshe dang*)   70
Russia   38, 54–56, 84, 100, 186, 227
– Anti-Russia Movement (*Ju'e yundong*)   56, 100

science   4, 25 f., 41, 82, 85 f., 90, 94 f., 113, 142, 148, 155 f., 160, 164, 166 f., 175, 178, 182, 185, 192 f., 198–200, 202, 208, 212, 221 f., 226
Second Civil War   180, 221
Second Revolution (*Erci geming*)   71, 130, 147 f., 152, 154, 158, 164
Second Sino-Japanese War   27, 178, 180, 209
Second World War   16
Self-Strengthening Movement   2, 7, 28, 39, 44 f., 61, 194
si (egoism, private)   116
Sino-Babylonianism   89, 91, 100, 110
social contract (*minyue*)   98 f., 112
social division of work   97, 112, 146, 149
social evolutionism   8, 99, 110, 168
socialism   2, 16, 28, 32, 71, 77 f., 94, 102 f., 121, 146, 163, 180, 195–197, 205 f., 225, 229
social organicism   94, 112
Society for Political Education (*Seikyōsha*)   20 f., 79, 82 f.
– Asia (*Ajia*)   79
– Japanese (*Nihonjin*)   79, 82
– national essence (*kokusui*)   20 f., 79–83

Society for the Mass Education (*Pingmin jiaoyu hui*)   177
Society for the Study of Socialism (*Shehui zhuyi jiangxihui*)   63f.
Society of People (*Heimin sha*)   64
Society of Potency (*Jishe*)   56
Society of Restoration (*Fushe*)   56
Southern Society   14, 22, 34, 54–56, 59–61, 65, 71, 90, 95f., 100, 103, 114, 119f., 125–132, 136, 139–141, 158f., 162, 181, 219
Subao Affair   59

Taiwan   1, 33, 39f., 45, 58, 64, 72, 82, 180, 218, 221f., 227
Three Dynasties (*sandai*)   113f., 116, 171
Three Principles of the People (*sanmin zhuyi*)   77, 103, 127
traditionalism   8, 28f., 34–44, 68, 72, 213f.
Tsinghua University   182, 201f., 209
– Tsinghua Institute of Research on National Learning   184
– Tsinghua School   182f.

United States   35, 38, 74, 117, 141, 148, 181–184, 188, 201, 209, 217
universal suffrage   141, 157, 176, 216f.

vernacular language (*baihua*)   18, 122–124, 136, 141, 180, 183, 185, 195
village self-government (*xiangzhi*)   83, 173f.

xincun (new village)   166
Xingzhonghui   53
Xinmin congbao (*Journal of New People*)   52, 56, 62, 94, 107, 114

Yellow Emperor   22f., 66, 68, 90f., 100–102, 109, 119
YMCA   177

zhongti xiyong (Chinese teaching as substance, Western teaching as application)   23, 87, 91f.

www.ingramcontent.com/pod-product-compliance
Lightning Source LLC
Chambersburg PA
CBHW031424150426
43191CB00006B/384